PERIODIC TABLES UNIFYING LIVING ORGANISMS AT THE MOLECULAR LEVEL

THE PREDICTIVE POWER OF THE LAW OF PERIODICITY

Prof. Dr. Antonio Lima-de-Faria
Lund University, Department of Cell and Organism Biology
Biology Building
Sölvegatan 35
SE-223 62 Lund
Sweden

Private Address:
Trumslagaregränden 10
SE-226 39 Lund
Sweden

E-mail: johan.essen-moller@comhem.se

PERIODIC TABLES UNIFYING LIVING ORGANISMS AT THE MOLECULAR LEVEL

THE PREDICTIVE POWER OF THE LAW OF PERIODICITY

Antonio Lima-de-Faria

Lund University, Sweden

World Scientific

NEW JERSEY • LONDON • SINGAPORE • BEIJING • SHANGHAI • HONG KONG • TAIPEI • CHENNAI • TOKYO

Published by

World Scientific Publishing Co. Pte. Ltd.

5 Toh Tuck Link, Singapore 596224

USA office: 27 Warren Street, Suite 401-402, Hackensack, NJ 07601

UK office: 57 Shelton Street, Covent Garden, London WC2H 9HE

Library of Congress Cataloging-in-Publication Data
Names: Lima-de-Faria, A., author.
Title: Periodic tables unifying living organisms at the molecular level : the
predictive power of the law of periodicity / Antonio Lima-de-Faria.
Description: New Jersey : World Scientific, 2017. | Includes bibliographical references.
Identifiers: LCCN 2017037719| ISBN 9789813227002 (hardcover : alk. paper) |
ISBN 9813227001 (hardcover : alk. paper)
Subjects: | MESH: Evolution, Molecular | Periodicity | Models, Molecular
Classification: LCC QH366.2 | NLM QU 475 | DDC 576.8--dc23
LC record available at https://lccn.loc.gov/2017037719

British Library Cataloguing-in-Publication Data
A catalogue record for this book is available from the British Library.

Printed in Singapore

Foreword

Evolution is a process inherent to matter and energy, since the simplest elementary particles already exhibit two fundamental properties: (1) They combine with each other in a rigid ordered way and (2) they show periodicity revealed by the emergence of similar patterns. As an obligatory consequence evolution extended to atoms, minerals, macromolecules, cells and organisms. The same holds for periodicity which became most evident in atoms where it takes its iconic form in the Periodic Table of the Chemical Elements.

Periodicity in living organisms could not be considered before, because the chaotic emergence of species, by random mutation and selection, precluded such an assumption as well as a search for a fundamental ordered process in biological phenomena. As a consequence the recurring similar structures and functions were called "Paradoxical Events". There are no "Paradoxical" or "Intriguing Events" in living processes. Every structure and function is the immediate product of a molecular cascade having its origin in atomic and ultimately in electronic events.

In this work an extensive analysis of the following structures and functions, was carried out at the molecular level. These are: luminescence, vision, regeneration, flight, plant carnivory, mental ability, placenta and penis. An impressive periodicity was revealed with the same essential properties that are present at the atomic level. (1) A given property re-emerges time and again (as much as over 40 times). (2) It suddenly appears in the simplest as well in the most complex organisms. (3) The sudden recurrence occurs in organisms that are not phylogenetically closely related, (4) the property resurfaces without previous announcement. (5) The novel structures and functions appear "ready made" as a surprise to the organism. This has been demonstrated by the manipulation of genes that led to the sudden production of additional complete and normal wings and legs in flies and birds.

These results are not surprising: (1) We tend to forget that the most diversified living organisms are constituted solely by the atoms of the Periodic Table, which moreover have not changed their properties when the cell was formed. (2) Most significant is that protozoa, plants and lower and higher animals (including humans) share common genes that now have been isolated and sequenced. Examples are: (1) The occurrence of the placenta in plants turned out to be directed by the same genes that form it in mammals. (2) The *Pax6* gene is responsible for the formation of all types of eyes found from the simplest invertebrates to humans.

The assembled molecular data led to the creation of a General Table of Biological Periodicity which exhibits remarkable similarities to the Periodic Table of the Chemical Elements. It also led to the formulation of a Law of Biological Periodicity with predictive power.

Periodicity is unifying biological phenomena introducing an unexpected general order in evolution that before could hardly be contemplated.

Lund University, Lund, Sweden.
August 24, 2017

Contents

Chapter 1

Biological Evolution is Now Being Studied at the Level of Elementary Particles

The Search for the Physical Rules that Predict the Atomic Behavior of DNA and of Proteins

There is at present a most detailed and coherent picture of the structure and function of DNA, RNA and proteins elucidating the biochemical activity of the cell. Replication, transcription and translation are well understood in chemical terms. Besides, a large number of DNA, RNA and protein molecules have been sequenced allowing the establishment of phylogenetic relationships between organisms at the molecular level. The overall result is impressive in every respect.

It is this enormous advancement that has led to the obligatory question: What are the physical rules that direct the evolution of these macromolecules? This question had been asked earlier for proteins because they are so many and diversified. Concerning DNA and RNA we are mostly in the dark but physicists are now helping in this direction.

The number of proteins in which the atomic structure was determined soon exceeded 300. However, by the 1990s it was not yet possible to formulate a set of rules that allowed the prediction of a protein's three-dimensional structure from the amino acids of its polypeptide chain. Proteins are gigantic atomic edifices that may consist of as many as 500 amino acid residues (bovine glutamate dehydrogenase, Lehninger, 1975).

Since then computer graphics and nuclear magnetic resonance have allowed a better resolution of their three-dimensional structure and function. But the picture is still far from satisfactory. The reason is simple. As C. Branden and J. Tooze (1991) already pointed out "We shall not unravel the chemistry of life in molecular detail without knowing at atomic or close to atomic resolution the structure of biological macromolecules, especially the proteins". Surprisingly this still applies to our present state of knowledge in proteins as well as in DNA.

The Present Creation of New Accelerators and Spallation Sources

New technologies are now allowing to go deeper than the atom, the research in this area being explosive.

Over 900 scientists have been performing experiments during 2013 using synchrotron radiation of the MAX II and MAX III accelerators from Lund University in Sweden. At present a still more efficient laboratory, MAX IV, has been completed (June 2016).

Synchrotron radiation is obtained when electrons are accelerated to a speed close to that of light. At the same time a magnetic field curves their trajectory.

The synchrotron radiation from the three storage rings of these accelerators covers the wavelength range from infrared through ultraviolet to hard X-ray range. Five independent beam lines have been used mainly for protein crystallography giving a better insight into their structure. The largest ring is 528 meters in circumference (Fig. 1.1).

X-rays may damage the materials studied. As a consequence new laboratories are being created to use instead neutron scattering to probe the structures but in addition the behavior of large molecules.

The special properties of neutrons make this new technology unique. For this purpose the European-Spallation-Source is now under construction at Lund University.

There are several fundamental reasons for using neutrons: (1) They probe structure, but most important, also motion. Measurement of structure extends from micrometres to one-hundred-thousandth of a micrometre and motion is recorded from milliseconds to ten-million-millionths of a millisecond. (2) Neutrons pass easily through most materials even when these are submitted to extreme conditions of high or low temperatures. (3) Neutrons are non-destructive and can image the inside of objects, being an extremely precise tool. (4) Since neutrons are scattered by atomic nuclei, they report which element and which isotope is present by substituting one isotope for another in key regions. (5) The nuclei of

Figure 1.1 Aerial view of Max IV accelerator laboratory which started operating in June 2016 at Lund University, Lund, Sweden.

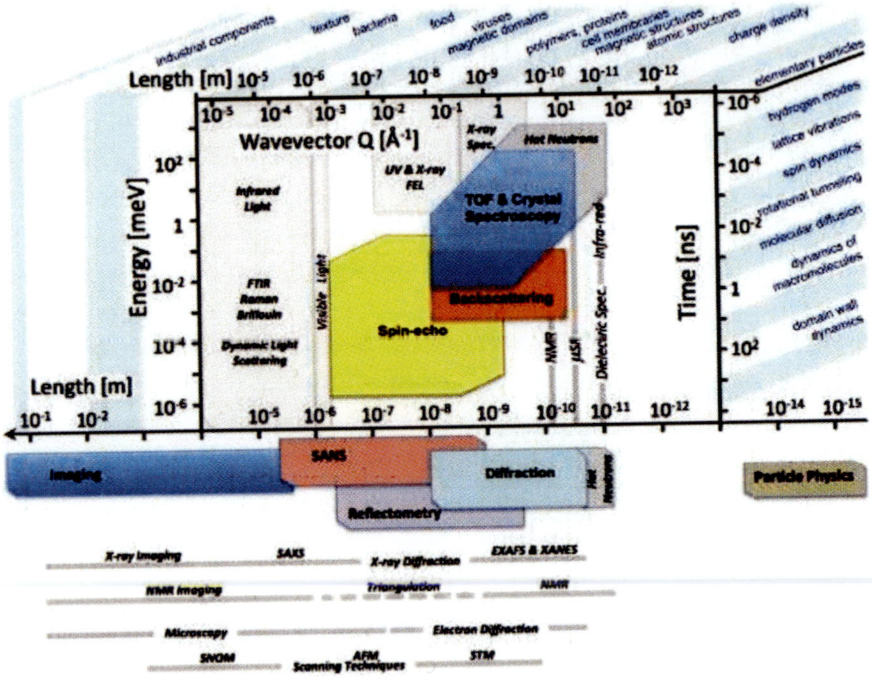

Figure 1.2 What do neutrons tell us. **European-Spallation-Source**. Graph showing the energy, time range and distance that is covered by neutron scattering. The type of structures as well as the technologies employed are also indicated. Neutron scattering provides a unique combination of structural and dynamic information.

ordinary hydrogen atoms contain only a single proton. If a neutron is added the hydrogen becomes deuterium. Deuterium substitution is used to study complex biological materials (Eriksson 2015) (Fig. 1.2).

Visualizing Individual Hydrogen Atoms and Tracing Atom Mobility within a Macromolecule

Single atoms have been photographed since 1979 and by 1992 pictures were obtained of iodine atoms where they could be seen to be regularly bound to each other. This spatial order allowed the detection of missing atoms since they left a cavity of corresponding size in the atomic layer (von Baeyer 1992) (Fig. 1.3).

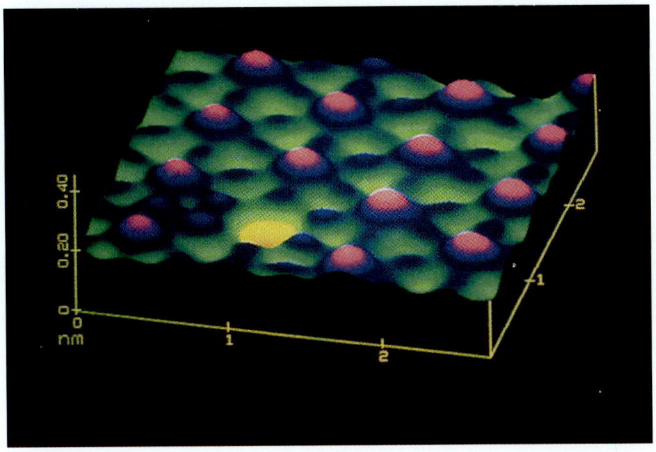

Figure 1.3 Iodine atoms are bonded to each other, but one is missing, leaving a gaping hole in the surface. Courtesy of Fran Heyl.

Since then better methods have been devised permitting the identification of atoms not only in single layers but in large complex structures.

Biological samples typically have a high hydrogen content but are easily damaged by X-rays or electrons. By contrast the use of neutrons in proteins and nucleic acids allows to visualize the hydrogen atoms. This is particularly useful for understanding enzyme mechanisms, protein-ligand interactions and proton transport across membranes. A combined neutron and X-ray crystallographic study of the enzyme aldose reductase elucidated the proton transfer step in the reaction mechanism involving a deprotonated tyrosine (Blakeley *et al.* 2008).

An even better result was obtained when deuterium, the isotope of hydrogen, was introduced.

Approximately one-half of all atoms in a protein are hydrogen atoms. These play a critical role in protein function including hydrogen bonding, electrostatic interactions and catalysis. In DNA hydrogen atoms have also critical functions connected with hydrogen bonding and electrostatic forces which are critical during replication and transcription. Unfortunately hydrogen atoms were difficult to visualize in a three-dimensional context, their locations being usually inferred from the position of their neighboring heavy atoms such as carbon, nitrogen, sulfur and others. But neutron crystallography has allowed the viewing of individual hydrogen atoms, even at medium resolution of around 2 Å, in the small plant protein *crambin* due to the unique scattering properties of hydrogen and deuterium (Chen *et al.* 2012).

Neutron diffraction has also been useful for determining orientations of molecules and the energy of rotation about a chemical bond. Moreover, the exchanging of labile hydrogens with deuterium has turned out to be a sensitive probe of dynamics and breathing motion in proteins (Kossiakoff 1982).

The technology available at present already allows the viewing of atoms and the tracing of their mobility, opening a door to a deeper knowledge of the physical evolution of DNA and other atomic complexes.

Chapter 2

The Unexpected Surge of Periodicity Among Plants and Animals is Anchored to that of Elementary Particles and Chemical Elements

Elementary Particles Show Already Evolution and Periodicity

Several generations of physicists worked at the European Nuclear Research Center (CERN) in Switzerland in an attempt to unravel the inner structure of protons and neutrons. Following decades of exhaustive research they had discovered several hundred particles. Confusion reigned, no one could make sense of this "proliferating zoo" as it was then called.

However, further research revealed that the initial plethora of particles could be reduced to two fundamental ones: quarks and leptons, which by combination produced all the others. Protons and neutrons appear to be composed of quarks. A proton is made of two up-quarks and one down-quark and a neutron consists of one up-quark and two down-quarks (Mulvey 1979). Elementary particles showed already a primeval form of evolution (Pagels 1982).

This result led also to the establishment of a Periodic Table of Elementary Particles. Murray Gell-Mann (1929–) succeeded in elucidating from the chaotic subnuclear landscape certain underlying patterns well hidden among the new properties. As Padamsee (2003) points out: "Order does exist, although at a deeper level than immediately recognizable" (Fig. 2.1).

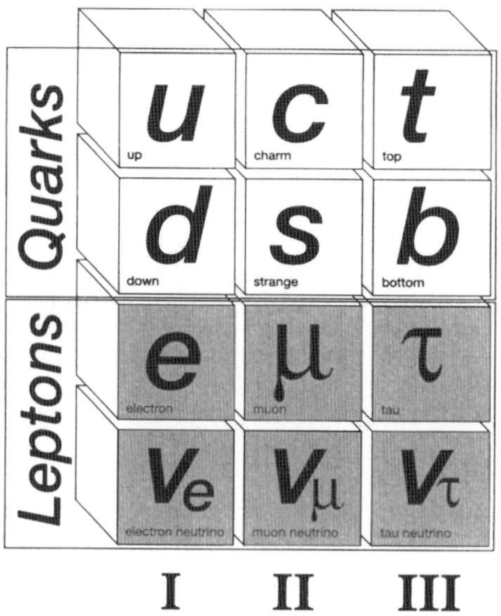

Figure 2.1 Modern Periodic Table of quarks and leptons showing the three generations of matter derived from leptons and quarks.

The Well Established Periodicity at the Atomic Level Contains Many Irregularities

By definition periodicity is "a tendency to recur at regular intervals". "In chemistry, the occurrence of similar properties in elements occupying similar positions in the periodic table" (Webster 1976).

As Scerri (2007) puts it "There is a fundamental relationship between the elements" and he adds: "The periodic law states that after regular but varying intervals the chemical elements show an approximate repetition of their properties". But he also points out that "Periodicity among the elements is neither constant nor exact".

This irregularity is usually not mentioned in order to ensure a straightforward picture, since the Periodic Table of the Elements is one of the icons of science (Sanderson 1967, Greenwood and Earnshaw 1989).

No phenomenon at the physical, chemical or biological level reveals total regularity. Already the periodicity of the movements of the planets in the Solar System contains irregularities. According to Kepler's first law of motion the planets, including the Earth, each go around the Sun on an elliptical orbit, but the Sun is not at the center of the ellipse, it is at an off-center position. Also the eccentricity of some planetary orbits, such as that of Mercury, is not negligible (Weinberg 2015).

The Periodic Table of the Elements is accordingly irregular. Hydrogen (H) can be placed close to lithium on the extreme left side of the table, but may also occupy a position at the extreme right at the side of the noble gases. Helium (He) is an element with an anomalous position in the periodic table. The same is true of europium (Eu) and ytterbium (Yb) that belong to the lanthanide series. Other irregularities include chromium (Cr) and copper (Cu) atoms that carry 1 rather than 2 electrons on their outer orbitals (Fig. 2.2).

Another anomaly is technetium (Tc) that is a radioactive element sitting right in the middle of the most stable range of the periodic table (the fifth period transition metals). "There is a clear line between radioactive and stable elements: everything below bismuth (No 83) is stable; everything above it is radioactive — except technetium and promethium (Pm), which stick out like sore thumbs" (Gray 2009).

Not less than 700 different graphic representations of the Periodic Table of the Elements, with quite different configurations, have been published exposing the variety of interpretations of periodicity among elements (Mazurs 1974) (Fig. 2.2).

As Gray points out "Chemistry is too complicated for any rule to be absolutely hard and fast".

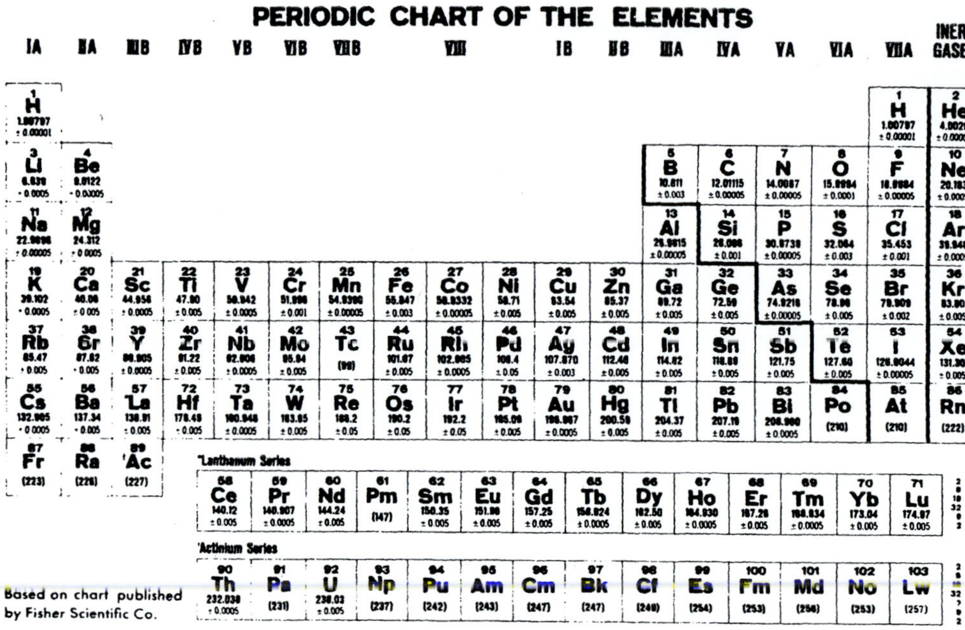

1

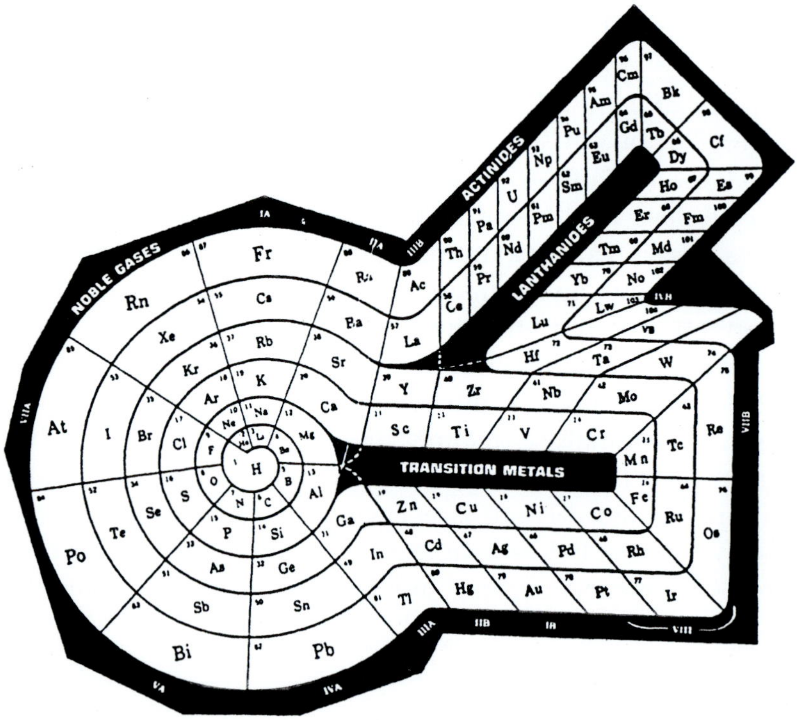

2

Figure 2.2 (1) The generally used graphic representation of the periodicity of the chemical elements. The periodic chart is based on the original table of Mendeleev from 1869. Since then many other chemical elements were added according to Mendeleev's prediction (in this chart 103). Initially the distribution of the atoms mainly followed their atomic weights. Later, as the mechanism of periodicity turned out to lie

Biology is still more complicated and as such rules are expected to be even less rigid. But this does not exclude, as is the case of the chemical elements, that there is a fundamental underlying order that pervades biological phenomena.

The Periodicity of the Chemical Elements has Determined that of Minerals and its Extension to the Biological Level Could not be Avoided

That the periodicity of the chemical elements has determined the periodicity of the minerals is evidenced by the fact that: arsenic, antimony and bismuth are considered as both minerals and chemical elements (As) (Sb) (Bi). They build covalent bonds between their atoms which results in their layered structure and common properties. Another example is the group of elements consisting of gold, silver and copper (Au, Ag and Cu) that belong to the same group in the periodic table and which are at the same time minerals. Their common properties are known to be the result of the metallic bond of their atoms. As Jaffe (1988) remarks: "Electrons govern the chemical behavior of atoms, and how they may be packed into minerals".

What obliged this periodicity to extend to the biological level? It simply could not be avoided because the macromolecules, that play the main role in the cell's function, DNA, RNA and proteins, do not consist of any other chemical elements than those that build the periodic table. Moreover, the phosphate backbone of DNA and RNA, their sugars and bases are all constituted by atoms that are not spread at random in the table but have a particular location. These are: phosphorus (P), hydrogen (H), carbon (C), nitrogen (N) and oxygen (O) all located at the extreme right building a small group (Lima-de-Faria 2003).

Significant, is that the chemical elements did not change their properties as they built the cell, a fact that we tend to overlook.

Figure 2.2 (*Continued*) in the proton and electron composition of each chemical element, every element became defined by its atomic number. This is the number of protons found in the nucleus of every atom of that element. The number of protons determines in turn the number of electrons that orbit around each nucleus. The vertical columns tend to have elements with the same basic properties. In this chart the heaviest element is Nr. 103 but in later ones new elements have been added. At present it finishes with Nr. 118 (Gray 2009). (**2**) In this spiral representation of Periodicity, hydrogen occupies the origin of the spiral, because all other atoms are known to be derived from this chemical element. Within the spiral there is full regularity but the *Transition Metals* stick out of it building a separate block. Besides, the *Actinides* and *Lanthanides* cannot be easily compressed into any representation: be it the conventional chart where they build two separate rows or this one where they protrude distinctly from the spiral. Inert gases are also called noble gases.

Biological Periodicity in 1995 and 20 Years Later. Novel Structures Appear "Ready Made"

The work "Biological Periodicity. Its Molecular Mechanism and Evolutionary Implications" (Lima-de-Faria 1995) was published two decades ago. This may seem to be a short time but the development of molecular biology during this period has been meteoric and revolutionary. Radical changes, in our way of interpreting biological phenomena, have occurred due to many unexpected results.

The work of 1995 established periodicity in living organisms. Against expectation, the same structure or function tended to occur following given regular intervals.

The intervals between the repetition of a given property, like those of the Periodic Table, varied in length, but that was to be expected since biological structures are known to be more complex than chemical ones.

Flight in animals turned out to be the best example. No invertebrate phylum is known to include animals with wings except the class Insects, yet there are as many as 30 highly diversified phyla of invertebrates from simple worms to complex mollusks. According to the fossil record the first insects had no wings. The capacity to fly arose from nowhere without any predecessors. The earliest dragonflies had, however, large wings spanning about 70 cm as evidenced by well preserved fossils. This size of a wing is larger than that of many birds and bats.

Flight, accompanied by its complex body structures and functions, developed independently not less than five times in evolution. As the vertebrates emerged the fish, amphibians and most reptiles had no flight ability but suddenly the pterosaurs, a group of reptiles not directly related to the insects, aquired large wings soaring over the oceans. Equally suddenly the birds started flying without having had a direct phylogenetic relation with the pterosaurs. Later, as evolution proceeded, mammals diversified into such different species as mice and whales, but none flew. Again, without immediate precedents, the bats, with their large transparent wings, became the acrobats of the air. In the meantime some teleost fish acquired this ability using either 2 or 4 of their fins to fly, circulating around boats with high dexterity.

Hence without immediate relation to one another wings were formed five times being in each case accompanied by a series of accessory structures and mental coordination, without which flight could not have occurred. What is remarkable is that contrary to expectation — instead of being the result of successive random mutations and selection — the wings appeared as a "surprise" to the organism. This has been demonstrated by the production of flies and birds with four wings instead of two. Following elaborated genetic manipulation four-winged *Drosophila* (Gehring 1998) and four-winged chicken were obtained in the laboratory without any selection or accidental mutations (Isaac et al. 1998, Logan et al. 1998, Ohuchi et al. 1998). These "ready made" wings are identical to those that the animal possessed initially, are accompanied by the same accessory structures and are functional. This is not extraordinary because insects with four wings exist in nature such as the butterflies and dragonflies. Birds with four wings were also found in captivity. One is simply enacting in the laboratory a genetic event that occurred naturally in another animal group or during domestication. The genetic change has immediate consequences for the organism and it occurred whether the animal liked it or not.

Other structures and functions showed periodicity such as luminescence. This is a key function because it does not start at the biological level but appears in minerals before DNA and the cell were formed.

The fact that luminescence in animals is preceded by the luminescence in minerals demonstrates that the phenomenon is a pure atomic event.

The development of molecular biology during the last 20 years has produced a large amount of results that allow to establish biological periodicity on a much firmer basis.

The Occurrence of Order at the Cell Level

Until recently, every molecular process in the cell seemed to be chaotic, but this view was modified by the demonstration of DNA repair, RNA repair and even protein repair. All three processes re-establish the original molecular order by quality control of defective molecules (Robertson et al. 2009, Cullen 2003, Cakmak 2009).

In addition the results available lately, from different areas of molecular biology, disclose that order dominated cell evolution. The trafficking of molecules in the cell is guided and controlled at every level. MicroRNAs shape the road map of the emerging embryo and cells have sensing mechanisms that are used by their chromosomes to adjust gene number (Matzke and Matzke 2003, Zoncu et al. 2011, Chan and Marshall 2012, Felsenfeld 2014).

As Brown (1999) put it "In view of the thousands of damage events that genomes suffer every day, coupled with the errors that occur when the genome replicates, it is essential that cells possess efficient repair systems. Without these repair systems a genome would not be able to maintain its essential cellular functions for more than a few hours before key genes became inactivated by DNA damage. Similarly, cell lineages would accumulate replication errors at such a rate that their genomes would become dysfunctional after a few cell generations".

Evolution Started with the Ordered Combination of Quarks and Leptons — DNA, and the Cell, were Later Arrivals in this Process that were Obliged to Follow the Energy States of their Atoms

DNA, and the cell, are not the primary events in evolution — as generally believed — but are terminal stages in this process.

The view that biological evolution started with DNA is an antiquated assumption. The results obtained in particle physics, chemical periodicity and crystallography reveal that three evolutions preceded the biological one (Lima-de-Faria 1988).

(1) Evolution of the elementary particles

Experiments carried out with atoms revealed that they are composed of elementary particles known as quarks and leptons. It turned out that elementary particles have their own ancestors. Combination of particles in pairs or in groups has been used in the formation of larger particles. Restrictions were imposed on this initial construction since the electrons can only move in permitted orbits and the number of types of neutrinos is limited (Fig. 2.1).

(2) Evolution of the chemical elements

As early as 1815 the chemist Prout (1785–1850) proposed that there had been an evolution of the chemical elements. The idea was mainly ignored even after the establishment of the validity of the periodic Table of the elements. But Prout's idea was confirmed in the last decades. All known chemical elements are now recognized as derived from hydrogen (Scerri 2007). An impressive simplicity is involved since a single, and the simplest element, is used in the construction of all the others. This is shown in the display of the periodic Table as a spiral, in which hydrogen is located at its origin (Fig. 2.2.2).

(3) Evolution of the minerals

Mineralogists did not dare to speak of any evolution in minerals since they appeared to them as static chemical combinations. After it was shown that minerals had also an evolution of their own (Lima-de-Faria 1988), mineralogists are now describing it as their own finding. Minerals occur in a limited number (about 4,000) and have been classified into species by mineralogists due to their well defined chemical composition. All obey a primary order that is evident in the six crystal systems to which they all belong. The simplest of these systems is the cubic, which appears as the primary one. The secondary order is evident in the formation of derived structures which obey atomic rules. These structures are also found in only limited numbers. The symmetries present in crystals were considered to be limited to values of 1, 2, 3, 4 and 6. The 5 symmetry, which is so common in living organisms, was absent in minerals. This finding was used as an argument against a connection between the two levels of organization. But the discovery of the 5 symmetry in quasicrystals (Nelson 1986) disposed of this argument (Fig. 11.3).

Like the minerals, every macromolecule (e.g. protein, RNA or DNA) is a prisoner of its atomic constitution and of the energy states of the electrons that build its atoms. These macromolecules were late arrivals in the evolutionary process. It is their atomic behavior that will now be elucidated by the new accelerators and enormous spallation sources.

Not only DNA, but RNA is now considered to direct evolution due to its atomic properties. In the human genome organism diversity arises through extensive RNA-processing and widespread RNA-directed rewriting of DNA (Herbert 2004).

It is time to ask how the intrinsic properties of atoms and electrons have guided not only the periodicity of the chemical elements but that of living organisms.

We are witnessing an impressive jump in this direction that leads to an unchartered but fundamental area of evolution.

The Graphic Display of the Periodicity of the Chemical Elements Fits into that of the Biological Properties

The problem arose of how biological periodicity could be compared graphically with that of the chemical elements.

The distribution of the atoms in the Table follows a main type of representation. Columns are built by atoms. These share the same basic property, being located vertically according to an increasing grade of complexity (from 1 to 6).

This type of periodicity distribution turned out to fit the graphic display of biological periodicity. As an example three of the most representative vertical columns of the Periodic Table of the Elements are displayed in Table 2.1.

Alkali metals, *Alkaline earths* and *Noble gases* build the three vertical columns that show the most regular recurrence of properties. The *alkali metals* share the common property of being soft metals, the *alkaline earths* are known for being less explosive than the alkali metals, and the *noble gases* have in common the property of almost never forming compounds with each other or with any other elements. An extra column was added to put in evidence the number of electrons in filled shells that build every chemical element in the noble gases. Neon, Argon, Krypton, Xenon and Radon all have 8 electrons in their outer-shell, in agreement with the general rule that every vertical column contains elements with the same number

Table 2.1 Periodicity of the Chemical Elements

Increasing grade of complexity	ALKALI METALS	ALKALINE EARTHS	NOBLE GASES	ELECTRONS IN FILLED SHELLS OF NOBLE GASES
1	LITHIUM 3 6.9	BERYLLIUM 4 9.0	HELIUM 2 4.0	2
2	SODIUM 11 22.9	MAGNESIUM 12 24,3	NEON 10 20.1	2 8
3	POTASSIUM 19 39.1	CALCIUM 20 40.0	ARGON 18 39.9	2 8 8
4	RUBIDIUM 37 85.4	STRONTIUM 38 87.6	KRYPTON 36 83.8	2 8 18 8
5	CESIUM 55 132.9	BARIUM 56 137.3	XENON 54 131.3	2 8 18 18 8
6	FRANCIUM 87 (223)	RADIUM 88 226.0	RADON 86 (222)	2 8 18 32 18 8

Each vertical column consists of atoms displaying the same basic property. This re-emerges independently of increasing grade of complexity of the atoms. The atomic number is the number of protons found in the nucleus of every atom (in red). This in turn determines the number of electrons that orbit around each nucleus. The number of electrons that fill the *outer shell* of each atom is responsible for the periodicity. Atomic weight is in black numbers. Compare with Fig. 2.2.1 (Original).

Table 2.2 Periodicity of the Chemical Elements

Increasing grade of complexity →

LANTHANIDE SERIES

LANTHANUM	CERIUM	PRAESEODYMIUM	NEODYMIUM	PROMETHIUM	SAMARIUM	EUROPIUM	GADOLINIUM	TERBIUM	DYSPROSIUM	HOLMIUM	ERBIUM	THULIUM	YTTERBIUM
57	58	59	60	61	62	63	64	65	66	67	68	69	70
138.9	140.1	140.9	144.2	(145)	150.4	151.9	157.2	158.9	162.5	164.9	167.2	168.9	173.0

ACTINIDE SERIES

ACTINIUM	THORIUM	PROTACTINIUM	URANIUM	NEPTUNIUM	PLUTONIUM	AMERICIUM	CURIUM	BERKELIUM	CALIFORNIUM	EINSTEINIUM	FERMIUM	MENDELEVIUM	NOBELIUM
89	90	91	92	93	94	95	96	97	98	99	100	101	102
(227)	232.0	231.0	238.0	237.0	(244)	(243)	(247)	(247)	(251)	(254)	(257)	(256)	(254)

The chemical elements with atomic numbers 57 to 70 and 89 to 102 are called *rare earths* due to their common properties (according to Dickerson and Geis, 1979). Other authors subtract or add atoms to the series. They build two parallel series: the *lanthanides* and the *actinides*, each series having particular properties. The atoms are distributed along each series following increased complexity (atomic number in red). Compare with Fig. 2.2.1 (Original).

of outer-shell electrons. But, *Helium* is exceptional since it has only 2 electrons in its outer-shell.

The *rare earths* (Table 2.2) build two parallel series with atomic numbers 57 to 70 and 89 to 102 due to their common properties. The first group, the *lanthanides*, are notorious for being chemically similar to each other. Their main differences are in magnetic properties some being strong others weak magnets. The same is the case with the second group, the *actinides*. They are all radioactive some being more radioactive than others.

When these two series are compared in parallel, atom by atom, they also show some similarities. This is why the rare earths are also called *inner transition metals* (Dickerson and Geis 1979).

These two parallel series also have their counterparts in biological periodicity (Table 12.1).

Chapter 3

Carnivory in Plants is Not a "Paradoxical Event" But is Due to the Expression of Specific Genes and Chemical Modifications of DNA

What is Carnivory

"Carnivorous" is the term opposed to "herbivorous" to denote flesh-eating instead of plant-eating. A whole group of mammals has been classified as the order *Carnivores* by zoologists. They are mainly flesh-eating, having powerful jaws and sharp teeth, like the dog or the lion, but some may at times use plants as their food such as the bear.

Within the mammals, but outside the Carnivores, are the toothed whales and the bats that eat also mainly flesh. Thus other anatomical traits have been considered more important in the creation of separate orders.

The term "carnivory" is used at present to describe the ability of animals and of plants to trap, ingest, and digest animals followed by the assimilation of the contents of their organs (McPherson 2010). Carnivorous plants have been described since Linnaeus time but only in the 20th century have their functions been established on a scientific basis.

Carnivory in Plants is a Recent Invention

Mammals have not inhabited the planet for a long time. The 5,411 species have evolved relatively rapidly occupying all continents and oceans. Mammals originated from reptilian ancestors in the Triassic period with the oldest recognized mammals appearing 225 million years ago. The real explosive phase of modern mammalian evolution did not begin until 55 million years ago. The Carnivores, like the major living groups, arose later (10 million years ago, mid-Eocene) and they are represented at present by 249 species (Wilson and Mittermeier 2009).

Equally, the carnivorous plants are not old. To start with all of them are flowering plants. There are no carnivorous algae, mosses, liverworts, ferns or conifers.

The flowering plants arose late in evolution but earlier than the Carnivores and the other major groups of mammals. Fossil pollen provides the earliest evidence of flowers at roughly 136 million years ago (Cretaceous) followed by a rapid major radiation 125 Myr ago (Frohlich and Chase 2007). But, according to McPherson (2010), all extant genera of carnivorous plants are found in recently derived groups and none belong to the basal geologically old angiosperms, and he adds that carnivory is a "relatively young evolutionary invention" among flowering plants.

Carnivory Evolved Independently at Least on Ten Occasions

The 20 extant genera of carnivorous plants have turned out not to be closely related. Their evolution led to multiple carnivorous plant lineages. The presently living genera belong to 12 different families of flowering plants. To these may be added two more families if *Philcoxia*, *Proboscidea* and *Ibicella* are proven to be fully carnivorous. Carnivory evolved at least on ten separate occasions (Albert *et al.* 1992, Williams *et al.* 1994) and may have evolved independently 12 times (McPherson 2010).

Carnivorous Plants are Derived from Five Botanical Orders and there are Proto-carnivores

Palaeo-evolutionary history and phylogeny based on molecular data of carnivorous plants (Bayer *et al.* 1996), advocates that all living genera are derived from five botanical orders as follows:

1 In Caryophyllales carnivory evolved only once, but sticky traps, snap traps and pitcher traps become established in this lineage. The four closely related families are descendants from a single carnivorous ancestor. These are: *Drosera*, *Dionaea*, *Aldrovanda* (Droseraceae), *Drosophyllum* (Drosophyllaceae), *Nepenthes* (Nepentaceae), *Triphyophyllum* (Dioncophyllaceae).
2 Carnivory evolved independently twice in the order Ericales. One in America: *Darlingtonia*, *Heliamphora*, *Sarracenia* (Fig. 3.1) and the other in Africa *Roridula* (Fig. 3.2).
3 In Lamiales it evolved at least twice independently. *Byblis* and *Pinguicula*, *Genlisea*, *Utricularia*.
4 The Australian pitcher plant *Cephalotus follicularis* is the sole representative in the Oxalidales (Fig. 3.3).
5 The Bromeliaceae are the only Monocotyledons represented. Carnivory appeared twice: *Brocchinia* and *Catopsis*.

In this classification *Roridula* and *Byblis* are considered to be carnivorous but they only have certain carnivorous traits. *Roridula gorgonias* has sticky leaves that trap insects but has no digestive enzymes (Ellis and Midgley 1996). Bugs live on *Byblis liniflora* but no digestive enzymes are present (Hartmeyer 1998). These two species are usually considered "proto-carnivores" since Plachno *et al.* (2006) found high activity of phosphatase in their tissues.

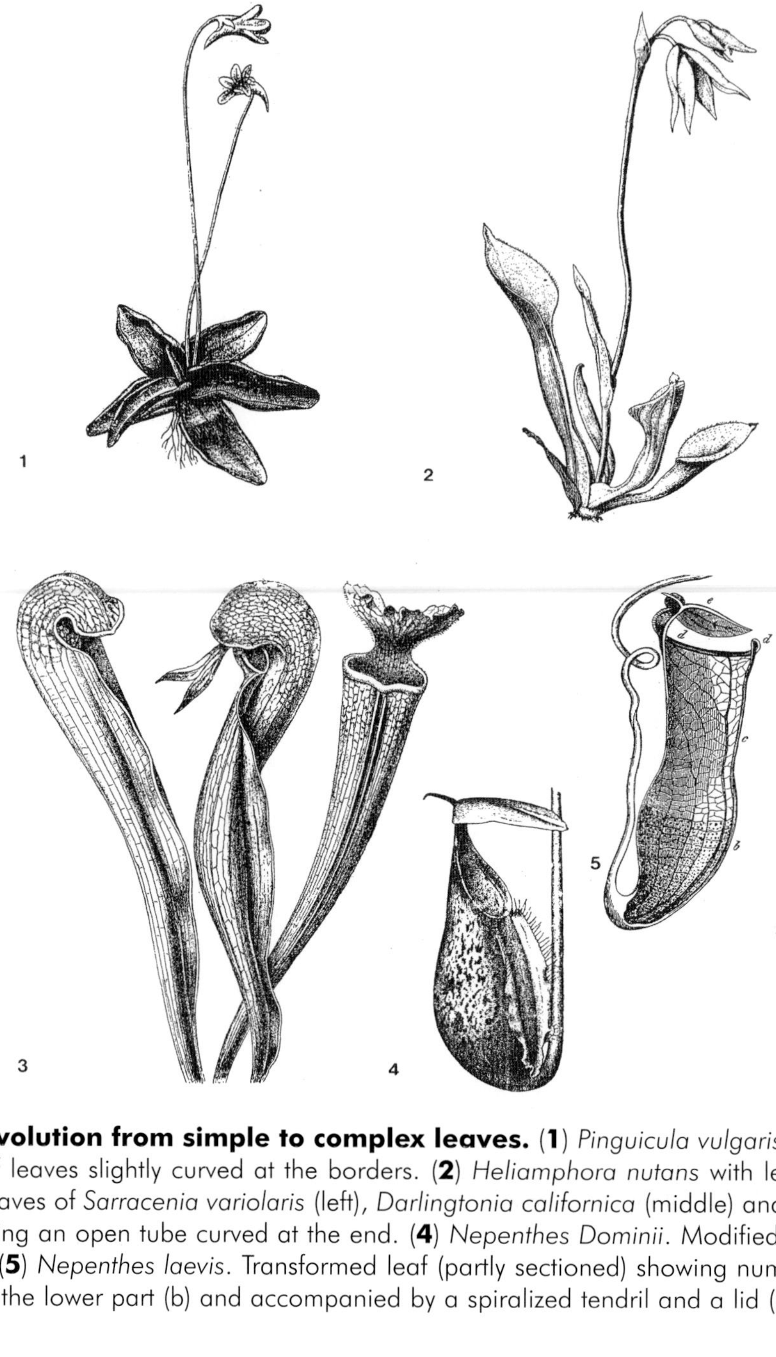

Figure 3.1 Evolution from simple to complex leaves. (**1**) *Pinguicula vulgaris*. Flowering plant with a rosette of leaves slightly curved at the borders. (**2**) *Heliamphora nutans* with leaves building an amphora. (**3**) Leaves of *Sarracenia variolaris* (left), *Darlingtonia californica* (middle) and *Sarracenia liciniata* (right) building an open tube curved at the end. (**4**) *Nepenthes Dominii*. Modified leaf as a pitcher with large lock. (**5**) *Nepenthes laevis*. Transformed leaf (partly sectioned) showing numerous glands (as black specks) in the lower part (b) and accompanied by a spiralized tendril and a lid (e).

Figure 3.2 Carnivorous plants and their glands. (**1**) *Byblis gigantea*. Flowering plant. (**2**) *Roridula dentata*. Whole plant. (**3**) *Drosera rotundifolia*. One leaf with closed tentacles. (**4**) and (**5**) *Drosera rotundifolia*. Two leaves: one seen at rest (from the side), the other with part of the tentacles encircling a victim. (**6**) *Drosera rotundifolia*. Section of the extremity of a leaf tentacle. (**7**) *Drosophyllum lusitanicum*. Cross section of leaf tentacle.

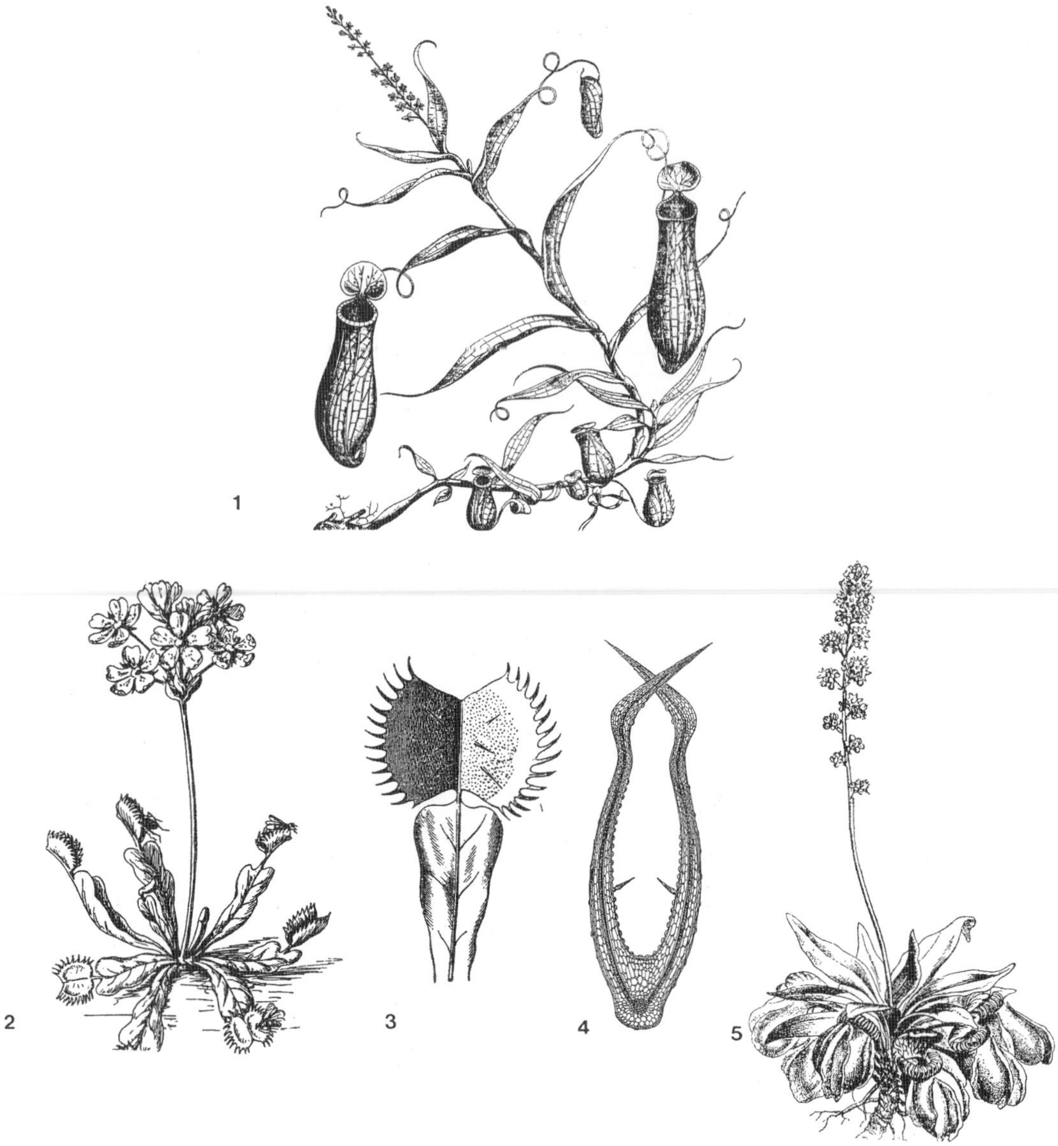

Figure 3.3 Modified leaves in carnivorous plants. (**1**) *Nepenthes* sp. Plant with two types of pitchers (small lower part and large upper part) besides it displays many tendrils. (**2**) *Dionaea* sp. Flowering plant with insects trapped in the leaves. (**3**) and (**4**) Open trap leaf and closed leaf (in cross section) of *Dionaea muscipula*. Note the 3 trigger hairs and glands on the open trap. (**5**) *Cephalotus follicularis* with two types of leaves in the same plant: flat leaves and pitchers (lower part).

More than 95% of the species displaying carnivory belong to the Lamiales and the Caryophyllales (Fig. 3.4).

Carnivorous Plants are Found on all Continents and Number Over 300 Species

The extant genera of carnivorous plants are spread over Europe, Asia, Africa, the American continent and Australia. The genus *Nepenthes* encompasses about 80 species in the islands of Borneo and Sumatra, but it also occurs in India. The number of species in *Drosera* exceeds 90, and is over 150 in *Utricularia*. *Genlisia* has 16 species spread through tropical areas from South America to Madagascar, and in Australia the genus *Cephalotus* is represented by a single species.

Carnivorous plants may be only a few centimeters long (*Pinguicula*) or may grow to 14 meters in height (*Nepenthes*) (Eilenberg *et al.* 2006).

A Carnivorous Organ is the Product of the Coherent Combination of an Impressive Number of Structures and Functions that Occur Separately in other Plant Families

Most significant for the origin of periodicity in carnivorous plants is the finding that the basic structures and functions participating in carnivory existed separately in multiple plant families without having anything to do with flesh digestion. It is only when they were suddenly combined coherently that carnivory emerged.

The traps have one origin, the enzymes are found elsewhere, the movements of leaves are present in most different families, the adhesive glands are found also in other genera and the sensitive hairs occur in many plants. The carnivorous organs have been put together by the assembly of traits that initially were not functionally related.

Leaf Transformations

In carnivory leaves are modified into the most unexpected shapes. These range from simple straight leaves to the most complex organs that rival those of vertebrates

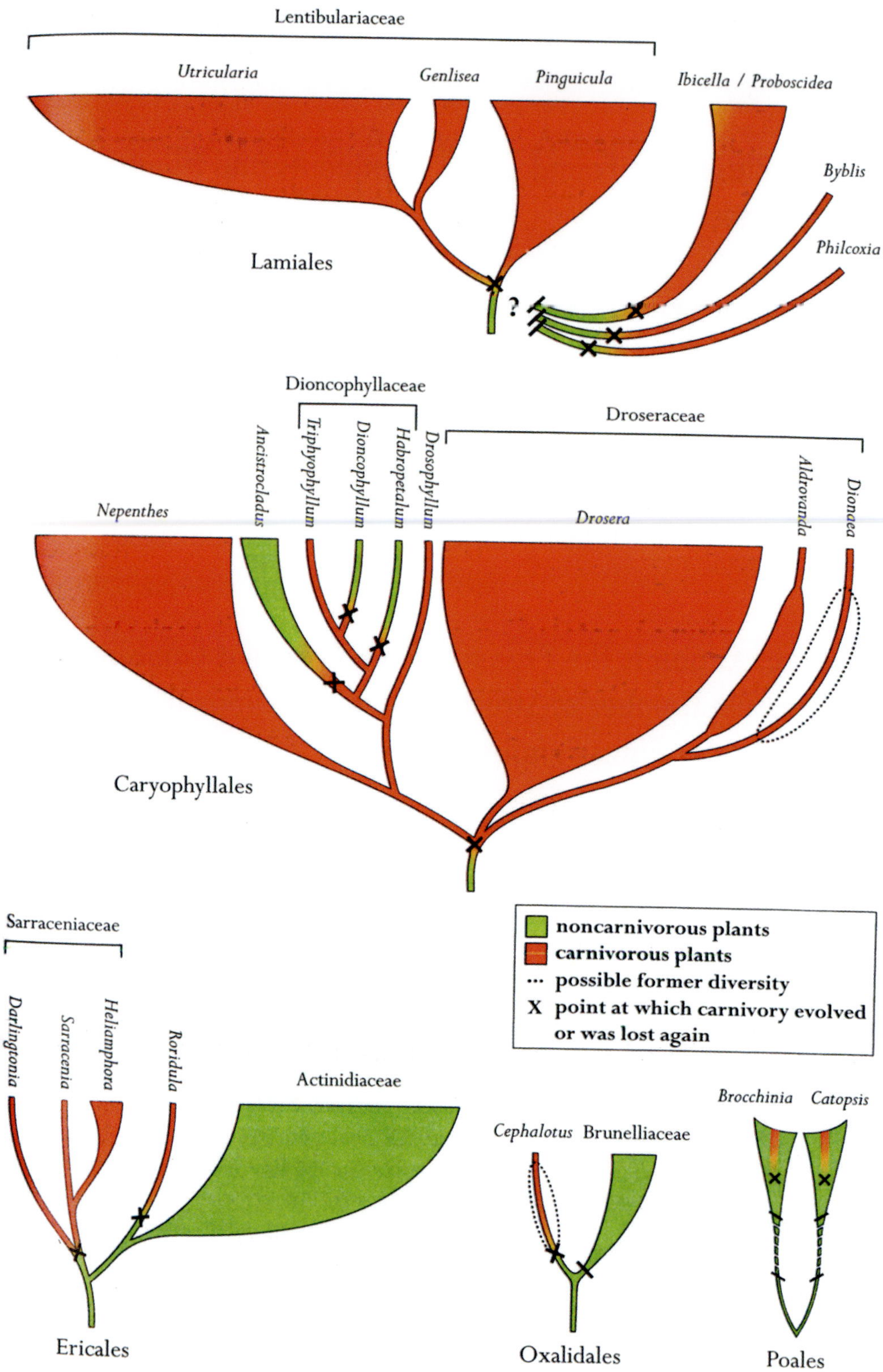

Lentibulariaceae
Utricularia
Genlisea
Pinguicula
Ibicella / Proboscidea
Byblis
Philcoxia
Lamiales
?
Dioncophyllaceae
Droseraceae
Nepenthes
Ancistrocladus
Triphyophyllum
Dioncophyllum
Habropetalum
Drosophyllum
Drosera
Aldrovanda
Dionaea
Caryophyllales
Sarraceniaceae
Darlingtonia
Sarracenia
Heliamphora
Roridula
Actinidiaceae
Ericales
noncarnivorous plants
carnivorous plants
··· possible former diversity
X point at which carnivory evolved or was lost again
Cephalotus
Brunelliaceae
Oxalidales
Brocchinia
Catopsis
Poales

Figure 3.4

(including humans) some of these structures with their accompanying functions are called "stomachs" by botanists.

In *Pinguicula* there is a rosette of leaves, at the base of the plant, which are flat having only the borders slightly curved (Fig. 3.1). In other species there is a subsequent increase in this curvature that finally leads to the formation of a vessel. A further development is present in *Heliamphora*, the leaf borders come so close to each other that they assume the shape of a funnel with their open upper parts resembling an amphora. In *Sarracenia* and *Darlingtonia* the leaves form trumpet-like tubes covered with hoods. These leaf vessels become partly closed by lids in *Nepenthes* (Fig. 3.1).

The extreme result of this process is that the leaf becomes transformed into a most sophisticated trap with a bladder form shut by doors (*Utricularia* and *Polypompholyx*). An opened flat surface is modified into a fully closed structure (Fig. 3.5).

Hairs and Bristles

Hairs and bristles are structures with mainly sensorial functions. In *Utricularia* the entrance to the bladder-like trap consists of stiff bristles near the door. When a small animal brushes against these bristles, the hairs distort the lower edge of the door causing it to spring open (Fig. 3.5). Other hairs with a sensorial function occur in the interior of leaves of *Dionaea* (Fig. 3.3).

The trapping mechanism is so specialized that it can distinguish between living prey and pebbles or sticks that touch the leaf. Moreover, the leaf will not close unless two of its hairs are touched in succession or one hair is touched twice (Raven *et al.* 1999).

The hairs, the bristles and spikes are not distributed at random. Their location, number and orientation are such that they maximize the efficiency of the organ.

The number of hairs increases sharply in *Darlingtonia* and *Sarracenia*. They are located inside the tube. As many as 50 hairs are located in the inside of the trap of *Genlisia*.

Figure 3.4 Diagram illustrating the evolution of carnivory in the five plant orders that comprise known carnivorous plant species according to McPherson (2010). This most valuable representation of the evolutionary relationships between carnivorous species exemplifies the dominating view of evolution applied to the plant kingdom. The comparison of this figure with Table 3.1 puts in evidence the sharp difference between this classical interpretation and the one based on molecular periodicity. They do not exclude but complement each other.

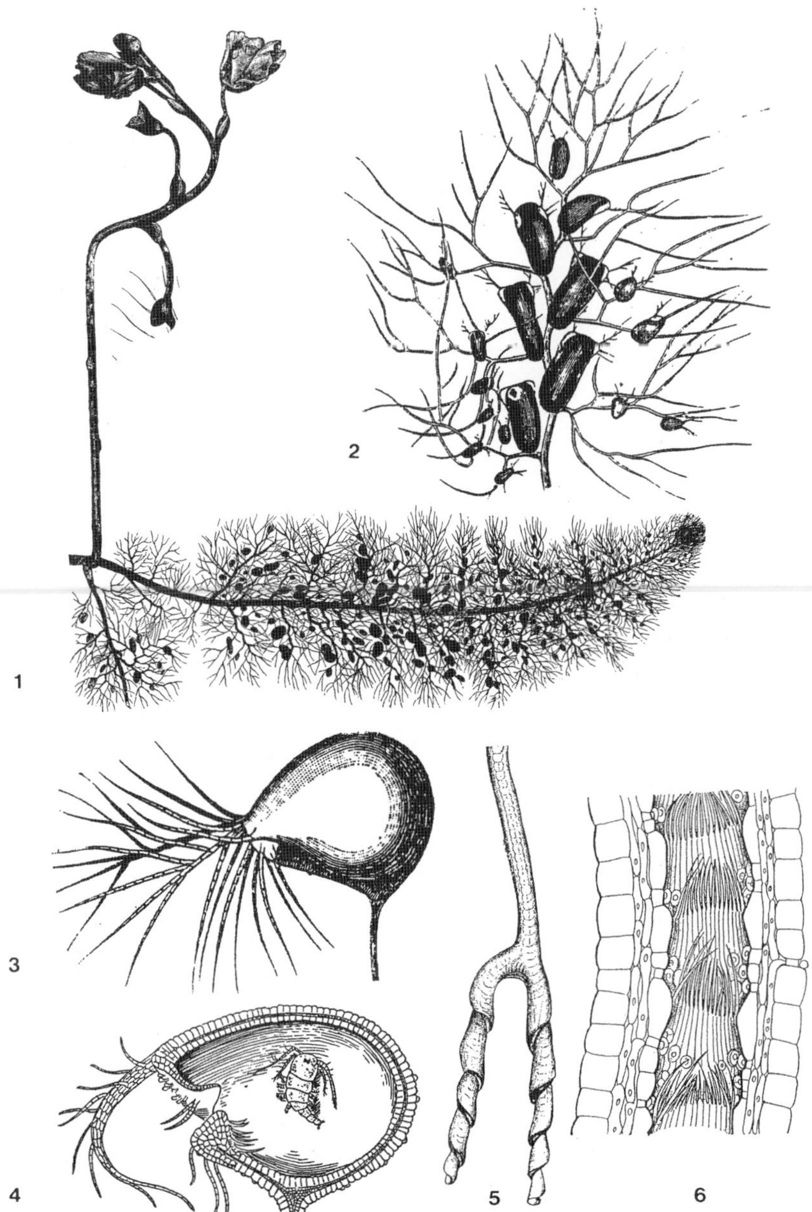

Figure 3.5 Extreme leaf modifications. (**1**) and (**2**) *Utricularia vulgaris*. Flowering plant and leaf fragment with bladders. (**3**) Isolated bladder showing long hairs at entrance. (**4**) The vesicle in cross section showing the mechanism that closes the opening and a trapped crustacean being digested. (**5**) and (**6**) *Genlisea ornata*. Leaf with its characteristic spiralized shape and its longitudinal section showing the internal distribution of the hairs.

Glands

Glands are independent structures covering the inner or outer part of the leaf in great numbers.

The upper part of the interior of the pitcher in *Nepenthes* has no glands and is waxy (about 1/3). The lower two thirds are covered by numerous microscopic glands which are more than 6,000 per cm^2. These glands have a double function. They not only secrete the digestive fluid but also absorb the products resulting from the digestion (Fig. 3.2).

In the Australian plant *Cephalotus* nectar glands are scattered over the entire surface of the tube excluding the rim, and they are also found on the external and internal surfaces of the lid.

On the walls of the digestive tube of *Genlisia* are many glands on the ventral and dorsal side. A second type of small glands occurs below the ridges. These have both digestive and absorptive functions.

Drosophyllum possesses stalked glands, which secrete a mucilaginous substance but also sessile glands which produce enzymes. Two-armed and four-armed glands are present within the bladder of *Utricularia*. Glands are empowered with movement as is the case in the tentacles of *Drosera*. Hence glands: (1) are microscopic or macroscopic, (2) can be of several types and shapes, (3) have four separate functions: secretion, digestion, absorption and mobility.

The glands are, like the hairs, not distributed at random. In most species nectar glands are located at the entrance of the funnel and digestive glands are present inside its lower part (e.g. *Sarracenia, Darlingtonia*).

Enzymes

No human stomach (or that of any other mammal) digests flesh without enzymes. The food is first processed by the enzyme amylase, excreted by the salivary glands, that begins to digest starch. The secretions in the stomach of vertebrates include hydrochloric acid and the enzymes: pepsin (splits proteins into amino acids), chitinase (hydrolyzes insect cuticules) and rennin (making milk susceptible to pepsin) (Wilson and Mittermeier 2009).

Proteases (also called proteinases) are proteolytic enzymes which cleave most proteins into small fragments. These are usually the digestive enzymes pepsin and trypsin. Other enzymes, such as phosphatases, catalyze the hydrolysis of esters and are specific for phosphate groups.

To start with, acid is also produced in the organs of carnivorous plants at the side of enzymes.

The digestive fluid of *Dionaea muscipula* contains a proteinase. This enzyme was secreted when the plant was fed small cubes of gelatin and its activity doubled over a 3-day period following feeding (Ryan 1973).

In *Drosophyllum* the secretion droplets held on the gland heads carry very active proteases, capable of digesting a hole in the emulsion of processed colour film within 30 minutes (Heslop-Harrison 1975). Two proteases are present, with molecular weights of 59,000 and 21,000, in *Nepenthes macferlanei*. The major enzyme turned out to be strikingly similar to animal pepsin. Lipase activity was also demonstrated (Tökés *et al.* 1974).

The acid proteases of bovine brain and other animal organs, including human liver, were compared with those of carnivorous plants (*Drosera* and *Nepenthes*) and found to have similar enzyme active sites. These plant and animal proteases were both inhibited by agent *DAN* (a methylester) in combination with cupric ions (Takahashi *et al.* 1974).

The presence of proteases, nucleases and phosphatases was demonstrated in *Sarracenia*. The expression of these enzymes was developmentally regulated during the pitcher opening stages (Gallie and Chang 1997).

Chitin is a major component of cell walls in insects. Some insects make up the bulk of the prey found in pitcher traps. Chitinase, found in *Nepenthes*, is the enzyme that digests chitin (Eilenberg *et al.* 2006).

Using fluorescence labelling, phosphatase activity was detected in digestive glands of 47 species of carnivorous plants (Plachno *et al.* 2006).

Digested animals are an additional source of nitrogen, phosphorus, sulfur, potassium and magnesium for carnivorous plants. Some can take up more than 50% of their sodium and phosphorus from their prey.

The digestive enzymes in plants are not only the same that participate in animal digestion but have the same active sites.

Movements

Carnivorous plants may use traps without or with active movements. The first type is found in *Drosophyllum, Darlingtonia, Sarracenia, Cephalotus* and other species. The second is mainly represented by *Pinguicula, Dionaea, Utricularia* and *Drosera*.

In the absence of movements the small animals become attached to the sticky leaves. But the addition of plant movements to the other functions, resulted in a most advanced activity. The edges of the leaves of *Pinguicula* gradually roll over the small insects that have adhered to their glands. This movement is fully coordinated with the arrival of an animal on the leaves. The speed of the movement may be impressive as in *Dionaea* where the leaves close within 0.5 of a second (Raven *et al.* 1999). The three sensitive hairs located on each leaf half are those that give the signal that an animal has landed on the surface. Even more remarkable is that, when the toothed edges are closed, the sharp teeth do not move against each other but get correctly interlocked blocking the escape of the animal — the organ gets fully closed.

The tentacles of *Drosera* move to trap the victim also in a coordinated way. The movement will start at a site depending on what part of the leaf the victim has landed. Either it will be held in the original position, the other tentacles moving over it; or the tentacles will bend inwards carrying the insect to a central position of the leaf where other tentacles will collaborate in the digestion. This movement is so well ordered that Slack (1979) compared it to that of the tentacles of sea anemones which also embrace their victims in a coherent wave-like fashion (Fig. 3.2).

Another type of directed movement is found in *Utricularia* where the door of the bladder opens and closes. *Polypompholyx* and *Aldrovanda* and *Dionaea* have also mobile trap systems.

The mechanism of the movements is at present explained by changes in cell water content, but this interpretation is considered inadequate by several authors.

Tendrils

Tendrils are thin and curled leaves that allow a plant to get attached to various supports. They are also part of the carnivorous organ in *Nepenthes* (Fig. 3.3). Within the same plant the pitchers can be located in the upper region or in the lower one close to the earth. Upper pitchers have the tendril attached at the back whereas lower pitchers have it at the front. The spirals in the tendril are absent in lower pitchers but are most pronounced in the upper pitchers.

The presence of tendrils in this carnivorous plant is significant. Tendrils are the main participants of movements in distant non-carnivorous plant families. Their association in this plant, with pitchers, shows how the formation of these organs is directly connected to a structure involved in movement. This situation contributes to elucidate the way carnivory and movement became coupled during evolution.

Covering Lids, Mobile Doors, Oriented Hairs and Transparent Windows are Incredible Structures that Convey Extreme Efficiency

In addition to the movements the most incredible refinements have been added to the main basic structures, resulting in an extreme efficiency in the trapping of prey.

During evolution not only both sides of a leaf have come together building a tube, or a pitcher, but the upper part of the leaf has bent itself backwards building a lid, which allows only a narrow entrance to the interior (*Nepenthes*) (Fig. 3.1). The lid is rich in nectar and it may be partly like a cap (*Sarracenia*). The lid is considered by botanists to diminish the amount of rainwater that would enter the digestion chamber diluting the digestive juices.

The lid in *Nepenthes* does not close upon the prey but in *Utricularia* the lid is transformed into a most ingeneous instrument. In this last species the trap has the shape of a bladder and its entrance is opened or closed by a door complete with a tripping device. Once a small aquatic creature brushes against the latter the trap is sprung — the door flies open, thus releasing a partial vacuum which sucks the victim inside, the door shuts imprisoning its prey, after which the vacuum is soon restored again and the trap is ready for further action (Slack 1979). Within 30 ms of tactile detection the door bursts open (Braam 2005). Trapdoors are also part of the bladder in the Australian species *Polypompholyx*. The carnivorous traps are a "place of no return".

The occurrence of hairs along the interior of the digesting tube is in itself a hindrance to the movements of the animal. But in a device developed in *Sarracenia* and *Genlisia*, there are within the pitcher tube, numerous hairs all oriented inwards not allowing the animal to move back to the entrance.

Transparent windows are another device, present in the vessels. In *Sarracenia*, the part of the pitcher opposite to the entrance is mottled with a series of white areolae or windows. In some species all sides are covered by these windows. Bright light comes from the windows. The insects see them as apparent exits, and make further escape attempts at other windows. The result is that the hairs become intercrossed building a "perfect valve" preventing any upward movement. Long and narrow, colorless, translucent windows are also present in *Cephalotus* which are similar to those of *Sarracenia* and *Darlingtonia*.

Extreme efficiency has been obtained by the addition of four types of devices.

DNA Evolution has Not Allowed Flowers to Become Carnivores

One could think that the different organs of a plant: flowers, leaves, stems and roots as well as fruits, would be able to develop structures and functions that would lead to carnivory.

Flowers, which have attractive colors and produce perfumes and seductive nectars, could easily have combined these functions with the digestion of insects that visit flowers constantly. Also many spiders hide between the petals to trap insects.

The spiders could as well become perfect victims since several species are flower dwellers. The digestion could be easily accomplished as flowers, such as the morning glory *Ipomoea purpurea*, turn out to have also proteolytic enzymes (Ryan 1973).

Flowers show even movements, which are an obligatory component of carnivory in several species. They open and close their petals at well defined hours of the day, as in *Calendula officinalis*, where they have a twelve-hour rhythm. Some open in the morning, whereas others in the evening (*Silene noctiflora*). This exact behavior allowed Linnaeus to construct a "flower clock" in his botanical garden (Guilliermond and Mangenot 1941, Denffer *et al.* 1971).

The large flowers of *Rafflesia* (up to 90 centimeters in diameter) exhale a putrid odor that attracts flies in great numbers and have a large central cavity where they could be trapped and digested, but this is not the case, *Rafflesia* is not carnivorous.

Stems are generally not involved in this process. Exceptions are *Byblis* and *Proboscidea* where the stems have glands. Also in the sepals, but not in the petals, of the flowers of *Drosophyllum* a few glands appear.

The absence of carnivory in flowers is usually not mentioned because it disagrees with the prevailing interpretation of evolution based on selection and random mutation.

Carnivory does not Occur in Roots Although they Produce Digestive Enzymes

Although it seems strange, roots could have developed carnivory to digest earthworms and other minute animals that they encounter as they penetrate into the soil. Corn (*Zea mays*) roots even produce enzymes, such as peptidases, like the leaves of carnivorous plants (Ryan 1973).

Yet, no carnivory is known in roots.

Fruits are Loaded with Digestive Enzymes but are Not Carnivorous

But the fruits? They could also become carnivorous because they may be filled with digestive enzymes. The papaya tree (*Carica papaya*), of tropical America, has a fleshy trunk, large palmately lobed leaves and flowers. The stem, leaves and flowers do not possess digestive enzymes but contain alkaloids. However, the fruits are loaded with papain. This is a proteolytic enzyme which is extracted in

1 2

Figure 3.6 Non-carnivorous plants with digestive enzymes but lacking modified leaves or flowers. (**1**) *Carica papaya*. Fruits of papaya tree. (**2**) *Ananas comosus*. Pineapple fruit.

crystalline form from the latex. Papain is 15 times more active than the crude product and is a protease with digestive as well as wound-healing properties (van Wyk and Wink 2004).

Pineapples (*Ananas comosus*) contain also digestive enzymes. Significant is that these are extracted not only from the fruits but also from the stem. They are a mixture of five proteolytic enzymes known as bromelain (bromelin *A* and *B* being the main compounds) (van Wyk and Wink 2004) (Fig. 3.6).

DNA evolution has not allowed flowers, fruits, roots or stems to become carnivorous although all of them produce digestive enzymes. Leaves were the main plant organ that was modified where the digestion of animals takes place.

Adhesive Glands Similar to those Found in Carnivorous Plants are Present Elsewhere without Being Accompanied by Carnivory

Plants covered by adhesive glands occur throughout the whole plant kingdom. These are capable of catching small insects but are non-carnivorous: three species of *Saxifraga*, a primrose (*Primula sinensis*), the common *Pelargonium zonale* and tobacco plants (*Nicotiana tabacum*); all bear sticky glands and belong to families that are not closely related.

Coordinated Movements also Occur in Orchids Where they Participate in Pollination

Again, the same functional solution that occurs in carnivorous plants, is present in a totally different plant family and participates in a different process.

The orchids, characterized by their complex flowers, are also able to produce coordinated movements.

The flowers of most plant species exhibit distinctly separated sexual organs but in orchids the anther and stigma are contained in one body, the column. One of the flower petals is also much larger than the others — it is called the lip. The lips turn out to be mechanical devices that trap the insects but only temporarily during the time they visit the flowers in search of pollen. The complex lips have the shape of a bucket filled with water (*Coryanthes maculatus*). "Petals and sepals grasp the pollinator and drop him into the bucket; he emerges from the flower through tunnel-like structures and is loaded with pollen". "The lip may also form a lid hinged with a claw-like appendage. When the flower opens the lip turns around and falls back, covering the column and enclosing the insect. The device remains closed while the insect is inside, but if the movement ceases, the lid soon opens again". "Many species within the genus *Pterostylis* have sensitive lips: when touched by the weight of even the smallest insect, the lip springs up to the column and imprisons the pollinator" (Kramer 1979).

There are sensitive hairs on the lip of the orchid *Bulbophyllum* that convey the information from a landing insect. This is followed by the immediate movement of the lip which is far from random since it is directed upwards. The lip is hinged like the door at the entrance of the bladder in the carnivorous plant *Utricularia*. Also like in this species the device closes and opens depending on the presence or absence of an entrapped animal.

The similarity of the whole process in orchids with the events in carnivory is impressive.

The Exquisite Pitcher Shapes that Leaves Assume in Carnivory are Found in Leaves and Flowers of Non-carnivorous Plants

There is another structural feature that is part of carnivory which exists independently in non-carnivorous species.

The transformation of the leaves into liquid containing vessels which may assume the forms of urns, tubes, pitchers or bladders is found in other plant families — some closely related others phylogenetically located faraway. *Dischidia Rafflesiana* (Asclepiadaceae) belongs to a totally different family than that of *Nepenthes* yet it has leaves of two different types: one with the regular shape common to most plants and another group of leaves that are nearly identical to those of the pitchers of *Nepenthes* (Fig. 3.7).

This similarity extends to three species of *Aristolochia* (Aristolochaceae) which are directly related to the family Nepenthaceae. In this case not the leaves but the flowers closely resemble the urns of *Sarracenia* and *Darlingtonia* (Sarraceniaceae) (Fig. 3.7).

Carnivorous Structures without Digestive Enzymes and Enzymes without Digestive Ability

Plants so closely related as belonging to the same family (Droseraceae) lack the ability to produce digestive enzymes although they have other accompanying carnivorous structures. *Roridula* has sticky leaves but no digestive enzymes, *Byblis* has no digestive enzymes but instead produces another enzyme: phosphatase.

Remarkable is that *Utricularia*, that has one of the most effective trapping organs, contains mainly acid phosphatases and only some protease. Phosphate uptake may be essential for many carnivorous plants according to Plachno *et al.* (2006).

Carnivory was Lost in Some Species or was Never Allowed to be Completed

According to McPherson (2010) *Triphyophyllum* has two closely related species which are non-carnivorous: *Dioncophyllum* and *Habropetalum* as a result of loss of carnivorous adaptation. Another case where carnivory was lost completely is *Ancistrocladus*. To these may be added *Ibicella* and *Proboscidea* (Martyniaceae) commonly supposed to be carnivorous, and the non-carnivorous *Philoxia* (Gratiolaceae) which has sticky leaves. The question arises whether these examples represent actual loss of carnivory or are cases in which the combination of most functions never was allowed to occur.

Figure 3.7 Non-carnivorous plants with modified leaves and flowers having shapes similar to those found in carnivory. (**1**) *Dischidia Rafflesiana*. (*a*) Plant with normal leaves (upper part) and others in the form of a pitcher. (*b*) Cross section of a pitcher. (**2**) *Aristolochia macrophylla*. Plant with flowers in the form of an urn and cross section of a flower. (**3**) *Aristolochia pallida*. Flower in the form of a pitcher with lock curved over entrance. (**4**) *Aristolochia rotunda*. Flower with large brown lock over the tube (left side of figure).

A Battery of Genes Decides the Fusion of Leaves — This is Accompanied by the Fusion of Flower Parts

Slack (1979) noticed that the carnivorous plants could be divided into two groups according to the arrangement of their flower parts. To the first group belonged: *Heliamphora*, *Sarracenia*, *Darlingtonia*, *Nepenthes*, *Cephalotus*, *Drosophyllum*, *Drosera*, *Dionaea* and *Aldrovanda*. In these 9 genera, which are closely related, the petals are separated from one another, the symmetry of the flower being radial.

The second group covers: *Genlisia*, *Pinguicula* and *Utricularia*. In these 3 genera the petals are not separate but are united to form a basal tube. The resulting symmetry is instead bilateral. This resemblance may seem superficial but it is not. The transformation of a flower into a tube form, as the result of the association of separate petals, is a well defined genetic event.

The genes that determine the sequential arrangement of the flower parts are well known. They are the homeotic genes which also decide the distribution of the body parts in insects and humans (Lu *et al.* 1996, Bender 2008). The transformation of bilateral symmetry to radial has been extensively studied in *Antirrhinum* and other plant species. These genes have been characterized in the utmost detail by Stubbe (1966) and Bradley *et al.* (1997).

Leaf fusion, that occurs in many carnivorous plants, leading to the formation of leaf tubes and pitchers, is directed by the same genes that determine the fusion of flower parts. The events occurring in the shoot meristem, when it develops into leaves and later into flowers, are orchestrated by a series of genes that have been isolated, cloned, sequenced, introduced into transgenic plants, and had their proteins isolated. A master gene is *LEAFY* (*LFY*) which ensures cell division within the developing embryo. Together with a battery of other genes, LEAFY decides both the formation of leaves and subsequently of flowers (Glover 2007).

Genes Decide the Emergence of Carnivory — Isolation and Characterization of its Genes and their Repression

Insects are the main victims of plant carnivory.

The pitchers of *Nepenthes khasiana* were used to identify plant chitinases during digestion. Chitinases are enzymes which hydrolyse insect chitin. Two different types of endochitinases were found in the liquid of closed traps. Moreover, injection of chitin into the closed sterile pitchers induced the appearance of additional endochitinase enzymes.

Four chitinase genes were isolated from the secretory region of the pitchers of this *Nepenthes* species. The evaluation of messenger RNA levels disclosed that the genes are expressed in the secretory cells. These results reveal, for the first time, the participation of genes encoding chitinases in prey-trap interaction and their differential expression during prey trapping (Eilenberg *et al.* 2006).

Proteins that inhibit proteolytic enzymes are found in virtually all animal tissues and fluids and have been the object of considerable research. On the other hand, plant proteinase inhibitors have only been researched recently. Kinetin (a plant growth hormone) was found to retard the production of proteinase in *Brassica* and oat leaves. Proteinase inhibitors accumulate in high concentrations in leaves. Immunological and immunoelectrophoretic analyses have shown that inhibitor proteins are synthesized in leaflets (Ryan 1973).

The isolation of specific genes responsible for enzyme production, the demonstration of their expression in pitcher plants as well as the occurrence of inhibition of proteolytic enzymes in leaves, enable a better understanding of the genetic events leading to the presence or absence of carnivory in closely related species where these genes may be expressed or repressed.

Within the Same Plant Carnivorous Leaves Emerge at the Side of Non-carnivorous Ones by Change in Gene Expression

The absence and the presence of carnivorous leaves within the same individual of several species of carnivorous plants is a key to the explanation of the occurrence of this phenomenon.

It demonstrates that the same DNA constitution can suddenly result in the exhibition of this property or inhibit its occurrence depending on gene expression. Carnivory emerges as an internal DNA event that appears "ready made" due to a change of gene expression during the organism's development. This is a well established phenomenon for many other properties. The differential expression of genes during development, called an epigenetic event, is a general phenomenon in plants (Matzke *et al.* 2015) and it is mainly due to RNA-directed methylation of DNA (Herbert 2004).

The examples are several:

Genlisia has non-carnivorous flat green leaves above ground and distinct subterranean carnivorous leaves which form corkscrew traps.

Triphyophyllum changes its leaf morphology over time and in response to habitat. The plant grows as a vine using hooks. Three distinct stages of its development are identified: (1) Juvenile, during which only non-carnivorous leaves are produced. (2) Carnivorous stage, characterized by the building of carnivorous leaves. (3) Mature stage, during which no carnivorous leaves are formed and instead flowers develop and the plant climbs. Significant is that in this plant carnivorous leaves are only produced occasionally (McPherson 2010).

Sarracenia has within the same individual plant elongated narrow leaves adapted to photosynthesis which are non-carnivorous (called phyllodia), and carnivorous funnel like leaves building a pitcher.

Nepenthes has also two types of leaves: those that are without pitchers and leaves with pitchers (Fig. 3.3).

Cephalotus produces two distinct types of leaves: ovate, flat, non-carnivorous ones, as well as hollow carnivorous pitchers. Occasionally most plants of this genus produce leaves of intermediate morphology that partly resemble the carnivorous foliage (McPherson 2010). This third type is relevant for it shows the partial expression of gene action in determining carnivory (Fig. 3.3).

Drosera represents also a valuable example of differential gene expression. *Drosera* covers 90 species, most of them produce solely carnivorous leaves. However, two species: *D. aff. arcturi* and *D. caduca* produce non-glandular, non-carnivorous leaves. Besides, *D. regia* has linear elongated leaves whereas *D. rotundifolia* has circular ones. In both cases they are covered with glands.

Pinguicula. From September to May the tropical plants produce flowers and then rapidly form a succession of large, glandular carnivorous leaves. As water becomes scarce (May to July) the carnivorous leaves die out and only small non-carnivorous foliage is produced. In temperate and tropical areas other species that resist periods of cold have only a single leaf type.

Utricularia has diversified into 228 species distributed worldwide which are both aquatic and terrestrial. Yet all species produce photosynthetic green leaves as well as highly modified carnivorous traps.

This means that: (1) Not less than 8 genera produce leaves of different types within the same individual plant. (2) The phenomenon extends over several hundred species (*Utricularia*). (3) Typical of genes showing an epigenetic behavior is the finding that the occurrence of carnivorous and non-carnivorous leaves is a product of the developmental process (*Triphyophyllum*, *Pinguicula*). (4) Different species within the same genus (*Drosera*) may have only one type of leaf or have two types.

Specific Genes are Responsible for the Directed Movements of Tendrils

Tendrils are modified leaves or stems. They occur in a wide number of plant families that are not directly related: Basellaceae (Climbing vines), Vitaceae

(Climbing shrubs), Tropaeolaceae (Climbing herbs) and others. Among carnivorous plants *Nepenthes* shows tendrils.

The movements of tendrils are: (1) directed to a support, (2) result from a reaction triggered by contact with a specific structure, (3) move in helices that are both left-handed and right-handed changing direction in the middle of the tendril, (4) adapt themselves to the shape and thickness of the support.

The best studied case is *Bryonia*. Its tendrils are coiled starting "a remarkable 'searching' movement, the tip describing a circle or an ellipse, and the whole tendril a conical surface. The axis of this cone is at first obliquely upwards, but later it may sink and fall below the horizontal." This movement is common to many climbing plants (Denffer *et al.* 1971). As a result of this circular movement the tendril strikes the support becoming concave at the site of contact. It embraces the support adapting itself to its shape by wrapping itself several times. The torsion of the tendril functions also like a spring adjusting to its distance from the support (Fig. 3.8).

If the tendril is touched by a matchstick, for only an instant, a concavity develops at the site of contact. The reaction takes less than 30 seconds. The concavity results from an extremely rapid growth of the side of the tendril opposite to that touched. Such growth would take at least 24 hours in normal cell extension.

Touching or striking with a glass rod that has been coated with damp gelatin is without effect; but curvature begins immediately if a thread of wool, weighing no more than 0.00025 mg, and incapable of being detected by human touch, is drawn along a sensitive tendril. There is a rapid conduction of the stimulus along the whole tendril — the speed being of the order of 4 mm/min, which is tenfold the conduction in phototropism (Denffer *et al.* 1971).

The movements of tendrils, of the leaves of *Mimosa pudica* and of carnivorous plants, have now been investigated at the molecular level. It is not surprising that they are found to be directed by the same agents since they have many features in common (Fig. 3.9).

The induction of tendril coiling by touch in *Bryonia* is accompanied by increased levels of octadecanoids. These function as a signal to the accumulation of the auxin indole-3-acetic acid (IAA) which is known to be an active inducer of tendril coiling. The central role of this molecular process in plant movements is well established (Friml 2003).

A number of inter- and intracellular signaling components, including hormones and potential second messengers, have been implicated in touch-induced alterations in plant development. Intracellular calcium (Ca^{2+}) has long been implicated as an important second messenger in mechano-signaling and response in both animal and plant cells. Transgenic plants expressing the jellyfish gene encoding aequorin, a Ca^{2+}-dependent protein, have demonstrated a rapid intracellular increase in Ca^{2+} in response to touch stimulus (Knight *et al.* 1991).

Control experiments in the plant *Arabidopsis* revealed the existence of touch-inducible genes called *TCH* genes. They were isolated by differential complementary DNA library screening. They are strongly and rapidly up-regulated in

Figure 3.8 Recurrence of prehensile organs occurring from plants to apes. (**1**) **Plants**. *Bryonia dioica*, Cucurbitaceae (Pumpkin family). Flowering plant with numerous tendrils attached to a stick and four successive stages of the growth of its tendrils around a stick. (**2**) **Molluscs**. *Octopus macropus*, Cephalopoda. (**3**) **Fishes**. *Hippocampus hippocampus*. Seahorse. (**4**) **Reptiles**. *Chamaeleon vulgaris*. Chameleon. (**5**) **Marsupials**. *Didelphis marsupialis*. Northern Black-eared opossum. (**6**) **Placentals**. *Alouatta* sp. Howler monkey from South America.

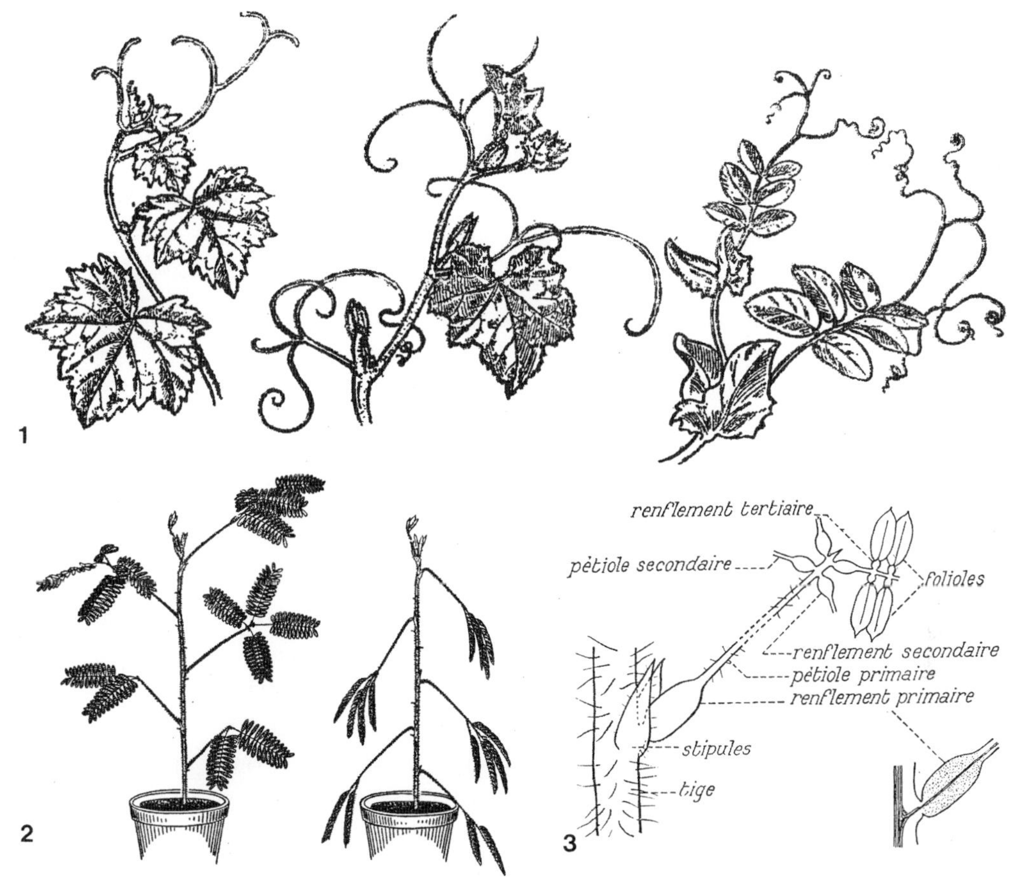

Figure 3.9 Plant movements. (**1**) Tendrils have different origins depending on the species. From left to right: Tendrils of grape vine *Vitis vinifera* (Vitaceae) which develop from side branches; tendrils of *Cucurbita* sp., a pumpkin, which grow from the leaves; tendrils of a pea plant (Leguminosae) which grow from the terminal part of a leaf.(**2**) *Mimosa pudica*, Leguminosae. Left: Plant with its leaves normally disposed. Right: The same plant following a slight touch. The leaves fall down and the leaflets close. (**3**) *Mimosa pudica*. The main morphological features of a leaf, showing the location of the primary, secondary and tertiary pulvini.

expression as in response to touch (Braam and Davis 1990). In the last two decades many other genes have been discovered, 2.5% of *Arabidopsis* genes are expressed in touch-stimulated plants (Braam 2005).

The Recurrence of Prehensile Organs from Plants to Apes

Prehensile tails in fishes, reptiles and marsupials may be seen as accidental events without relationship. Still worse, their comparison with the helical tendrils of plants and the prehensile tail of placental mammals, so close to humans as the apes, seems even less acceptable. However, there are no fortuitous events in nature, every genetic trait is the product of a well defined molecular cascade originating and directed by well defined segments of DNA.

A few decades ago similar structural and functional events, in plants and humans, would seem to be beyond any comparison, as they were considered fully unrelated. The genetic scenario has changed radically. Who could think that the genes that direct the formation and segmentation of our vertebral column would be the same that decide the formation and segmentation of the floral organs into sepals, petals, anthers and stigma? These are the homeotic genes that have been localized in chromosomes and DNA sequenced (McGinnis *et al.* a and b 1984, Lu *et al.* 1996).

This homology was preceded by the establishment of other evidence on their genetic similarity. The fusion between plant and animal cells, by means of Sendai viruses, and accompanied by radioactive markers, disclosed that human chromosomes divided in the plant cytoplasm (Lima-de-Faria *et al.* 1977, 1983).

Another source of evidence had also been collected earlier when the distribution of the ribosomal RNA genes along the chromosomes was found to be the same in plants and humans. The ribosomal genes, were found to be located in over 506 species close to the telomere DNA sequences. Their location was defined by a mathematical equation (Lima-de-Faria 1973). Thus, the results from molecular cytogenetics have placed humans in a more modest position and as an integral part of the other living organisms, being the simple invertebrates or plants.

Prehensile organs are found in invertebrates (*Octopus*), in fishes (sea horses) and reptiles (chameleons). They roll their arms (*Octopus*) or their tails (sea horse) firmly around a prey or a support respectively. The spider monkeys (*Ateles*) are so called because they use their tail as an extra arm which allows them to attach to tree branches making them the acrobats of the forest canopy.

An additional resemblance is displayed by the marsupial *Didelphis marsupialis*. It rolls its tail firmly round tree branches like the placental ape *Ateles belzebuth* (Fig. 3.8).

The molecular events behind the prehensile movements of these animal organs are not yet known because they have not been before considered relevant. It is thus expected that molecular mechanisms similar to those found in tendrils participate in the prehensile capacity of the tails of fishes, reptiles and apes.

Protection Against Own Digesting Enzymes and Protection Against Own Venom are Directly Related

How is it that the tissues of the traps of carnivorous plants are not digested by the enzymes that accumulate in concentrated form in their pitchers? How is it that the human stomach is not also digested by the corroding hydrochloric acid and the enzymes that it uses in the digestion of food? This last question was asked and explained in 1972 by Davenport in an article entitled: "Why the Stomach Does Not Digest Itself." As he pointed out this organ secretes an acid that at the concentration present can dissolve metals such as zinc and kill living cells. The stomach walls resist the attacks of the enzyme pepsin and of the acid as if they were "made of porcelain" as Claude Bernhard (1813–1878) already observed in his pioneer work of experimental medicine and physiological chemistry. The corrosive juice is prevented from attacking the stomach tissues by a complex physical-chemical barrier.

Like in carnivorous plants movements participate in the process. The digestion is accompanied by vigorous contraction movements occurring in the strong muscular wall of the stomach pouch. The lining of the interior of this organ is a glandular mucosa. The barrier to digestion lies mainly in the columnar epthelial cells of the mucosa. They block the passage of hydrogen atoms (originating from the hydrochloric acid, HCl) and facilitate the passage of sodium ions (from various sources) out of the mucosa.

A comparable protective situation occurs in carnivorous and non-carnivorous plants. *Drosera* plants encircle insects with their tentacles loaded with juices (Fig. 3.2), and digest them with great efficiency. The epithelial cells of the plant tissues face the same chemical attack encountered by the epithelial cells of the stomach. Cellulose is many times accompanied by lignin and wax in the cells of leaves. These may function as a protective system against the penetration of the enzymes into their own cells. But this may be only a major protective barrier, other processes may accompany these structures as indicated by the study of *Colchicum* plants. In them the toxic alkaloid colchicine is produced in the interior of the cells and acts in their cytoplasm directly on the chromosomes. *Colchicum automnale* is wild in Europe. The cells of this non-carnivorous plant produce colchicine in concentrations of 0.3 to 1.2% of the dry weight of the stem, flowers or seeds.

Cytologists found that colchicine interfered with cell division, inhibiting the normal separation of chromosomes, by disrupting their attachment to the tubulin molecules of the spindle. The result was that the daughter chromosomes remained within the same cell, duplicating its chromosome number. Although the colchicine concentration is high, and it is active within the cell, *Colchicum* does not double its chromosome numbers during thousands of cell divisions that are part of the development of the plant. Moreover, generation after generation the wild plants have maintained their 38 chromosomes, a number which is equal to that of their close relatives that are not known to produce this alkaloid (Fig. 3.10).

These results lead into the development of venom defense in vertebrates ranging from amphibians and reptiles to mammals. The relationship is not fortuitous since the biochemical origins of venoms have been traced to digestive plant and animal juices. Venoms participate in the predigestion of the prey (Halliday and Adler 2004).

Dendrobates auratus, a brilliantly coloured, tropical American frog, is one of the most poisonous amphibians. The poison is considered to be a defense against predators (Burnie 2004). Other frogs and toads are known to produce poisons from glands in their skin. The Cane toad can squirt toxic secretions from its parotoid glands up to 1 meter into its predators. *Dendrobates* and other American frogs secrete one of the most lethal poisons known — *batrachotoxin*. A mere 0.00001 g is enough to kill a human (Halliday and Adler 2004). Yet these frogs do not kill each other although their skin contact is most intimate during mating between males and females.

This situation recurs in reptiles. Approximately 200 genera of snakes have specialized glands that produce toxic venoms. The glands of viperids produce powerful destroying enzymes notably phospholipase A_2. Within 24 hours the venom of a rattlesnake digests all the organs of its prey. However, the rattlesnakes do not kill each other. They are immune to their own venom, containing antibodies against various venom components. The same is the case with the poisonous frogs where the sodium channels of their nerve membranes are protected from the action of batrachotoxin, otherwise they would be immediately paralyzed (Mebs 2002) (Fig. 3.10).

As one reaches the mammals, unrelated groups of animals have developed during their evolution defense against snake venom. Indian gray mongooses are renowned for their ability to kill snakes. They turned out to be unaffected by snake venom. This is alpha-neurotoxin which works by attaching itself to acetylcholine receptor molecules on the surface of muscle cells. The neurotoxin paralyzes and kills the victim, but in mongooses and in snakes the receptors are shaped in a way that the venom cannot attach to them (Hedges 1997).

European moles, hedgehogs, pygmy shrews and American squirrels have all developed antisnake defenses (Poran and Coss 1990) (Fig. 3.11).

Hence, we are approaching a position where the mechanisms responsible for protection against digestive enzymes in plants will be better elucidated. Not only protective structural details participate, but immune processes appear as an obligatory component in this process be it in animals or plants.

Coherence is Explicit in Carnivory. Plant-animal Similarities

The coherence is so explicit that can hardly be overlooked. It becomes most evident when the structures that occur in animals and plants are found not only to

Figure 3.10 Defense in plants and animals. (**1**) **Effect of colchicine**. Difference between normal chromosome division and division in colchicine treated cells. Instead of cells with 4 chromosomes, cells with the double number are formed. The chromosomes become unable to move to the spindle poles. (**2**) *Colchicum autumnale*. The flowers, corm or seeds contain up to 1.2% of colchicine (of dry weight). Yet no disturbance occurs in cell divisions. (**3**) **Poisonous frog**, *Dendrobates auratus*. Poisonous frogs

Figure 3.10 (*Continued*) protect themselves from Batrachotoxin. (**4**) **Highly venomous snakes**, *Naja sp*. Venomous snakes are immune to their own venom. (**5**) *Nepenthes sp*. Cells of carnivorous plants are not digested by own enzymes. (**6**) *Homo sapiens*. The human stomach cells are not digested by the enzymes produced by the stomach.

Figure 3.11 Defense in animals. (**1**) **Indian gray mongoose**, *Herpestes edwardsii*. Unaffected by snake venom. (**2**) **Mole**, *Talpa europaea*. Protected from snake venom. (**3**) **Hedgehog**, *Erinaceus europaeus*. Protected from snake venom. (**4**) **Pygmy shrew**. Shrews are protected from snake venom. (**5**) **Squirrel**, *Spermophilus lateralis*. Squirrels developed antisnake defense.

be similar but also the functions are equally located and activated according to a similar time sequence.

1 The presence of prey in vertebrates, including humans is recognized by the organs of smell. Their sensorial epithelial cells respond to molecules from the

environment (Abercrombie *et al.* 1990). Prey in plants is recognized by epithelial cells located on sensorial hairs that are at the entrance of the pitcher or urn.

2 As the prey is recognized the mouth of animals opens widely engulfing the victim and closing subsequently. In plants the door of the trap opens widely when the victim approaches and, as this is caught inside, it closes rapidly (*Utricularia*).

3 The vertebrate stomach is a large urn with an upper and a lower cavity (called the antrum). It is in the antrum that the food is mixed with digestive enzymes (Davenport 1972). Equally in plants the digestion also takes place in the lower part of the pitcher. The upper 1/3 does not contain glands, but thousands of them cover the inner walls in the lower 2/3 of the vessel (*Nepenthes*).

4 Digestion in both animals and plants includes the same enzymes but also the presence of acid which breaks the covalent bonds of ingested chemicals.

5 The nutrient materials are assimilated by the cells of the intestine. In carnivorous plants, it has been demonstrated, by using *Drosophila* flies labeled with radioisotopes, that the cells of the sticky leaves assimilate as well the nutrients (*Roridula*) (Ellis and Midgley 1996).

6 The debris resulting from animal digestion are expelled by the anal opening or may be regorged. Plants lack alimentary canals, but they have cells that can digest contained material (e.g. polysaccharides, lipids). The lysosomes present in plant cells also contribute to the digestion of foreign bodies (Abercrombie *et al.* 1990).

7 Movements are an obligatory part of the alimentary canal in vertebrates involving the mouth, tongue, throat, stomach, and intestines following a well established time sequence. In plants the glands encircle and press the victim in an ordered way (*Drosera*). The two lobes of the leaves of *Dionaea* correctly intertwine their spikes closing the vessel completely and as they move the victim is crushed.

There is nothing mystic about coherence. This phenomenon is at present understood as a result of atomic and molecular assembly. Every molecule or structure in the cell results from the self-assembly of well known atoms. The cell is built from inside out in which every atom has been tested against each other. Only the coherent electronic structures have been accepted initially and in future molecular reactions. These coherent interactions have resulted in the cell having its own repair mechanisms (Robertson *et al.* 2009, Fasken and Corbett 2005, Ogé *et al.* 2008), regulating signals (Terry *et al.* 2007) and even the ability to measure of the size of its own organelles (Chan and Marshall 2012).

The coherence is also evident in the non-random distribution of glands and hairs. The nectar glands that attract insects are located close to the entrance of the pitchers, whereas the digestive glands that consume the victim line the interior walls of the vessels. The same is the case with the hairs. The sensitive ones that release movements are also located near the entrance doors or other strategic sites. Other types of hairs are also conveniently located, these impede the escape of the prey by filling the interior of the trap and being directed downwards.

Carnivory — An Event Decided Mainly by DNA Transformations

The evidence gathered above may be summarized as follows:

1. Carnivory is not an "intriguing" phenomenon, but is genetically determined.
2. The fusion of leaves — a primary event in carnivory — is determined by a battery of genes that have been isolated, cloned and sequenced.
3. Another primary event is prey digestion. Four chitinase genes were isolated from the secretory region of pitchers and were proven to be expressed in the secretory cells.
4. Expression is also accompanied by repression. Inhibition of genes determining digestive enzymes takes place in leaves.
5. Movements of tendrils, which are also part of carnivory, are decided by touch-inducible genes. As much as 2.5% of the genes of the plant *Arabidopsis* are expressed in touch-stimulated plants.
6. Carnivory is not only a genetic phenomenon but also an epigenetic one. All cells of an organism carry the same DNA but the formation of different structures or functions, within the same individual, is called an epigenetic event. This is due to the modification of its own DNA or to transformations of the proteins that DNA produces during the development of the organism (Felsenfeld 2014). Within the same individual plant the leaf morphology changes during development resulting in the production side by side of non-carnivorous and carnivorous leaves. This epigenetic event occurs in individual plants of eight different genera (Fig. 3.3).
7. This means that carnivory appears as a "ready made" event decided mainly by atomic interactions occurring within DNA and is not necessarily the product of mutations taking place over longer periods of time.
8. Evolution has not allowed flowers to become carnivorous although their colors, perfumes and nectars are highly seductive to insects. Carnivory is a phenomenon limited to leaves with only a few species in which sepals and stems have glands.
9. The successive modification of leaves resulted in highly modified trapping organs. These assumed quite different shapes from spiral branches to funnels, pitchers and bladders.

10. Sensorial hairs were added to these traps that convey molecular signals resulting in leaf movements.
11. Both nectar glands that attract insects and digestive glands that both digest and absorb nutrients were also added.
12. Most enzymes present in plant vessels are strikingly similar to those participating in animal digestion.
13. Movements are an obligatory component in several genera attaining an impressive speed.
14. The distribution of the hairs and of the glands is non-random occupying positions where their efficiency is most evident. The movements are not either chaotic but are fully coordinated with the other processes.
15. Extreme efficiency is obtained by the inclusion of covering lids, mobile doors and transparent windows.
16. The exquisite shapes that leaves assume in carnivory are found in leaves and flowers of non-carnivorous plants. Besides, some plant genera have sticky leaves but no digestive enzymes whereas others have digestive enzymes but no sticky leaves.
17. Some genera have only certain carnivorous traits. Moreover, carnivory was lost in some species or was never allowed to become complete.

The Periodicity of Plant Carnivory

Periodicity is revealed by the following:

1. Carnivory in plants is a recent invention. The flowering plants arose late in evolution and carnivory arose even later, being present in only recently derived groups.
2. Carnivory evolved independently on at least ten separate occasions. The present living genera belong to 12 different families of flowering plants.
3. There has been an explosive evolution in some genera, which contain over 150 species displaying carnivory, whereas other genera are represented by a single species (Fig. 3.4).
4. Not a single trait, but a coherent package of structures and functions reappears in the different families resulting in extreme efficiency.
5. The reappearance occurs in related, but equally in unrelated families.
6. The package of structures and functions starts with simple forms but these become successively highly complex in other species.

The Construction of the Periodic Table of Plant Carnivory

The plant genera included in the Columns of the Periodic Table cover the best studied species found among the currently recognized carnivorous plants of the world. Genera, where traits are a source of debate, were left out.

Each vertical column contains the genera that exhibit the same basic property namely: (1) Leaf transformation. (2) Gland secretion. (3) Sensitive hairs. (4) Digestive enzymes. (5) Directed movements (Table 3.1).

The genera are located along the column, starting at the top, with those that exhibit the simplest traits and finishing with those having the highest complexity.

Column 1 Leaf Transformations

Drosophyllum — Leaves straight and slightly curved at top. Family Drosophyllaceae. Europe (Iberian peninsula and Morocco). *Pinguicula* — Leaves slightly curved. Family Lentibulariaceae. Eurasia and Americas. *Drosera* — Leaves linear and others slightly curved in *D. regia*. Other species have flat circular leaves (*D. rotundifolia*). Family Droseraceae. Worldwide distribution. *Heliamphora* — Leaves like a funnel with a small lid. Family Sarraceniaceae. Tropical South America. *Sarracenia* — Leaves forming funnel-shaped tube. Pitchers with a lid-like wood which curls slightly backwards. Family Sarraceniaceae. North America. *Darlingtonia* — Leaves form long narrow pitchers which are curved funnels covered by a lid with the shape of a dome. Family Sarraceniaceae. North America. *Cephalotus* — Leaves forming a pitcher. The mouth has a well developed rim with three ridges. There is a lid that overhangs the mouth. Family Cephalotaceae. Australia. *Nepenthes* — Two types of pitchers. Upper and lower pitchers vary markedly in shape some being cylindrical and bulbous with a large lid. Family Nepenthaceae. Borneo, Australia, Madagascar. *Aldrovanda* — Leaves building bilobed trap with pointed bristles. Family Droseraceae. Eurasia, Australia, Africa. Worldwide. *Dionaea* — Leaves formed by kidney-shaped halves which come together like a trap. Family Droseraceae. North America. *Genlisia* — Trap leaves consist of spirally twisted branches with claw like cells. Family Lentibulariaceae. South and Central America, Africa. *Polypompholyx* — The trap is a bladder with a wing and a door. Family Lentibulariaceae. Australia. *Utricularia* — The traps are bladders consisting of a hollow bag at the end of a stalk. The entrance is sealed by a valve and the door opens and closes sucking the prey. Family Lentibulariaceae. Australia, North and South America, Europe, Asia.

Table 3.1 Periodic Table of Plant Carnivory

Increasing grade of complexity	LEAF TRANS-FORMATIONS	GLAND SECRETION	SENSITIVE HAIRS	DIGESTIVE ENZYMES	DIRECTED MOVEMENTS
1	*Drosophyllum* *Pinguicula* *Drosera* Leaves slightly curved	*Dionaea* *Genlisia* Microscopic and small glands	*Drosophyllum* *Drosera* No hairs	*Darlingtonia* No enzymes	*Drosophyllum* *Heliamphora* *Sarracenia* No movements
2	*Heliamphora* *Sarracenia* *Darlingtonia* *Cephalotus* Leaves forming tube with a lid	*Cephalotus* *Nepenthes* Nectar and diges-tive sessile glands	*Nepenthes* *Cephalotus* Hairs outside pitcher	*Utricularia* Some proteases plus phosphatases	*Darlingtonia* *Nepenthes* *Cephalotous* *Genlisia* No movements
3	*Nepenthes* Leaves forming two types of pitchers	*Pinguicula* *Drosophyllum* Mucous and sessile digestive glands	*Polypompholyx* *Utricularia* Hairs at entrance of bladder	*Aldrovanda* Proteases	*Pinguicula* Leaf rolls over prey *Drosera* Tentacles move
4	*Aldrovanda* *Dionaea* Leaves bilobed building trap	*Drosera* Mucous tentacles and digestive glands	*Dionaea* *Aldrovanda* Spines and trigger hairs	*Drosera* *Drosophyllum* *Dionaea* Proteases plus phosphatases	*Polypompholyx* Trap door opens and closes
5	*Genlisia* Leaves as spi-ralled branches	*Polypompholyx* Two-armed and four-armed glands	*Darlingtonia* *Sarracenia* Downward point-ing hairs	*Sarracenia* Proteases nucleases and phosphatases	*Dionaea* Trap closes within 0.5 seconds *Aldrovanda* Trap closes within 0.01 seconds
6	*Polypompholyx* *Utricularia* Leaves building traps as a bladder	*Utricularia* Three types of large glands	*Genlisia* Complex system of hairs forming valves	*Pinguicula* *Nepenthes* Proteases lipases chitinases and others	*Utricularia* Trap door closes in 30 milliseconds

This property is divided into the 5 main traits that result in the ability that plants have to trap and digest animal prey. Remarkable is that all the 5 traits start with simple structures (or their absence) and successively the complexity increases leading to extreme forms of exquisite refinements. These traits are decided by specific genes (see text). Their evolution in parallel follows a well known genetic event characterized by genes with similar functions being closely associated along chromosomes. Carnivory occurs solely in flowering plants. These arrived late in evolution and most species have about the same degree of organization. The increase in complexity refers in this case to the trait exhibited by the plant. The periodicity is evidenced, among other features, by the same trait reemerging in unrelated families. Besides each trait starts with a simple structure or function that becomes successively more complex in species with similar body complexity since all are flowering plants (Original).

Column 2 Gland Secretion

Dionaea — Microscopic digestive glands over entire surface of trap. Near marginal teeth are nectar glands. *Genlisia* — On the walls of the bulbs there are glands crowded on ridges. There is in addition a smaller type of gland. *Cephalotus* — Nectar glands on collar of pitcher and digestive glands inside pitcher. *Nepenthes* — Lid with nectar-secreting glands and digestive zone of pitcher covered with small glands. *Pinguicula* — Numerous stalked mucus-secreting glands and digestive sessile. *Drosophyllum* — Numerous stalked mucus producing glands and sessile glands. *Drosera* — Sessile digestive glands and stalked mucus glands building tentacles. *Polypompholyx* — Bladder trap with two types of glands: two-armed glands and four-armed glands. *Utricularia* — Large glands of different types. There are two-armed glands and four-armed glands with digestive functions. In addition globe-shaped glands and glandular hairs occur in the bladder.

Column 3 Sensitive Hairs

Drosophyllum — No hairs. *Drosera* — No hairs. *Nepenthes* — No hairs inside pitcher. The wings of the pitchers are fringed with hairs or bristle-like teeth. Teeth of the rim are sharp. *Cephalotus* — No hairs inside pitcher. Frontal ridge of pitcher with hairs. *Polypompholyx* — The trap contains stiff pointed hairs close to the wing. Interior of bladder is also covered with hairs at entrance. *Utricularia* — Trigger hairs at entrance of trap and glandular hairs on both sides. *Dionaea* — Stiff spine-like teeth on lobes. Inner surface of each lobe with sensitive trigger hairs. *Aldrovanda* — Each leaf lobe bears pointed sword-like teeth. Has also microscopic trigger hairs and four-armed hairs with mucus. *Darlingtonia* — Large number of downward pointing hairs inside pitcher. *Sarracenia* — Hairs form valve which only allows advance in one direction. *Genlisia* — The entrance of tube has tiny claw-like cells. Beyond them are rows of sharp inward-pointing hairs. The rows consist of as many as 50 hairs. They form valves that exclude movement of animals in opposite direction. The lips possess claw-like processes followed by rows of downpointing hairs similar to those of the tube.

Column 4 Digestive Enzymes

Darlingtonia — No enzymes. *Utricularia* — Only some proteases but presence of phosphatases. *Aldrovanda* — Proteases. *Drosera* — Proteases plus phosphatases. *Drosophyllum* — Proteases plus phosphatases. *Dionaea* — Proteases plus phosphatases. *Sarracenia* — Proteases, nucleases and phosphatases. *Pinguicula* — Proteases, phosphatases, amylases, esterases and ribonucleases (Roccia *et al.* 2016). *Nepenthes* — Proteases, lipases and chitinases.

Column 5 Directed Movements

None of the following genera display movements:
Drosophyllum, Byblis, Heliamphora, Sarracenia, Darlingtonia, Nepenthes (no movements of pitcher but presence of tendrils), *Cephalotus, Genlisia.*

Pinguicula — Margins of leaf role over prey. *Drosera* — Leaf blade rolls over victim. Tentacles bend inwards. *Polypompholyx* — Trap door opens and closes efficiently. *Dionaea* — Two or three stimuli are needed before the trap will close but it is extremely rapid (0.5 of a second) (Raven *et al.* 1999). *Aldrovanda* — One touch of trigger hair will suffice to close trap. It is considered one of the the fastest moving structures in the plant kingdom (0.01 seconds) (McPherson 2010). *Utricularia* — Trap door opens and closes sucking in the victim in 30 milliseconds (Braam 2005).

Chapter 4

Luminescence Occurs from Minerals to Fish but not Beyond. It is Both an Electronic and a Genetic Event

Luminescence is an Electronic Process Resulting from Changes in Atomic Energy States

Luminescence does not start at the biological level but is evident in minerals which have no DNA or RNA. Besides, it goes still deeper. Only a restricted number of specific chemical elements in the minerals is responsible for the emission of light. This process can be traced to still lower levels reaching the elementary particles. It is the electron which is the main actor in luminescence.

To start with, the ability of a compound to absorb photons (the quanta of light) depends on its atomic structure, particularly on the arrangement of electrons surrounding its atomic nuclei. When a photon strikes an atom, or molecule, capable of absorbing light at a given wavelength, energy is absorbed by some of the electrons which are thus boosted to higher energy levels. The atom, or the molecule, reach an "excited state" (energy rich) and can return to a "ground state" (low-energy), with the simultaneous emission of energy in the form of light. It is this emission of light by excited molecules that is called *fluorescence* (Lehninger 1975).

Fluorescence and *phosphorescence* differ only in the amount of time it takes for electrons to return to their ground states. Fluorescence takes place in fractions of a second, in phosphorescence materials continue emitting light for hours after the exciting radiation has been turned on. The term fluorescence derives originally from the finding that the mineral fluorite, when irradiated with ultraviolet radiation re-emits light in the visible range.

Chemical reactions, radioactive decay and electric currents can likewise lead to electronic energy transitions resulting in the emission of light. These, together with fluorescence (caused by the absorption of light) are different forms of *luminescence*.

In living organisms, that emit light (bioluminescence) enzymatic oxidoreductions take place in which a free-energy change is utilized to excite a molecule to a high-energy state. This is followed by the emission of light as the molecule returns to its ground state.

The chemical reaction involved in bioluminescence consists of two components a protein, luciferin and an enzyme, luciferase (molecular weight circa 100,000). Luciferin reacts with *ATP* (adenosine triphosphate) forming an intermediate compound that bounds to the catalytic site of luciferase. When in this form, the enzyme is exposed to free oxygen and it is oxidized, emitting light when its electrons return to the ground state. One quantum of light is emitted for each molecule of luciferin oxidized (Lehninger 1975).

Luminescence was Widespread in Minerals Before it Arose in Living Organisms and it was Found Even Earlier in Chemical Elements

Initially luminescence among minerals was found only in fluorite. The finding was considered so exceptional that it led to the creation of the term fluorescence. Soon other minerals turned out to re-emit light when previously irradiated with ultraviolet.

Not less than 21 minerals are luminescent, emitting light of different colors, depending on the chemical element present in the mineral that functions as the energy activator. In scheelite ($CaWO_4$) it is the atom W (tungsten), which is a regular component of the molecular structure of this mineral, that is the source of luminescence. Uvarovite ($Ca_3Cr_2Si_3O_{12}$) contains the atom Cr (chromium) responsible for the emission of red-colored light. Most colors of the rainbow occur: red, orange, yellow, green and blue (Fig. 4.1).

What is remarkable is that the chemical elements that appear in these minerals are not of all kinds. There is order already at this stage. These trace elements have been considered to be "defects" or "mistakes", but at present there is a simple atomic explanation (Wenk and Bulakh 2004). Crystallization is a highly dynamic process in which atoms not only assemble at tremendous speeds but are obliged to change their relative positions during their movements. At high temperatures minerals are structurally disordered and chemically homogeneous. Upon cooling the atoms may order themselves to a structure with a different configuration (*e.g.* quartz (SiO_2) changes from hexagonal to trigonal symmetry).

Several types of modifications occur in the regular position of atoms during crystallization: (1) an atom is missing creating a vacancy (Fig. 1.3), (2) an interstitial atom is introduced, (3) dislocations occur over a group of atoms causing distortions.

Table 4.1 lists the luminescent minerals in which the chemical elements assumed to be color activators are included. Following short and long wave ultraviolet radiation the minerals are divided into two groups. In all but 10 of them the chemical elements considered to be responsible for color are known. In 16 cases Mn (manganese) is present, followed by Eu (europium) and U (uranium) in 7 and 6 cases respectively. Chromium, copper, sulfur, tungsten and silver occur only in a few minerals. When these 8 elements are considered: 5 turned out to be "transition metals" and 2 are "rare earths" whereas only 1 is a "nonmetal". All of them have an atomic number higher than 16. Hence their distribution in the periodic table is far from random.

Of significance is that while these chemical elements function as activators, other elements: iron, cobalt and nickel suppress luminescence in minerals. This ability to activate and to repress reappears later in DNA. Since the early days of

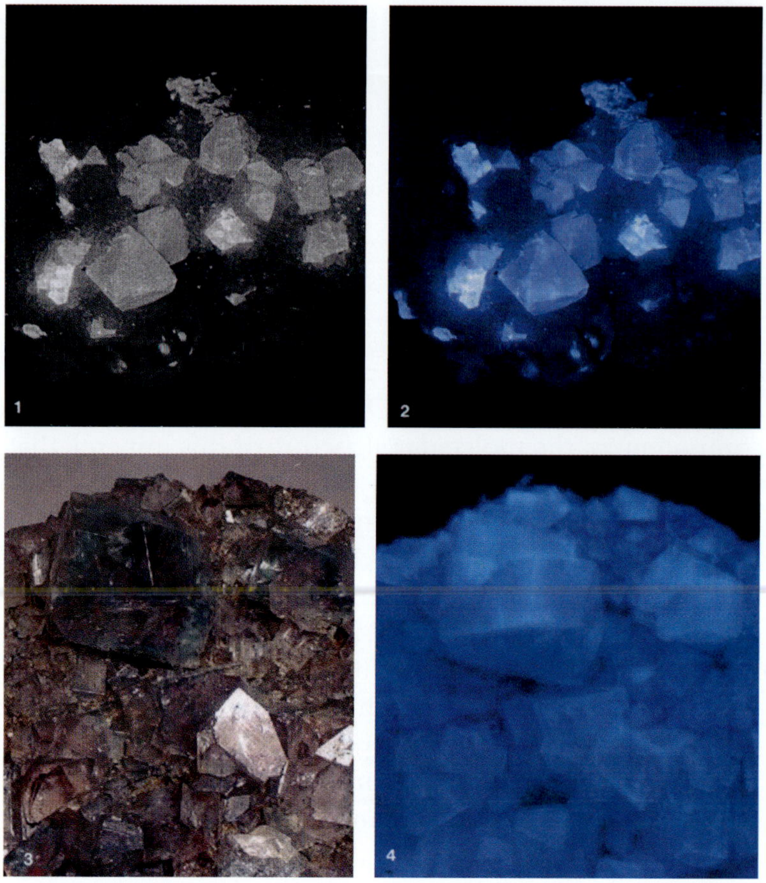

Figure 4.1 Luminescence in minerals. (**1**) Scheelite, $CaWO_4$, Tetragonal system. Crystals of scheelite seen under daylight. (**2**) Scheelite, $CaWO_4$. Luminescence in crystals of scheelite submitted to ultraviolet radiation (**3**) Fluorite, CaF_2, Cubic. Crystals of fluorite seen under daylight. Large cubes are accompanied by smaller ones. The large cubes have a blue-green color dispersion (natural fluorescence). (**4**) Fluorite, CaF_2. Luminescence in crystals of fluorite submitted to ultraviolet radiation.

genetics it is known that particular DNA segments ("activators") enhance the activity of genes, located in the same or other chromosomes, whereas "repressors" inhibit the function of other sequences. The activation and suppression of light is a regular feature in bioluminescence. Insects do it as they produce intermittent flahes of light. Their luminescent organs contain nerves directly connected to the photocytes, that are considered to regulate the flash emission (Shimomura 2006).

Not only atoms that get inserted as trace elements into minerals contribute to luminescence. This phenomenon appears already in single chemical elements.

Table 4.1 Luminescence in Minerals

Mineral	Short wave ultraviolet	Long wave ultraviolet
Halite	Red (Mn)	Red (w) (Mn)
Fluorite (P)	White/yellow (org), blue (Eu)	White/yellow (org), blue (Eu)
Calcite (P)	White, red (Mn), yellow, Green (U), blue (Eu)	White, red (Mn), yellow Green (U), blue (Eu)
Aragonite (P)	White, yellow, green (U)	White, yellow, green (U)
Barite	White, red (w), yellow	White, red (w), yellow
Apatite	Orange-yellow (Mn), blue (Eu)	Orange-yellow (Mn)
Autunite	Green (U)	Green (U)
Scheelite	White, yellow, blue (W)	Brown, yellow
Corundum	Red (w) (Cr)	Red (Cr)
Anthophyllite	Red (Mn)	Red (Mn)
Benitoite	Blue	
Tremolite	Orange-yellow (Mn)	Orange (Mn)
Wollastonite	White, orange (Mn), yellow, blue	White, orange (Mn), yellow, blue
Willemite (P)	Green (Mn), yellow (Cu)	Green (Mn)
Microcline	Red, blue (Eu)	Blue (Eu)
Albite	Blue	White, blue
Sodalite	Orange (S)	Orange (S)
Sphalerite	Orange (Mn)	Red, orange (Mn), blue (Cu, Ag)
Uvarovite	Red (Cr)	
Witherite	White, yellow, blue	White, yellow, blue
Selenite	White, yellow, blue	White, yellow, blue

When minerals are irradiated with ultraviolet radiation of short or long wave lengths they re-emit light in the visible range; either with a duration of fractions of a second (fluorescence) or for hours (phosphorescence, *P*). Assumed activator elements for specific colors are indicated. Mn = Manganese, Eu = Europium, U = Uranium, W = Tungsten, Cr = Chromium, Cu = Copper, S = Sulfur, Ag = Silver, *w*, weak fluorescence; *org*, organic traces. From: Wenk and Bulakh (2004).

Strontium (Sr, atomic number 38) in the form of strontium aluminate produces extremely bright luminous paints that glow in the dark, due to the capacity to absorb light from the surroundings and then slowly releasing it over the course of hours (Gray 2009).

Luminescence in minerals can also be caused by heating them to high temperatures (100°C). Thermoluminescence is best known in nonmetallic minerals that contain impurity ions as activators such as fluorite, calcite and apatite. Other minerals also become luminous when crushed as a result of the disruption of their atomic configuration. Fluorite is one of them (Klein and Hurlbut 1985).

The Colors Emitted by Living Organisms are the Same that are Emitted by Minerals

All luminescent animals produce an array of colors that build a rainbow of hues covering the full visible range and beyond: blue, green, yellow, orange and red. These turn out to be the main colors displayed by minerals (Table 4.1 and Fig. 4.2).

Most bioluminescence in the oceans is blue. Green is the next most common. Violet, yellow and orange occur rarely. In minerals blue and yellow are the most common colors.

The light flux ranges from 10^3 photons per second (bioluminescent bacterium) to more than 10^{12} photons per second (fish). The glow may be persistent (bacteria) but flashes of 43 ms are released from lantern fish light organs. The waveband of the light emitted may be adjusted by means of muscles and by complex optical parts that reflect, refract or filter the light.

The Absence of Luminescence in Later Groups of Organisms

The chemical generation of light in living organisms is found from bacteria to fish.

Most significant is that among the vertebrates the fish were the first and the last to emit light. No amphibians, reptiles, birds or mammals are known to irradiate light. Plants did not either become luminous while many species of fungi are luminous.

One is confronted with a recurrence of this phenomenon, as well as its repression or loss.

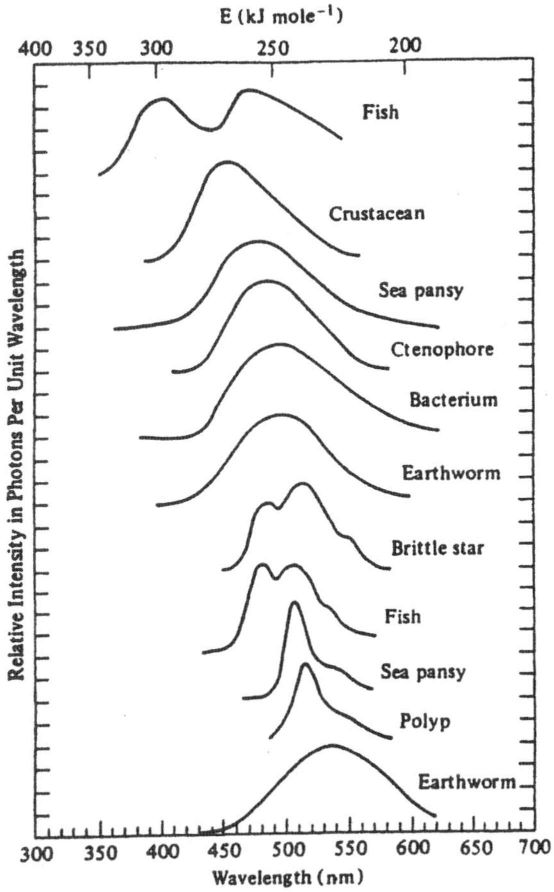

Figure 4.2 Wavelength (lambda; nm) and energy (kJ mole photons-1) emitted by bioluminescent organisms. Searsid fish exudate, crustacean *Cypridina* reaction *in vitro*, sea pansy *Renilla* reaction *in vitro*, ctenophore *Mnemiopsin* emission *in vivo*, photobacterium emission *in vivo*, earthworm *Diplocardia* emission *in vivo*, brittle star emission *in vivo*, fish *Porichtys* emission *in vivo*, sea pansy *Renilla* emission *in vivo*, hydropolyp *Clytia* emission *in vivo*, earthworm *Diplotrema* emission *in vivo*.

Luminous Bacteria have a Common DNA Sequence

Luminous bacteria may be terrestrial and free-living (which grow on fecal pellets) while others are symbiotic (forming specific associations with marine fishes and squid). The bacteria provide the host with light while the host provides the bacteria with an ideal growth environment.

Two single substrates participate in luminescence: A reduced flavin mononucleotide and a long-chain aliphatic aldehyde. These are oxidized by molecular oxygen and luciferase. The aldehyde is continously synthesized in the reaction, resulting in a persistent glow (Fig. 4.3).

The distribution of bioluminescence among bacteria is not even. All species in the terrestrial genus *Photorhabdus* are luminescent, but marine genera with bioluminescent species, such as *Photobacterium* and *Vibrio*, include many closely related non-luminous species.

Comparative DNA analysis has revealed that all luminous bacteria share a common gene sequence — the *lux operon* that encodes for the biosynthesis of luciferase and its substrates. This highly conserved DNA sequence appears in bacteria from very different ecological localities. However, bacteria isolated from squid families and teleost fishes disclosed deep divergences among the hosts which are not reflected in the symbionts, pointing to evolutionary independent origins (Guerrero-Ferreira and Nishiguchi 2007, Dunlap *et al.* 2007).

The maximum wavelength emission observed in bacteria is ca. 490 nm (Airth 1961).

Fungi Derive their Luminescence from at Least Three Different Lineages

Initially luminescence was discovered in fungi when rotten wood was found to glow in the dark.

Luciferin and luciferase participate in the chemical reaction that occurs in fungi leading to constant light emission with maximum intensity in the range 520–530 nm. The light emission is continuous and persistent.

Several features characterize their luminescence (Desjardin *et al.* 2008): (1) In some species only the mycelium is luminous, in others it is only the fruit bodies. (2) A luminescent species is sister to a non-luminescent one (an event occurring both in *Omphalotus* and *Armillaria*). (3) In the *Armillaria* lineage different tissues: mycelium, mycelial fans and rhizomorphs are luminescent, whereas the fruit bodies are not luminous. (4) The mycelia can sustain luminescence for up to 10 weeks. (5) Most known luminescent fungi were described in the genus *Mycena*. These mushrooms have lamellae and spore bearing tubes. Most species have only a luminous mycelium, in others both the mycelium and the fruit bodies emit light. Only the spores, or only the lamellae, or only the stipes, glow in related species of *Mycena*. (6) Of the 500 species of the genus *Mycena* only 35 have been reported to be luminescent. (7) Recent multi-gene molecular sequences data sets suggest at least 3 independent origins of bioluminescence belonging to three distinct evolutionary lineages.

The Myxomycota fungi are not luminous (Campbell 1988).

"Not All Dinoflagellates are Bioluminescent and Luminescent and Nonluminescent Strains of the Same Species are Common", Besides All Other Algae are not Luminous

Botanists divide the Algae into different classes but differ greatly in their arrangement due to the variation in complexity. The main criterion used in division is the dominant type of cellular pigment (chlorophylls, phycobilins, carotenes and xanthophylls). Seven classes are usually distinguished (here accompanied by an example): (1) Euglenophyceae (*Euglena gracilis*, unicellular). (2) Dinoflagellatae (*Noctiluca miliaris*, most unicellular). (3) Chrysophyceae (Diatoms, predominantly unicellular). (4) Xanthophyceae (*Vaucheria*, all levels of organization). (5) Green algae (*Ulva lactuca*, increased complexity, both freshwater and marine). (6) Brown algae (*Laminaria saccharina*, remarkably diverse in form, most marine). (7) Red algae (*Delesseria sanguinea*, predominantly marine).

Of all the 7 classes in only one of them, the Dinoflagellates, did luminescence emerge. There is no indication of this phenomenon in the complex green, brown and red algae which also inhabit the oceans. Like the fish, that live in this habitat, this group of algae could be as well luminous. It is not a question of depth. Some algae are found at 200 meters below sea level and the luminous dinoflagellates are part of the marine plankton covering the surface of the seas. It is this location that allows to observe the well known luminescence of *Noctiluca miliaris* — a spectacular display of blue sparkling light as waves break on beaches (Fig. 4.3).

Dinoflagellates are the only photosynthetic organisms capable of this behavior and they account for most of the planktonic bioluminescence in the ocean. The molecular biology is well understood. The luminous system consists of the enzyme luciferase, its substrate luciferin, and a protein that binds luciferin. The process is compartmentalized in discrete organelles called "*scintillons*" found within the cell. The peak of wave length is near 490 nm like that of most marine animals.

Hence, the dinoflagellates appear as an independent outburst of luminescence among the thousands of species of algae. Besides *Noctiluca*, various species of *Ceratium*, *Gonyaulax* and *Peridinium*, produce luminescence. But as Hackett *et al*. (2004a) write "Not all dinoflagellates are bioluminescent, however, and luminescent and nonluminescent strains of the same species are common". This evidence indicates that suppressor and activator DNA sequences ought to be involved in the sudden re-emergence of this phenomenon, since it occurs within the same species.

There is an extensive fossil record that locates the first dinoflagellates 245-208 million years ago. Molecular analysis of DNA and proteins allowed the building of a phylogenetic tree of this group. They are also known for their exceptional

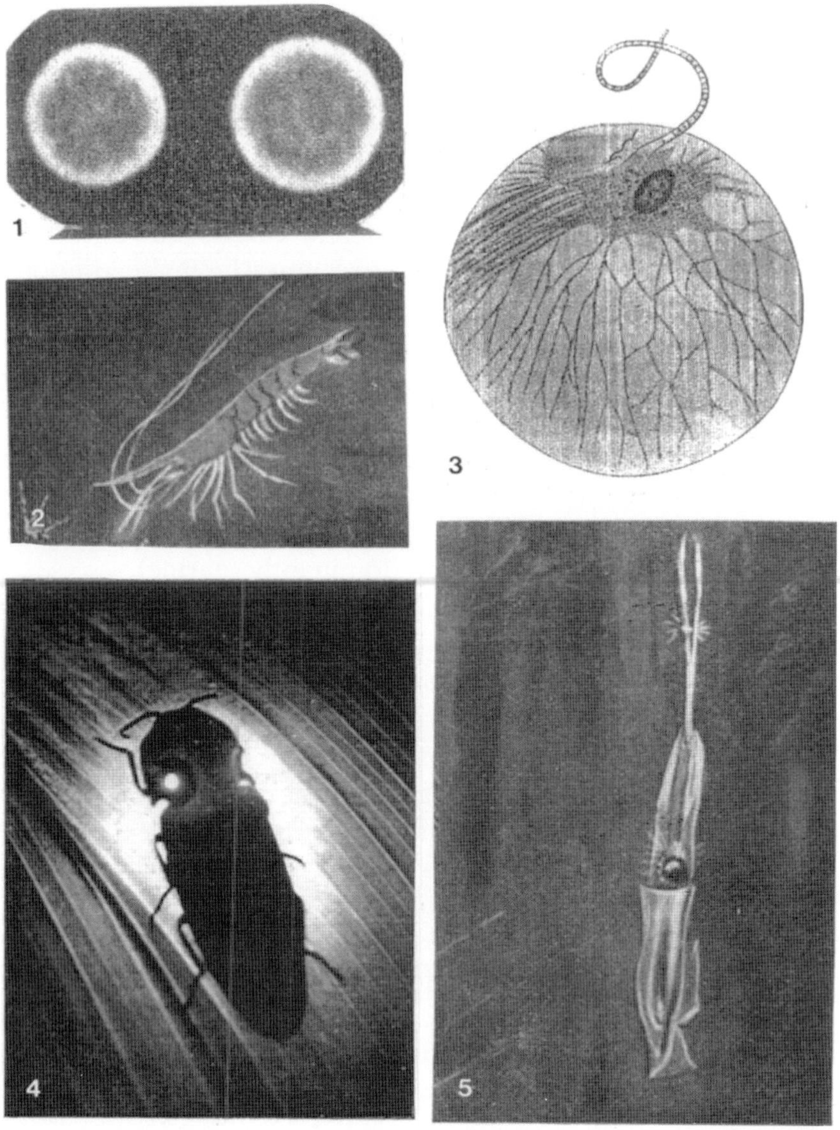

Figure 4.3 Bioluminescence. I. (**1**) **Photobacteria**. Two colonies of *Photobacterium* photographed in obscurity. (**2**) **Shrimp**. A luminous crustacean species. (**3**) **Noctiluca**. Adult stage of *Noctiluca sp*. (**4**) **Insect**, Coleoptera. The strong light emitted by *Pyrophorus noctilucus*. (**5**) **Squid**, Mollusks. A luminous species.

amount of DNA. Most algae have 0.54 pg DNA/cell whereas dinoflagellate DNA content ranges from 3-250 pg/cell corresponding to approximately 3000-215000 million bases (The haploid human genome is 3180 million bases and hexaploid *Triticum* wheat is 16000 million bases) (Spector 1984).

An Impressive Number of Invertebrate Groups were Not Allowed to Develop Luminosity

Vertebrates with their differentiation from fishes to mammals, and such a great shape variation within the mammals (from bats to whales), seem to represent the greatest diversification in evolution found among living organisms. But this is not the case. Their basic pattern decided by the skeleton has been fully preserved, and all species are variants of this rigid initial structure. On the contrary the less conspicuous invertebrates have differentiated in a much wider way. They comprise 97% of all animal species. The chordates (which include the vertebrates) occupy only one phylum and the invertebrates extend into 31 phyla, from the simple Placozoa to the complex Echinoderms.

Some 700 genera from 16 major phyla are luminous. There are thus many thousands of individual species of living light. But within these phyla there are equally thousands of species that do not exhibit this phenomenon.

The number of phyla with no known luminous species is equally impressive (Campbell 1988). Among them are: Mesozoa (minute marine parasites), Platyhelmintha (flat worms), Rotifera, Gastrotricha, Kinorhyncha, Nematoda, Acanthocephala, Endoprocta, Priapulida, Sipuncula, Echiura, Tardigrada, Pentastomida, Onychophora, Phoronida, Gnathostomulida, Loricifera, Brachiopoda, Chaetognatha and Pogonophora (beard worms).

Some of these phyla also harbor thousands of species such as the Platyhelmintha (about 13,000) and the Nematoda (about 20,000). Although there has been an enormous diversification in these invertebrates most of them were not allowed to develop luminosity.

The Class Insects Comprises at Least 1 Million Species Distributed Over Seven Hundred Families. However, In Only a Few Families Does Luminescence Occur but it Attains Extreme Intensity

Chapman (1969) wrote an extensive treatise covering most aspects of the structures and functions of insects. According to him a number of insects appear to be luminescent but this is due to the presence of bacteria covering their bodies. Self-luminescence, not involving bacteria, is only known to occur in a highly limited number of insects: Collembola (e.g. *Onychiurus armatus*), Homoptera (*e.g. Fulgora lanternaria*), larval Diptera (families Platyuridae and Bolitophilidae) and in a number of species in the Coleoptera (primarily in the families Lampyridae, Elateridae and Drilidae).

The Collembola, or springtails, are considered the most primitive insects but comprise as many as 6,500 species. Yet only a few are luminous.

Fulgora lanternaria, called lanternfly is not a fly but a Homoptera (family Fulgoridae), a group to which belong the cicadas and treehoppers. The Fulgoridae have eye-spots on the wings that can be flashed. This is the only species in which luminescence is well established (Romoser 1973).

The Diptera build the fourth largest order of insects with 130 families and 122,000 species. Yet the only true examples of luminescent Diptera are the larvae of the New Zealand "glowworm" (*Bolitophila luminosus*) and of the North American *Platyura fultoni* (these have also been called *Arachnocampa luminosa* and *Orfelia fultoni* respectively) (Meyer-Rochow 2007). As pointed out by Oba *et al.* (2011) in their study of the luminescent animals of Japan: "Despite the huge diversity, only a small number of luminous species are known from a single family of fungus gnats".

The Coleoptera have colonized every sort of habitat on land and in freshwater. The light emitted at night by some Coleoptera is so intense that can substitute an electric lampbulb to read a book. The name of this insect is *Pyrophorus noctilucus* a designation that explicitly describes this phenomenon (from Greek *pyr* fire and *pherein* to bear) (Francé 1943) (Fig. 4.3).

Luminous insects are popularly known as "fireflies" a name generally used as well by scientists. But "fireflies" are not flies (order Diptera) but beetles which build the order Coleoptera. This is one of the largest group of insects containing 370,000 species in 166 families. Yet luminescence is restricted to only a few families. It has been extensively studied in: Lampyridae (with three subfamilies Lampyrinae, Luciolinae and Photurinae), Phengodidae and Elateridae. The most extensive study of the molecular organization of the luciferase gene covers the investigation of 32 species of these families (Oba *et al.* 2010).

The light organs of these beetles appear in males and females differing in size or position, or may be lacking altogether usually in the male. In a number of genera there is also a sexual difference in the body form. The females are wingless and larviform contrasting with the flying males (Herring 2007).

The phenomenon is extremely punctual since it emerges in the most primitive insects (Collembola) as well as in the most advanced (Coleoptera). Between these two extremes, a tremendous diversification led to the occurrence of a huge number of species in which luminescence was obliterated.

Among the 100,000 Species of Molluscs Only Certain Families and Species Became Luminous

The phylum Molluscs is divided by taxonimists into 232 families that extend from the simple bivalves to the complex cephalopods. These harbor: nautiluses, squids, so called cattlefish and octopuses. Among the molluscs light emission has been recorded mainly in cephalopods and dorids. These last ones are marine slugs that have lost the protective shell of most gastropods. They exhibit exuberant colors and forms.

A great difficulty that remained unexplained was the presence in the same genus of luminous and non-luminous species. Two species of the squid *Sepiola* are examples of this situation. The first species gives visible light. In the second the luminous gland was completely lacking in 1/3 to 1/2 of the individuals.

Among the Cephalopods, all species within a genus are luminous but there are exceptions: *Loligo, Gonatus* and *Grimalditeuthis* (Fig. 4.3).

Great Variation in Number, Location and Organization of Light Glands in Squids, Octopuses and Slugs

Among the molluscs, the Cephalopods are covered by hundreds of luminescent organs that glow in the ocean waters (Fig. 4.3). The females of *Watasenia scintillans* have as many as 566 to 687 photophores (males 450 to 543) located in the ventral mantle.

Characteristic of this group is the great variation of the luminous organs in: location, size, morphology and sexual difference. The male of *Lycotenthis diadema* differs from females in having additional photophores on the arms. *Ctenopteryx siculus* males have a large visceral light organ which females lack.

On the contrary light emission by special additional organs is present only in mature females among seven of the 13 genera of the Chranchiidae. The special photophore is located at the arm tip and is structurally quite different from the other photophores on the eye or the mantle. Males lack these organs.

The organization of the luminous organs is complex and the association of the different tissues and chemicals is fully coherent. In *Charybditeuthis maculata* they consist of a reflector built by modified muscles which occupies the bottom of a cupola filled with luminescent matter. The whole is covered by a convergent lens having its origin in conjunctive tissue (Fig. 4.4).

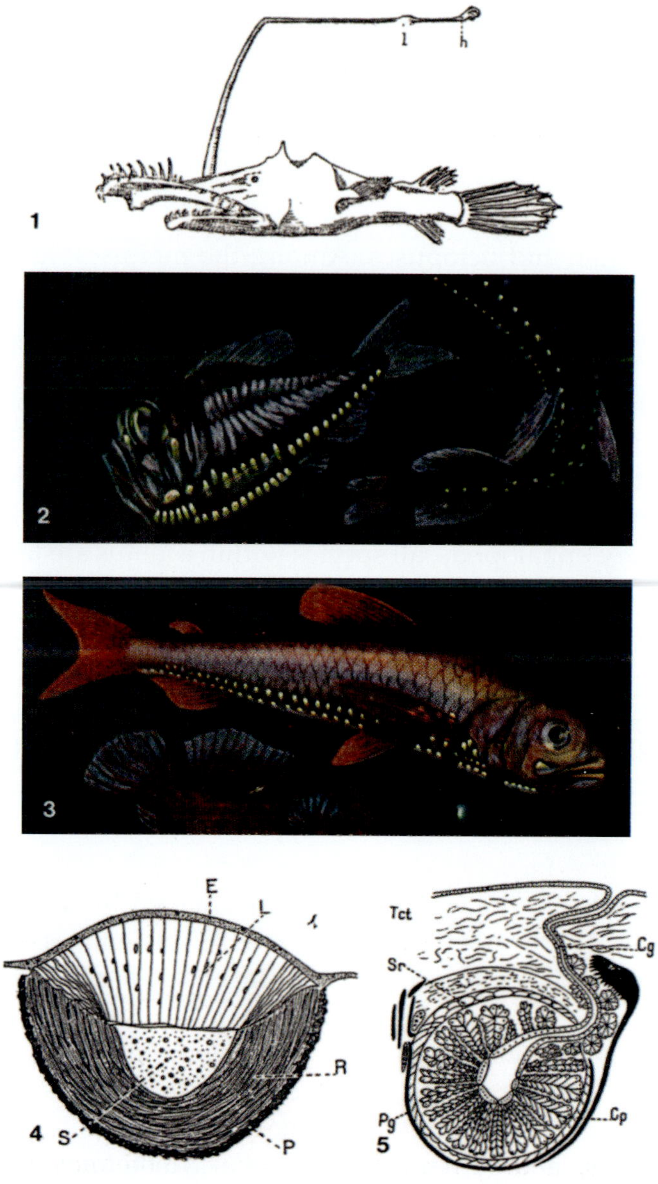

Figure 4.4 Bioluminescence. II. (**1**) **Abyssal fish**. *Lasiognathus saccostoma* having a tentacle with a luminous bulb (l). (**2**) **Abyssal fish**. *Neoscopelus sp*. With numerous luminous organs along the body. (**3**) **Abyssal fish**. *Melanocetus sp*. Also with luminous bodies located on lower ventral region. (**4**) **Luminous organ** of a Cephalopod (Molluscs). *Charybditeuthis maculata*. *L* lens, *R* reflector, *S* luminous substance. (**5**) **Luminous gland** of abyssal fish. *Gonostoma*. Light emitting cells (*Cp*), refringent substance comparable to a lens (*Sr*).

Another group of molluscs are the slugs. One family consists of seven genera of which two are luminescent. In both cases the luminosity is intrinsic (not due to bacteria association). There are differences between the two genera: in one, the luminescence is intracellular; in the other, there is a discharge of the luminous chemicals outside the cell where the chemical reaction occurs (Vallès and Gosliner 2006).

Fishes Look Like Christmas Trees with their Long Lines of Luminous Organs that "Are Extraordinarily Elaborate"

Ocean living fishes such as *Neoscopelus* sp. and *Melanocetus* sp. have their bodies covered by over 40 and over 60 luminous glands respectively (Fig. 4.4).

Herring (2002 and 2007) has described in detail the organization of light organs in abyssal and near-shore fishes.

In all but one of the 11 families and 35 recognized genera of anglerfishes the females have a bulbous luminous lure which is lacking in the dwarf males. This organ is located near the end of a long filament that protrudes from the head. The lure is movable and can even rotate and produce either a glow or a flash. As Herring (2002) pointed out these light organs "are extraordinarily elaborate". Besides, the different parts are coherently put together. These consist of: sensory filaments, papillae, light pipes and shutters.

Lynophryne is one genus in which the female has, in addition to the bacterial lure in the head, a multibranched barbel, hanging from the lower jaw, which is covered by luminous organs. In these the light is not bacterial but intrinsic showing that these fishes possess two independent luminous systems.

Their lure is also complex: it has a reflector, the bulb opens to the exterior by a pore and the light emission is under the fish's control. Males lack all these structures.

Lanternfishes, as their name tells us, provide the most extensive evidence of bioluminescence in oceanic animals. They have supra and infracaudal light organs which are formed from scales. Three of the 33 genera of this family lack these organs altogether, whereas in other genera the caudal organs are substituted by a large head photogland.

The fish *Malacosteus* has photophores that emit both blue and red light located near its eyes. The luminous organ with a sub-orbital location is like a sac filled with photocytes. The red light produced by these photocytes is further modified by a brown filter with an emission maximum still in the red. On the other hand the postorbital light organ emits blue light only.

The wavelengths of the emitted light have been well established for most organisms. Wampler (1978) measured the relative intensity in photons per unit wavelength. It can be seen that it is weakest in some earthworms and strongest in some fishes (Fig. 4.2).

Light emission in photoproteins is considered to be maximally only a single photon to be emitted per protein molecule. On the other hand photon fluxes emit more than 10^{12} photons per second in fish (Widder 2010).

Marine Mammals are not Luminous Yet they Live at Depths Exceeding Those of Luminous Fish

It is usually thought that the luminous fishes live at the bottom of the sea or in the abyssal regions. But luminescence in the oceans is most common at 200-1000 meters in depth. It is in this area that 80% of the fish and crustaceans that emit light are found and is also the region populated by cephalopods and gelatinous animal plankton that are luminous (Cronin *et al.* 2014).

Close to the bottom of the ocean, at abyssal regions, bioluminescence is less common, being circumscribed to ophiuroids, sea pens, polychaeta worms and some fish.

Marine mammals comprise many diversified species that also inhabit the bottom of the ocean and deep rivers.

The manatee (of the Amazonas), the common delphin and the Harp seal are found at depths of 250 m, the Humpback whale at 500 m and the Sowerby's beaked whale at 750 m (Cronin *et al.* 2014). Elephant seals may dive as far as 1,500 m or more (Macdonald 2002).

But female Sperm whales (*Physeter macrocephalus*) and their dependent young generally inhabit tropical, subtropical and temperate waters deeper than 1,000 meters. These whales are estimated to consume 750 squid per day, most of them being giant squids (*Architeuthis* sp.) which inhabit the same ocean depths. At times their dives may last for more than an hour and reach to depths of 2,000 and 3,000 meters. No whale species, like other mammals, are known to be luminous (Wilson and Mittermeier 2014 and Macdonald 2002).

There is no obvious biological or genetic reason, that would prevent these mammals from developing luminous organs, since they have preserved most of the genes that are found in fishes and the earliest vertebrates.

The structure and gene content of the DNA of the primitive marine amphioxus (lancelet animals) was compared with tunicates and humans (Putnam *et al.* 2008). It revealed that 17 ancestral chordate groups of genes, having their location in the same chromosome, are conserved in the amphioxus and modern vertebrate genomes. As much as 90% of the human genome is encompassed within these gene groups, revealing a tremendous permanence of gene organization and function from fish to humans. Even the position of microRNAs embedded in these clusters of genes were preserved (King *et al.* 2011).

Non-luminous Molluscs Occur Also at Ocean Depths from 200 to 7,000 Meters and Freshwater Lakes with Depths of 1,600 Meters do not Harbor Luminous Species

The Monoplacophorans which look like limpets but build a class of their own among the molluscs occur in the Atlantic, Pacific and Indian Oceans at depths of 200 to 7,000 meters, yet there is no indication that they are luminous although they are more than 300 million years old.

Also among the Molluscs (in the class Cephalopods) the Spirulas are relatives of cuttlefish (*Sepia*). Their bodies contain an internal coiled and chambered shell that acts as a float. They are found at depths of 2,000 meters but also in this case there is no indication that they are luminous (Burnie 2004).

It should be mentioned that there are lakes with comparable depths to those of the seas but "bioluminescence is nearly non-existent in freshwater for unknown reasons" (Cronin *et al.* 2014). Lake Bajkal, in eastern Russia, is the world's deepest and largest natural freshwater reserve. Its depth is 1,637 meters. It has also impressive dimensions: length 630 kilometers and breadth 80 kilometers, containing more than 1/5 of all the world's freshwater (Coenraads and Koivula 2007).

The Bats Living in Darkness Did not Become Luminous

The bats which fly actively in the darkness of caves and among trees during the night could as well have become luminous like the fish that swim in the darkness of the ocean. Yet this was not the case. The bats use an echolocation system based on sound to detect prey, but light flashes could also have added to the efficiency in finding their food sources, which in several species are not insects, but fruits from which they suck their juices (Burnie 2004). A luminous lantern would have been most useful but none of the 977 species of bats has luminous organs. Actually the Fulgoridae (luminous insects) have eye-spots on wings that even flash (Romoser 1973). Hence, bats could have lanterns on their heads like fish, or on their wings like insects, but none of these solutions arose.

Chemical Evolution of Luciferins and their Distribution Among Marine Animals

The chemical structures of the four best-known luciferins used by marine organisms are well known, and turn out to be as diverse as their phylogenetic distribution: (1) Bacterial luciferin may occur in free-living or symbiont bacteria that are part of luminescent organs in squid and fish. (2) Dinoflagellate luciferin is limited to this group and to some crustacean species. (3) Coelenterazine is on the contrary used by many different marine groups including: radiolarians, cnidarians, ctenophores, squids, ostracods, copepods, decapods, chaetognaths and fish. (4) Cypridina luciferin is found so far in only two groups: ostracods and some fish (Widder 2010). The chemical composition of luciferin in fireflies and earthworms is also well established (Shimomura 2006).

Supplemental proteins serve as secondary emitters in some bioluminescent systems. Best known is the jellyfish green fluorescent protein. This is an "accessory protein" that provides luciferase with luciferin. Others are "antenna proteins" which modulate the color by means of energy transfer (Titushin *et al.* 2011).

DNA Sequencing of 10 Full-length Photoprotein Genes Elucidates their Origin. Moreover Light Production and Light Reception Seem to be Connected as they are in Minerals

Jellyfish (phylum Cnidaria) and comb jellies (phylum Ctenophora) have calcium-activated photoproteins. This term was coined for proteins that would emit light upon the addition of a substance other than a luciferin (Hastings 2011).

The ctenophore *Mnemiopsis leidyi* is a representative of the earliest branch of animals that emit light. Its genome has 10 full-length photoprotein genes situated within two genomic clusters with high sequence conservation.

Photoprotein genes are co-expressed with three opsin genes which are involved in light sensing. One opsin functions as a photopigment *in vitro*, absorbing light at wavelengths that overlap peak photoprotein light emission. This indicates that light production and light reception may be functionally connected in ctenophore photocytes, a result that links mineral luminescence with that of the earliest luminescent animals. In minerals photoreception is a component of their light emission (Schnitzler *et al.* 2012).

Complete DNA Sequencing of Luciferase Genes Demonstrates that they have Been Highly Conserved

The luciferase gene produces the sole enzyme responsible for bioluminescence. The complete nucleotide sequence and the exon-intron structure of the luciferase gene has been determined for the *Hotaria* group of fireflies (Choi *et al.* 2003). The DNA sequence spans 1950 base pairs and consists of six introns and seven exons coding for 548 amino acid residues. It turns out that it is a highly conserved structure among the various species of this beetle group since the divergence of the amino acids has been minimal. Moreover, this gene was identical in length and exon-intron structure in the whole group.

At present more than 10 luciferase genes have been isolated from fireflies. These include several species of *Luciola*, *Photuris*, Luciolinae, and *Hotaria*. The Luciolinae are composed of seven genera including about 400 species, which are widely scattered throughout the world.

Flowering Plants and Mammals can Become Luminous. Expression of the Luciferase Gene in Whole Plants, Monkey Cells and Whole Rabbits

The firefly luciferase gene was used to make whole plants luminous. Initially naked DNA was introduced by electroporation, later a plasmid was added when the gene became part of carrot cells. In a final step a bacterial plasmid was used when *Nicotiana tabacum* plants were inoculated. Both plant cells and whole plants became luminous confirming the stable expression of the luciferase gene in all the plant tissues (Ow *et al.* 1986) (Fig. 4.5).

The nucleotide sequence of the luciferase gene of the luminous beetle *Photinus* was determined from the analysis of complementary DNA and genomic clones. The gene contains six introns (all less than 60 bases in length). The full-length luciferase gene was inserted into mammalian expression vectors and introduced into monkey cells in which luciferase was expressed (de Wet *et al.* 1987).

Lately transgenic rabbits have been made to luminesce over the whole body by inoculation of the luciferase gene (Stewart 2006). In the animal every tissue is luminescent as is the case in transgenic tobacco plants.

The genetic manipulation shows that both flowering plants and mammals that are not luminescent in nature can be made to be so. This finding reinforces the evidence that indicates that DNA evolution did not allow the emergence of this property in these distantly connected groups of living organisms.

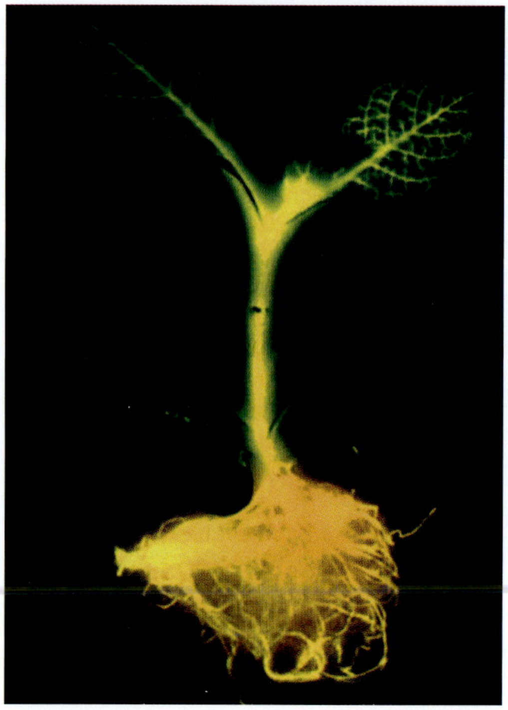

Figure 4.5 Stable expression of the luciferase firefly gene in the transgenic plant Nicotiana tabacum. All cells of the plant are luminescent.

Molecules Involved in Luminescence have Had their Own Evolution

Organisms have evolved, cells have evolved, but we usually tend to think that molecules are static. Their well defined atomic composition seems to be a permanent feature.

But it is the evolution of molecules that has obliged cells and organisms to evolve. Luciferins and luciferases, like any other macromolecules, are immersed in a huge molecular environment — the cell — which is a super dynamic living unit.

The chemical structure of dinoflagellate luciferin bears a striking similarity to chlorophyll, which suggests that it may have originated in photosynthetic species. But photosynthetic plants are not luminous.

In Dinoflagellates 8 luciferase genes have been identified. Phylogenetic analysis indicates that the most primitive luciferase of these eight genes is from *Noctiluca scintillans* since this luciferase gene codes for a single catalytic domain. However, in the other seven sequenced luciferases (all from photosynthetic dinoflagellates) there are three homologous catalytic sites. These sites are highly conserved across this group. There are, however, differences: (1) In some species the luciferase gene may have a large unique noncoding region. (2) Other species lack luciferin binding proteins.

Bioluminescence changes at the molecular level are considered to have resulted from specific enzyme mutations — the enzymes are unique and independently derived (Widder 2010).

Communication in Bacteria Deciding Collective Luminosity and the Release of Luminous Chemicals by Body Glands in Invertebrates

A new phenomenon, called autoinduction, was discovered in luminescent bacteria (Hastings 2011).

At concentrations less than 10^6 cells ml^{-1} in liquid culture, bacterial cells grow but do not synthesize luciferase. Luminescence only happens when the cells reach a higher growth density as the result of a substance released by the bacteria themselves into the medium which influences the transcription of luciferase genes. This is an impressive example of communication in bacteria by means of a diffusible chemical initiating luminosity in a cell community.

Luminous crustacea such as *Cypridina* (now called *Vargula*) injects its luciferin and luciferase into the sea water from two separate glands. These ejections form small luminous spots that hang in the water and that are luminous for minutes.

Another release of luminous chemicals is part of the behavior of *Pholas dactylus*. It emits a cloud of luminescence when disturbed. The same emission is found in the tube-worm *Chaetopterus*.

Single cells (bacteria) or complex organs (glands) are used in the emission of substances that cause the emergence of luminescence.

Luminescence May be Spread Over the Whole Body but in Most Cases Appears Localized in the form of Complex Organs

When certain species of springtails (Collembola, insects) such as *Achorutes muscorum* are stimulated, luminescence occurs over the whole body (McElroy 1964). This is significant because it indicates that most cells of this organism have the property of emitting light irrespective of the tissue or organ to which they belong.

One could expect that this situation would be the general form in which luminosity would be displayed since it is a simple one, but it turns out that, throughout evolution, luminosity has become confined not only to specific tissues and body locations but it even resulted in the building of utmost complex and specialized organs.

Each light organ of *Photuris* (insects) consists of large cells, the photocytes, lying just beneath the epidermis followed by several layers of dorsal cells. Covering the epidermis is a transparent cuticle. This is critical otherwise no light would be transmitted.

The photocytes are orderly arranged in the form of cylinders running at right angles to the cuticle and within each cylinder there are tracheae which convey oxygen necessary for the chemical reactions. Also of importance is that nerves enter the photocyte cylinder and build vesicles that have neurosecretory functions. Light production is not dispersed in the cell, but takes place in numerous granules that are packed within each photocyte.

In *Photinus* (Coleoptera) the two lanterns contain about 15,000 photocytes forming 6,000 cylinders (Chapman 1969).

The nerves control light emission by stimulating the release of acetylcholine at the nerve ending. This initiates the molecular ATP cascade whose products diffuse to the photocyte granules and leads to the production of light by removing the inhibition of luciferase.

The Coherence Involved in Luminescence is most Evident in the Building of Photophores

Single cells in photobacteria are considered to be the initial form of *photocytes*. They also occur in Hydrozoa and Ctenophores.

But the complexity already starts in the symbiotic bacteria which reside in a pouch. The bacteria continually produce light and the animal which harbors them has added a shutter that allows to control the exit of light.

The progressive assembly of more tissues led to the building of *photophores* (with various degrees of complexity) found in teleosts, insects and polychaeta.

It is in the formation of luminous glands that the complexity reaches its maximum of coherence.

The photocytes become "part of beautifully complex organs that contain filters, mirrors, lenses and other apparatus for controlling the color and direction of light" (Cronin *et al.* 2014). The filters modify the wave length of the radiation, the mirrors focus it on targets, the lenses increase the intensity. The coherence is revealed by the fact that all the different parts have properties which are in full agreement with the main function of the organ. An additional demonstration of coherence is revealed by the different parts of the gland not being located randomly but occupying a position related to each other that results in maximal effect (Fig. 4.4).

Bioluminescence was Originally Seen as a Magic Event and Lately as a Random and Spasmodic Phenomenon — Selection has also Been Evoked and Denied

When confronted with the fluorescence of minerals and of beetles in the middle 1800s the emission of light was seen as magic. It was then called "cold light" since no physical or chemical explanation of the phenomenon was possible.

It was only by the end of the XIX century that the chemical mechanism became understood and one had to await until the middle 1900s for an atomic explanation which elucidated the electronic changes that lead to photon emission.

Even when the DNA sequence of the genes is now known an antiquated approach continues to prevail. Like most phenomena in biology, bioluminescence is seen as a random and chaotic event. Shimonura (2006) expressed it in this way: "There is no obvious rule or reason in the distribution of luminous species among microbes, protists, plants and animals". And Harvey (1952) was still more explicit: "It is as the various groups have been written on a blackboard and a handful of damp sand cast over the names. Where each grain of sand strikes, a luminous species appears. It is an extraordinary fact that one species in a genus may be luminous and another closely allied species contain no trace of luminosity". Hence, "there is no obvious rule" and the phenomenon is "extraordinary".

As expected selection has been also used by some authors as the explanation, but denied by others.

Schnitzler *et al.* (2012) speaks of a "strong purifying selection" to explain the high DNA sequence conservation of photoprotein genes. The addition of the term "purifying" to selection makes it still more abstract.

Other authors, studying luminescence in fungi, have pointed out that "luminescence may not confer selective advantage because there are both luminescent and non-luminescent strains of the same species" (Weitz 2004).

The functions of bioluminescence have been classified as including: luring, defense, camouflage, communication and illumination. However, as Cronin *et al.* (2014) point out: "direct behavioral confirmation of many of the proposed functions is lacking and most hypotheses are based on morphology alone".

This view is reinforced by Campbell (1988) who states: "The apparent spasmodic occurrence of bioluminescence between, and within, different phyla has long been a puzzle", and is "still far from compatible from the modern Darwinian-Mendelian concepts applicable at the level of whole species and groups. If only we could ask the right questions!"

"Conventional Evolution Theory" Does not Explain Luminescence's Occurrence and Distribution. Instead "Intrinsic Chemistry" is Considered a Better Explanation by Various Authors

Treating at length most aspects of luminescence Herring (1978) remained perplexed by all the collected observations and experiments carried out on this subject.

He was obliged to conclude that what he called the "Conventional evolution theory" (Darwinism) could not be used to explain luminescence: (1) Competitive advantage conferred by luminosity could not be confirmed due to the high diversity and location of its organs. (2) Avoidance of predation leading to selection was not evident. (3) Coordinated behavior of luminous animals did not result in any particular advantage. (4) The occurrence of luminescence in totally unrelated animals was incomprehensible because it could neither be considered to occur at random nor to have an adaptive advantage.

Instead Herring concluded that an "intrinsic drive" toward diversification explained better the evolution of this phenomenon. Widder (2010) also speaks of an "intrinsic chemistry" directing enzyme evolution in bioluminescence.

Last, but not least, is the fact that luminescence occurred in single chemical elements and minerals before DNA and the cell existed, precluding old-fashioned interpretations of evolution.

The Periodicity of Luminescence. "Bioluminescence is Estimated to Have Evolved Independently at Least 40 Times"

1 All authors agree that luminescence has had independent origins throughout its evolution. Campbell in his extensive study of bioluminescence (1988) was obliged to conclude that it had arisen independently more than 30 times during evolution; and he added that the "spasmodic occurrence between and within phyla has long been a puzzle". Hastings (2011) has increased the number of independent origin of luminescence to 40 or more times. Reviewing the origins of

bioluminescence in the ocean Widder (2010) came to the same conclusion: "bioluminescence is estimated to have evolved independently at least 40 times".

2 The punctuated emergence of luminescence is shown by many examples. Among the 7 classes of Algae in only one of them, the Dinoflagellates, does luminescence emerge. Of the 500 species of the genus *Mycena* (Fungi) only 35 have been reported to be luminescent (Desjardin *et al.* 2008). The Coleoptera is one of the largest group of insects. Again, only a few families are luminous (Oba *et al.* 2010). Among the Dorids (Molluscs) one family consists of seven genera of which two are luminescent (Vallès and Gosliner 2006). Lanternfishes are known for their strong luminescence, but 3 of the 33 genera of this family lack luminous organs (Herring 2007).

3 Another feature, which is typical of periodicity, is that closely related species of insects do not necessarily exhibit luminescence (Shimomura 2006). The same is the case in fungi (Weitz 2004) and in dinoflagellates (Hackett *et al.* 2004a).

4 Some bioluminescent organisms are capable of making their own luciferin *de novo* from simple common metabolites. As Day *et al.* (2004) state: "In these cases the bioluminescent organism must encode all the genes required to code for the necessary biosynthetic enzymes required for the entire biosynthetic pathway of their luciferin". This was demonstrated in *Vibrio* which encodes the genes on the *lux* operon necessary to produce its luciferin *de novo* as well as its luciferase (Boylan *et al.* 1989). Moreover, other photoproteins than luciferin have been isolated and identified chemically such as Aequorin (Hastings 2011).

5 There is an increase in complexity of the photoemitters. These start with crystals and become successively more elaborate as simple photocytes are transformed into glands consisting of many additional structures.

6 High light intensity emerges irrespective of animal phylogenetic location. Invertebrates such as insects and vertebrates (fishes) emit the strongest light.

The Periodicity of Luminescence is the Result of Electronic and DNA Events

First, the emergence of periodicity is based on electronic energy changes, since both at the mineral and biological levels, it is due to electronic energy transitions resulting in the emission of light.

Second, it is essentially a DNA process since all bacteria that are luminescent share a common gene sequence — the *lux* operon.

Table 4.2 Periodic Table of Luminescence

Increasing grade of complexity	ABSENCE OF LUMINESCENCE	PRESENCE OF LUMINESCENCE	ORGANIZATION OF PHOTOEMITTERS
1	MINERALS Most non-luminous	MINERALS Over 20 are luminescent	CRYSTALS Fluorite Calcite
2	CYANOBACTERIA ALGAE Most non-luminous	EUBACTERIA *Photobacterium* *Vibrio* DINOFLAGELLATES Luminous	PHOTOCYTES Bacteria Hydrozoa Ctenophores
3	FUNGI (MYXOMYCOTA) Not luminous	FUNGI *Mycena* Luminous	LIGHT SECRETION Bivalvia Annelids Crustaceans
4	INVERTEBRATES Of 31 phyla 20 have no luminous species	INVERTEBRATES POLYCHAETA STARFISH JELLYFISH Weak luminosity	PHOTOPHORES Molluscs Insects Polychaeta
5	FLOWERING PLANTS Non-luminous FISH Most non-luminous AMPHIBIA Non-luminous	INVERTEBRATES MOLLUSCS INSECTS Extreme luminosity	GLANDS Cephalopods Fishes
6	REPTILES Non-luminous BIRDS Non-luminous MAMMALS Non-luminous	FISH Near-shore and abyssal Highly luminous	ELABORATE REFLECTORS LENSES SHUTTERS AND MIRRORS Molluscs Fishes

The notorious feature of luminescence is that it was totally precluded in later evolution although it was most strong and elaborated in simpler organisms. Yet there is a well defined increase of complexity among these organisms. Luminescence became maximal in fishes. The occurrence of luminescence, before the arrival of DNA and of the cell, anchors this phenomenon to atomic and electronic processes. The increase in complexity refers to the phylogenetic position of the organism that exhibits the property studied. The increase in complexity of the organ responsible for

Third, luminosity is an epigenetic phenomenon. Within the same species of fungi different tissues differ in luminosity: the mycelium is luminescent whereas the fruit bodies are not.

Fourth, the photoprotein genes have been DNA sequenced and the luciferase genes have also been DNA sequenced.

Fifth, genetic manipulation, shows that the luciferase gene can be isolated and inoculated into other non-luminous organisms in which every cell of the body becomes suddenly luminescent (Fig. 4.5).

The Periodic Table of Luminescence

Table 4.2 consists of three columns: (1) Absence of luminescence. (2) Its punctuated presence. (3) The organization of the structures in which it emerges.

Column 1 Absence of luminescence

At present more than 4,000 mineral species are currently known (Wenk and Bulakh 2004) and most of them are non-luminous. The same is the case of the chemical elements that are over 112, also most of them are non-luminous. As cells arrived in evolution the Cyanobacteria and most algae did not acquire luminosity. The same was the case with the Myxomycota fungi. Later an impressive long list of invertebrates was not either allowed to become luminous. Most fishes do not possess light organs, but the elimination of this phenomenon was total in flowering plants, amphibians, reptiles, birds and mammals.

Column 2 Presence of luminescence

A few chemical elements, such as strontium produce bright light. The luminescent minerals are no less than 21 (Table 4.1). Among the Eubacteria the genera *Photobacterium*, *Photorhabdus* and *Vibrio* are known for their luminosity. Some Dinoflagellates (*Noctiluca*) are luminous. Most luminescent fungi are described in

---

Table 4.2 (*Continued*) luminescence follows also mainly the organism's complexity. The reappearance of luminescence is extremely punctuated and high light intensity emerges irrespective of the animal's phylogenetic location. Another feature of the periodicity of bioluminescence is that it arose independently at least 40 times (Original).

the genus *Mycena*. Some of the invertebrate phyla suddenly display extreme luminosity such as the molluscs and insects. On the other hand, polychaeta, starfish and jellyfish are less luminous. Suddenly as the vertebrates emerge, among the 24,500 species of fish, some become highly luminous.

Column 3 Organization of photoemitters

Crystals of calcite and fluorite have a simple but well defined atomic structure that makes them luminous. Later single cells acquired luminosity (called photocytes) which appeared in Bacteria, Hydrozoa and Ctenophores. A more elaborated process then emerged in which light became secreted by the organism: Bivalvia, Annelids and Crustaceans. The complexity increased as several tissues were combined in the building of simple photophores. They appeared in molluscs, insects (Collembola) and in some fish (around the eyes). The next step were the glands of Cephalopods and some fish species which varied in complexity. Finally, reflectors, shutters, mirrors, filters and lenses were added to these structures in molluscs and fishes.

Chapter 5

Placenta in Plants and in Animals. Its Punctuated Emergence is Decided by Common Genes

The Placenta Characterizes a Whole Animal Group

The placenta tends to be ignored because it is a transient organ. It is discarded after the foetus is born, no longer being part of the many organs that are permanent and obligatory constituents of the body.

Yet it is so important anatomically and physiologically that it is the dominant trait that separates a whole group of vertebrates from all the others: the placental mammals.

Depending on the way they reproduce, mammals are divided into three groups. (1) Monotremes (duck-billed platypus and echidnas) lay eggs. (2) Marsupials have no true placenta, in most species the young are kept in a pouch outside the mother's body. (3) In placentals the unborn young develop in the mother's uterus.

The placental mammals are also called eutherians. They have an allantoic placenta, and a long period of gestation, after which the young are born as immature adults (viviparity). This term is used to distinguish from egg laying (oviparity). During pregnancy, food and oxygen pass from mother to foetus through an organ known as the placenta, while waste substances move in the opposite direction.

Definition of the Placenta

Mossman (1937) defined the placenta in a way that has been widely accepted since then: "as any intimate apposition or fusion of the fetal organs to the maternal [or paternal] tissues for physiological exchange". Later, Holmes (1979) gave the following definition that applies to eutherian mammals: "a double vascular spongy structure formed by interlocking of fetal and maternal tissue in the uterus, and in which maternal and fetal blood vessels are in close proximity, allowing nutritive, respiratory and other exchange" and the same author adds that the placenta in plants is "the ovule part of the carpel and a sporangium bearing area".

The Structure and Function of the Placenta in Mammals

The placenta consists of both embryonic and maternal tissues. The foetal components derive initially from the trophoblast (a ring of cells) connected with the embryonic bloodstream through its contact with other tissues. This greatly increases the surface area for exchange of solute molecules.

The functions of the placenta include: (1) Allowing passage of small molecules from uterine capillaries into the capillaries of the umbilical vein leading to the foetus which are mainly: oxygen, salts, glucose, amino acids, small peptides, fats, antibodies and vitamins. (2) The removal of embryonic excretory molecules, notably carbon dioxide and urea, by diffusion from the umbilical artery into the maternal circulation. In this respect the placenta functions like a kidney in its ability to be an excretory organ. (3) Storage of glycogen and its conversion to glucose. (4) Hormone production, substituting for maternal ovaries in production of the sex hormones. (5) Production of endorphins (peptide neurotransmitters having morphine-like effects). (6) Prevention of red blood cell exchange between mother and foetus avoiding agglutination of these cells (Abercrombie *et al.* 1990).

The great significance of the placenta is that without it the mother's body would reject the baby like any other foreign body (Macdonald 2002).

The Placenta has an Anatomical Ancestry and has Varied Considerably in Mammals

The placenta does not arise in evolution from nowhere. Like all other body organs it has an ancestry. The mammalian placenta "replaces the embryonic lung of reptiles and birds". The trophoblast and the mesodermal tissues of higher mammals are equivalent to the chorion, amnion and yolk sac found in reptiles and birds (Romer and Parsons 1978).

Most vertebrates, such as fishes, lay eggs that are usually fertilized and develop outside the maternal body in water environments. The appearance of an egg enclosed in an impervious shell, as in birds, was a major evolutionary step, since it freed land-living vertebrates for their dependence on water for breeding. Within the egg the embryo could develop in a private pond enclosed in a thin membrane, the amnion. In addition to yolk sac and amnion, the egg contained two additional membranes: the allantois and the chorion. It is from these four extra-embryonic membranes that the mammalian placenta arises (Fig. 5.1).

In mammals the *maternal component* of the chorioallantoic placenta with its blood vessels comprises several tissue layers. Together they form a membrane that varies in thickness. Electron microscopy led to the division of this structure into three types: (1) epitheliochorial placenta (ungulates, whales, dolphins, lower primates), (2) endotheliochorial placenta (carnivores, elephants) and (3) haemochorial placenta (rodents, higher primates including humans) (Mess and Carter 2007) (Fig. 5.2).

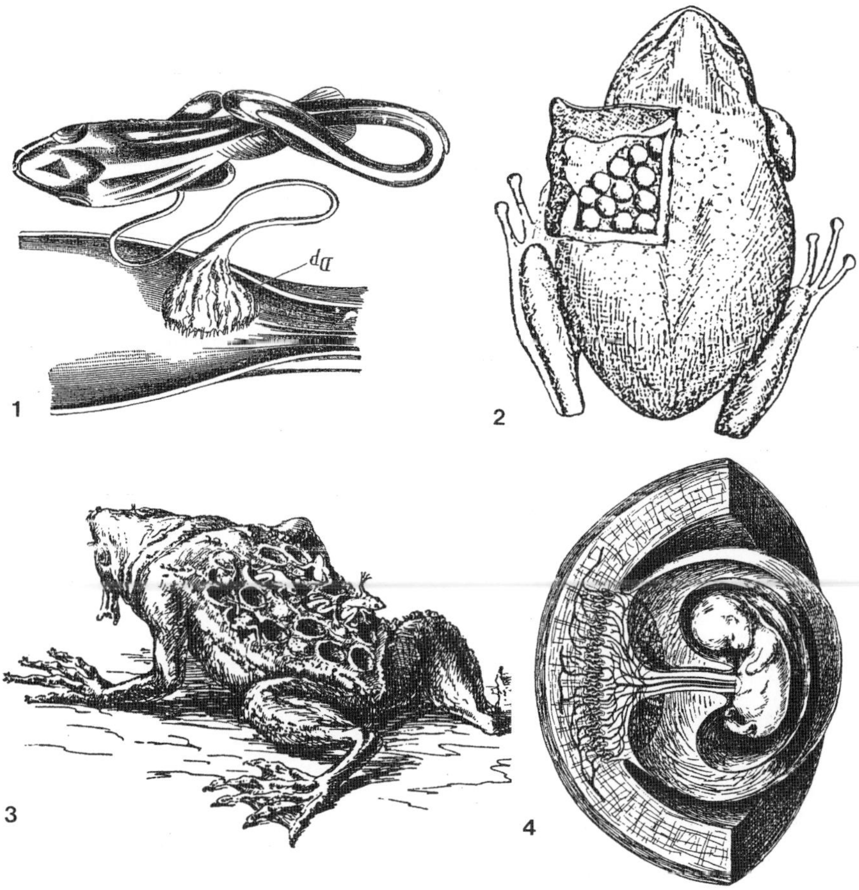

Figure 5.1 Placenta in fishes, amphibians, and humans. (**1**) *Mustelus laevis*, Cartilaginous fish, Shark. Embryo showing the placenta in contact with the uterus. (**2**) *Nototrema marsupiatum*, Amphibian, Frog. The dorsal pouch with part of the skin lifted to show the developing eggs. (**3**) *Pipa pipa*, Amphibian, Surinam toad. Young can be seen emerging from pits on the female's back. (**4**) *Homo sapiens*. Human embryo in the uterus showing the placenta to the left.

The Marsupial Solution Shows also Variation

Marsupials give birth at a very early stage of the embryo's development and nourish the newborn by milk instead of by a placenta. Their offspring are born in an almost embryonic state after a very short gestation (only 12 days in some bandicoot species).

In all marsupials investigated so far, the yolk sac is where the foetus develops until birth. The shell coat surrounding every foetus during early development

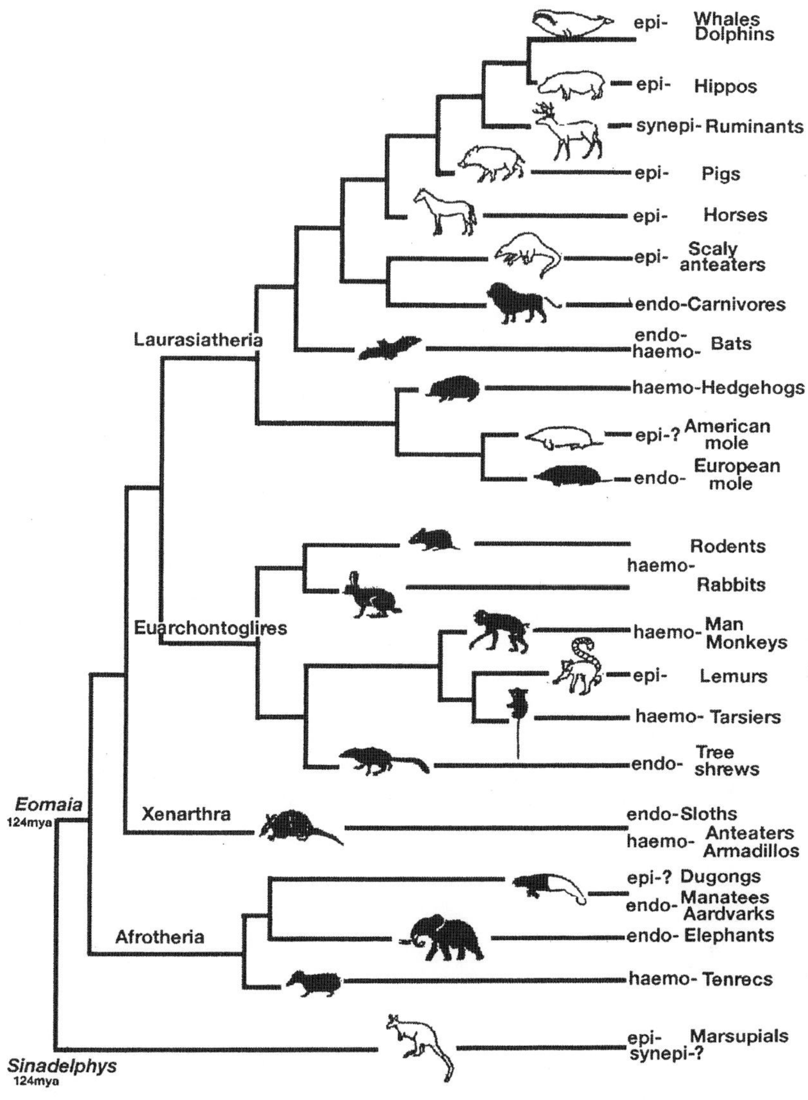

Figure 5.2 Evolution of the placenta. Placental mammals. Phylogenetic tree of eutherian mammals, showing the placental types of the various orders. Modified from Vogel (2005) and Wildman *et al.* (2006).

prevents direct contact between the yolk sac and the uterine (maternal) epithelium. However, it is permeable to gases and nutrients. According to Freyer *et al.* (2003) several morphological and biochemical features are consistent with the placental function of the yolk sac.

There are differences in placentation among the marsupials. In addition to the yolk sac placenta there is in some species a type of placenta which brings them closer to placental mammals. This occurs in bandicoots, animals with rabbit- and rat-like features inhabiting New Guinea and Australia which have an additional chorioallantoic placenta (Mess and Carter 2007) (Fig. 5.2).

The Placenta Occurs in the Most Complex as well as in the Simplest Living Organisms Being they Animals or Plants

The placenta is formed across the evolutionary scale. It has been described in detail in many invertebrate phyla such as plathelminthes (Flatworms), nemathelminthes (Roundworms), molluscs, echinoderms, arthropods, onychophorans, entoprocts (sessile animals similar to bryozoans), bryozoans and others (Moosbrugger *et al.* 2012). To these are to be added: sponges (Porifera), cnidarians (Corals, Jellyfish), nemerteans (Ribbon worms) and annelids (Segmented worms) (Ostrovsky 2013). Among lower chordates it appears in ascidians and all salps (Moosbrugger *et al.* 2012). It is most distinct in vertebrates such as fishes, amphibians, reptiles and placental mammals. In plants it is formed in simple mosses as well as in the most complex flowering plants.

The Plant Placenta and its Functional and Genetic Similarity with the Mammalian Placenta

The placenta has been extensively described in flowering plants by generations of botanists and plant physiologists. The placenta is the place in the ovary where the ovules develop. Ovule formation starts as a hemispherical protuberance of cells from the placenta (Greulach 1973). "The ovules are provided with nutrients through vascular bundles from their point of origin on the placenta" (Denffer *et al.* 1971).

Plants have sex organs but one thinks of them as structurally and functionally very different from those of humans. Such a view is not necessarily correct. In flowering plants the stigma and the ovary (with its egg cells) maintain a configuration which resembles the vagina, the uterus and the ovary of humans. But this does not end the similarity. Nagl (1973, 1976) has pointed out that the suspensor of several plants is most similar to the trophoblast of mammals and humans. The trophoblast is the tissue surrounding the embryo, from which the placenta is formed. Both the suspensor and the trophoblast have comparable anatomical locations in relation to their embryos. Both are involved in the transport of nutritional substances necessary for the embryo and both show DNA amplification and polytene chromosomes (Fig. 5.3). The study of the placenta of rodents has reinforced

Figure 5.3 Placental processes occur in plants, onychophorans and insects. (**1**) *Geranium phaeum*, Geraniaceae. Section of a seed showing the embryo (Em) and the suspensor (S). (**2**) *Peripatus capensis*, Onychophoran. (**3**) *Abedus herberti*, Belostomatidae. Male carrying eggs. (**4**) *Lethocerus americanus*, Belostomatidae. Giant water bug.

this similarity. A large increase in chromosome number (polyploidization) and endoreduplication of chromosomes occur regularly in their placenta. As a result the trophoblast cells become gigantic (Zybina and Zybina 2005). The structural and functional similarity between plant and animal placentas was reinforced by the study of Spielman *et al.* (2001) who carried out a comparative analysis of the genetic basis of gender in flowering plants and mammals (Fig. 5.4).

In mammals many gene loci that show genetic imprint are involved in foetal and placental growth. Parental imprinting is an epigenetic effect in which the

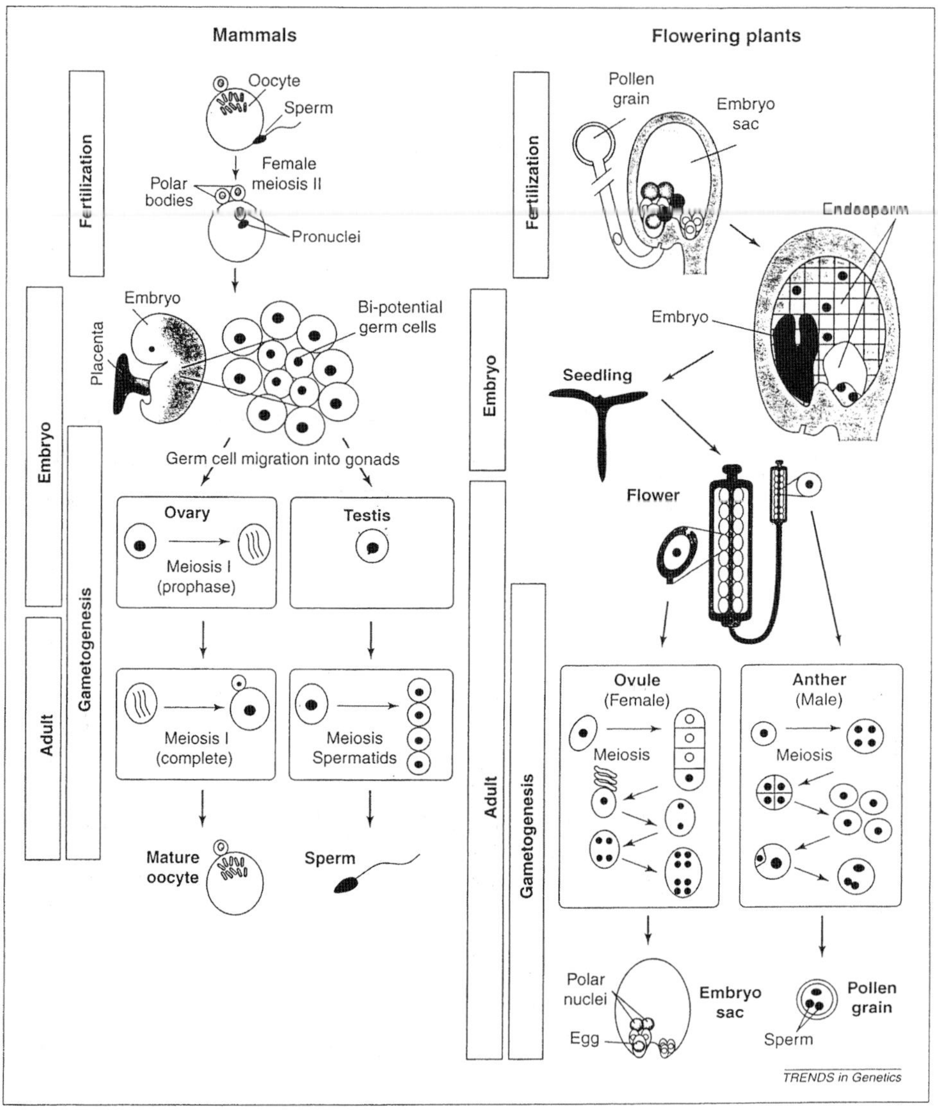

Figure 5.4 Comparison of the development of the placenta and other organs in mammals and flowering plants.

genes contributed to the next generation will have different expression patterns depending on whether they were maternally or paternally transmitted.

Imprinting in flowering plants disproportionately affects the endosperm (Fig. 5.4), a structure that transfers resources between the plant parent and embryo. When the dosage, of active copies of paternally expressed imprinted genes, is increased, the result is a promotion of the growth of the placenta and endosperm, whereas increasing maternal doses has the opposite effect. Besides, the suspensor, like the human placenta, is a terminal tissue (Spielman *et al.* 2001).

Similarly to mammals, flowering plants are sensitive to the balance of parental gene sets. Crosses between *Arabidopsis* plants with different numbers of chromosomes resulted in either a large or a small endosperm depending on whether the high number of chromosomes was carried by the mother or the father. These results are quite comparable to what happens in mice in which parent-of-origin effects reflect a change of active alleles of imprinted genes.

Placenta in Mosses, Ferns and Gymnosperms

The placenta of mosses has been described in detail in *Oedipodium griffithianum*. It consists of cells with highly convolute wall labyrinths on both the sporophytic and the gametophytic side (Ligrone and Duckett 2011).

In the great majority of native ferns (Polypodiaceae), the sporangia arise from a projecting mound of leaf tissue, the placenta or receptacle. In many species the sporangia are covered by a protective membrane and at a later stage a cohesion mechanism opens the sporangium and ejects the spores. These, like the fertilized ovules of flowering plants and the fertilized eggs of mammals, are a source of a new progeny (Denffer *et al.* 1971). The combination of light microscopy with ultrastructural evaluation elucidated the ontogeny and development of the placental cells in the fern *Ceratopteris richardii* (Johnson and Renzaglia 2008).

In Cycadales no indication of a placenta could be found but it may be very well present. The Conifers have a placenta that has also been less studied (Zenkteler and Relska-Roszak 2003).

Among the Simplest Invertebrates there is a well Developed Placenta — The Onychophorans

Due to their simplicity this phylum of invertebrates has been sometimes described as consisting of "half-worm, half-arthropod" animals.

They are the only survivors of a group of worms that were abundant in the Cambrian period. They have cylindrical bodies up to 15 cm long, 14 to 43 pairs of fleshy legs each terminating in a pair of claws and an antennae bearing head.

Velvet worms, as they are also called, include Peripatus (Fig. 5.3) which is known for its advanced system of birth. Each embryo is surrounded by an amniotic membrane and connected to the oviduct by a placenta and an umbilical cord. These two organs have the same functions as those present in vertebrates.

The Placenta in Bryozoans

This organ's structure has been studied using the electron microscope to establish, on a firm basis, its structural details.

Bryozoans are tiny aquatic animals less than 1 mm long that live inside box-like cases joined together forming colonies. They have diversified into 4,300 species being one of the major animal phyla. They have no circulatory or excretory organs. The class Gymnolaemata is almost entirely marine and includes the majority of species. The freshwater bryozoans (that belong to another class) consist of only about 50 species.

The calcified brood chambers have: (1) a double-walled hood, (2) a space used in embryonic incubation and (3) a closing structure. In placental species a vesicle is part of the process of nutrient delivery to the developing embryo. The placenta is formed by the distal wall of this vesicle. Its cells undergo a drastic increase in size and number. The cells also change in shape and in organelle composition. The embryo also increases in size during incubation (Moosbrugger *et al.* 2012).

Insects have Highly Efficient Placentas. The Genes Involved in Milk Proteins have Been DNA Sequenced. The Acquisition of a Novel Function Does not Need Changes in DNA

Free living insects develop by the embryo consuming the maternally derived nutrients accumulated in the yolk of the egg.

Wasps, such as *Aphidius ervi*, are efficient parasites of several other insects, namely aphid species. The wasp establishes a profound interaction with the host during the embryological development of its eggs: (1) The egg is yolk-less and contains very few energetic reserves. (2) The wasp induces an inhibition of the aphid progeny. (3) It redirects nutrients from the host's body in favour of its eggs.

In this process the wasp develops a placenta like structure that recalls the mammalian placenta. A rapid cell growth leads to the building of a membrane that wraps the embryo until the first larval stage and invades the host tissues by microvilli projections building a placenta of simpler dimensions than that found in vertebrates but that is equally effective. Of importance is that phylogenetically very close insects lack this organ (Sabri *et al.* 2011).

A more advanced situation is found in another insect, the cockroach *Diploptera punctata* (Williford *et al.* 2004) which is known to be viviparous. There are also intermediate cases in which species of cockroaches retain embryos within the female's body throughout their development but do not provide nourishment in addition to that present in the egg.

In *Diploptera punctata*, the brood sac epithelium also functions as the source of nutritive secretion ingested by the embryos during most of the gestation time. As gestation progresses the yolk is replaced by proteins (called *Milk* proteins). These are deposited in large amounts leading to a 60% increase in embryo protein content. These built a multigene family that encodes water-soluble proteins. Twenty five distinct complementary DNAs of these *Milk* proteins were cloned as well as the exon-intron structure of one *Milk* locus. The gene expression of *Milk* proteins during the gestation period was confirmed by molecular analysis.

As these authors stress, the acquisition of a novel function does not need to be associated with the change of the coding sequence, or the protein expression pattern, provided that a new interacting component appears, which in this case is the nourishment of embryos.

Care of eggs by adult insects is common but tending solely by male insects is very unusual. This behavior is best known in the giant water bugs, the Nepoidea comprising the families Nepidae and Belostomatidae. In this last family egg tending involves the males "back-brooding" — carrying the eggs on their backs, in a behavior shared by over a hundred species in five genera (Gullan and Cranston 2005). Examples are *Abedus herberti* and *Lethoceros americanus* (Fig. 5.3).

In Diptera there are 61 Independent Origins of Viviparity

The Diptera (Flies) with their 122,000 species divided into 130 families have had the opportunity to create different types of reproduction. There are species in which: (1) The eggs deposited at an advanced stage of embryonic development emerge immediately as larvae (a transition between egg laying and viviparity). (2) Full viviparity, the larvae hatch inside the female before deposition. (3) The offspring is deposited as pupae.

There is also a transitional stage of placental formation (called pseudo-placental). In it the larvae are nourished by glandular secretions of the mother.

True viviparity has evolved on many separate occasions in Diptera. It is recorded in 22 families and Meier *et al.* (1999) identified at least 61 independent origins of viviparity. Six families appear to have it in their ground plan. In true viviparity the chorion is shed within the uterus and is expelled with the larva.

Like in cockroaches, there are in the Diptera *milk* producing glands. The uterus releases a yellow fluid which participates in intrauterine nourishment of the larvae. The number of protein gene families present in the genome is estimated to be 674 in *Drosophila* (Flies).

Scorpions have Several Solutions for Nourishing their Embryos

Scorpions are the most ancient group of all Arachnids comprising about 1,400 species. They show a large diversity of maternal structures from close to ancestral egg laying to placental formation.

Electron microscopy of the embryonic tissues of the desert scorpion (*Paruroctonus mesaensis*) revealed the presence of two types of placenta. One type, which is considered primitive, releases nutrients that are absorbed through the embryo epidermis.

In the second type, called follicular placenta, cells (near the oocyte) form a trophic network. Each embryo develops within a specialized region of the tubules and the embryo borders a mass of follicle cells that transfer nutrients and gases (Farley 1998).

Simple Molluscs, Like Freshwater Clams, have Unique Brooding Strategies. Transition from Egg Laying to Viviparity May be Rapid

Brooding, *i.e.* the incubation of embryos, larvae, or both, within the body of a parental animal is represented in molluscs by a wide spectrum of reproductive modes. The majority of marine bivalves are egg laying but some freshwater species combine egg laying with viviparity or have viviparity alone.

Five species of *Corbicula* from Indonesia, have a unique feature. They use different ways of brooding. There is a transition from reproduction by releasing free-swimming larva to reproduction in which the gills are utilized, being modified into brood sacs. The females release their young as juveniles after a prolonged incubation. This brooding is characterized by considerable anatomical and histological changes leading to the appearance of high cylindrical mucous cells as is the case in the placentation of other animals.

The evolutionary transition from egg laying to viviparity may take place rather rapidly and therefore can also be observed in invertebrates among species of the same genus or even within a single species (Korniushin and Glaubrecht 2003).

Other Indonesian freshwater molluscs (*Tylomelania*) were investigated by molecular techniques including mitochondrial 16S ribosomal RNA genes to access their phylogenetic relationships. The reproductive system of *Tylomelania* is fundamentally different from that of its relatives. Its oviduct is modified to become a brood pouch which is embedded in nutritive material contained within the egg capsule. There is no direct transfer of nutrients to the embryos during their growth period. Instead a huge albuminous gland provides the nutrients for development within the egg capsules.

Closely related species have other structural solutions since an albumin gland has not been found in them. In all cases no secretive tissue could be observed (Rintelen and Glaubrecht 2005).

In Sharks the Placenta Evolved to a Degree that Rivals Some Placental Mammals and Emerged Independently in Over 11 Occasions

Scanning electron microscopy allowed to disclose the highly diversified placental features of the shark *Scoliodon laticaudus* (a cartilaginous fish): (1) The eggs are minute and nearly yolk free. (2) Maternal nutrients are acquired by placental transport and by absorption of uterine fluid. (3) There is a unique uterine implantation site in which maternal blood is in contact with the embryonic yolk sac placenta. (4) The latter is solid containing a network of capillaries. (5) The umbilical stalk surface is amplified by a massive capillary network. This is thought to participate in gas exchanges and tranfer of small molecules. (6) Some of the placental features resemble those of viviparous rays (Wourms 1993).

This type of yolk sac placenta is estimated to have evolved independently on 11 to 20 occasions and is now found in 21 genera within 5 families of sharks.

More than half of the extant cartilaginous fish are viviparous including all the rays, and 69% of the sharks (Fig. 5.1).

It is the Male Sea Horse that Develops a Placenta

Remarkable is that the placenta does not develop obligatorily in females but is formed in males having the same structural organization and function.

The male sea horse (a teleost fish) carries the eggs and young in special pouches and nourishes them as well. The placenta does not develop in a uterus but in the pouch of the male. The female deposits her eggs into the male's pouch where they become embedded in a spongy structure irrigated by blood vessels. Oxygen, carbon dioxide and nutrients are exchanged across the placenta.

This is the same functional solution appearing in a different sex and connected with a different organ.

The Use of other Body Organs as Placentas in Toads and Insects

Both vertebrates and invertebrates, use the back of their bodies to locate the placenta.

The Surinam toad *Pipa pipa* carries its eggs in pockets on the back of the female and each embryo develops into a tadpole. The tail of the tadpole is used as a placenta through which nutrients from the mother's body are absorbed (Fig. 5.1). The black salamander of the Alps is viviparous. In this case its tadpoles develop within the mother but the gills of the tadpoles function as a placenta.

It makes no difference if you are a toad or an insect. The solutions emerge with great similarity.

Abedus herberti is called a back-brooding insect because it is like a copy of *Pipa pipa*. It carries all its eggs cemented to its back. Besides, like in the sea horse, it is the male *Abedus* that carries the eggs. This is also the case in a related species of *Lethocerus*, a giant water bug in which males tend the eggs (Fig. 5.3).

All the species in the Belostomatidae (insects) are considered to be back brooders, where females glue their eggs to the dorsum of their mates. Each egg is adhered to the male's back with a mucilaginous glue, which collectively forms a gelatinous layer supporting the clutch (Costa 2006, Gullan and Cranston 2005).

94 Human Genes are Highly Expressed in Placenta

The anatomy of the placenta varies greatly across mammals. Moreover, the placenta is considered to be a dynamic organ that has evolved rapidly (Wildman 2011).

Many of the anatomical features of the human placenta are conserved. Despite this conservation there are many recently evolved placenta-specific genes that are important in the function of the human placenta.

The question arose as whether the same genetic elements (*e.g.* orthologous genes and retrotransposons) participated in the formation of the placenta in different vertebrates.

Several studies have demonstrated that genes participating in placenta formation and function have evolved during mammalian descent especially in primates and humans. Uddin *et al.* (2008) determined that of 1,240 human genes that show adaptive evolution in primates, 70 are highly expressed in placenta. Subsequently Hou *et al.* (2009) identified 94 human genes that are highly expressed in placenta. Some of these genes participate in pregnancy complications and result in aberrant placenta formation in mice. Primate genes build placenta-specific gene families and are assembled in clusters with placenta-specific expression. They code mainly for cathepsins (a group of proteolytic enzymes) and prolactins (protein gonadotrophic hormones, involved in pregnancy and lactation) (Wildman 2011) (Fig. 5.2).

Genes seldom work alone in the genome of animals or plants but are dependent on the intervention of regulatory elements (sequences of DNA located in the same or other chromosomes) for their expression. The glycoprotein hormone α-subunit gene is expressed in the pituitary of all mammals. Its expression in the placenta is limited to only primates and horses. In humans two different elements are required for placenta-specific expression of the α-subunit gene. The *upstream regulatory element* binds a protein unique to placenta, whereas the *response element* binds a ubiquitous protein. The first regulatory element has been retained in a number of mammals, but lacks expression in the placentas of rodents and cattle. This difference was found to derive from a single nucleotide change (Nilson *et al.* 1991).

The role of retrotransposons in the evolution of placentation in mammals has been studied by Kaneko-Ishino and Ishino (2010). They concluded that in mice these genetic elements were silenced by DNA methylation. Additionally they identified and localized, both in human and mice chromosomes, genes responsible for placentation. The distal part of the mouse chromosome 12 is an important imprinted region for placental function and the *PEG10* gene on human chromosome 7 is responsible for an early embryonic lethal condition and placental abnormality (Kaneko-Ishino and Ishino 2010).

Lipocalin Milk Proteins are Similar in Plants, Insects and Humans

The existence of similarity of genes between plants, invertebrates and humans seemed before to be unacceptable. But the discovery of the *Hox* genes disposed of this assumption overnight. *Hox* genes are responsible for the sequence of floral organs in plants, for the segmentation of the body in insects and for the formation of the vertebral column in humans (Carrasco *et al.* 1984, Lu *et al.* 1996, Bender 2008). The plant placenta is a part of the floral organs.

The DNA sequencing of the genome of the plant *Arabidopsis thaliana* confirmed the similarity between plant and human genes. Comparison with the sequences of the human genome revealed a high percentage of genes implicated in normal human functions, as well as diseases (such as Alzheimer's), to be present in this plant. The divergence of plants from humans is estimated to have taken place 1.6 billion years ago. This work disclosed a remarkable conservation of protein functions between plants and humans (Xu and Møller 2011).

Milk production, which is a part of lactation and placenta formation, also turns out to be decided by closely similar genes found in insects and humans.

Two milk proteins, produced by the tsetse fly *Glossina*, were molecularly characterized and found to constitute 50% of the secretions from which the intrauterine larva derives its nourishment. These proteins are small, contain a high proportion of hydrophobic amino acids and are located in the *lipocalin* family of proteins. Lipocalins are an important component in the milk secretions of viviparous cockroaches, marsupials and placental mammals. The milk proteins of the insects and mammals are considered to be dependent on similar genes (Yang *et al.* 2010). An unexpected finding is that lipocalins are also present in wheat as well as *Arabidopsis*. The molecular analysis of their genes disclosed that they share homologies with mammalian and insect lipocalins (Charron *et al.* 2002).

Coherence in Placenta Formation is Guided by Hormones

Studies in humans and other mammals have shown that the formation of the placenta is the result of the action of two cascades of hormonal signals. The first is produced by the follicle stimulating hormone which increases the intracellular concentration of cyclic AMP. This, in turn, leads to the production of phosphorylated proteins that affect transcription of DNA into RNA. The follicle stimulating hormone promotes the maturation of the egg follicles, which secrete a second hormone, estrogen. This initiates the second cascade. The estrogen prepares the walls of the uterus, where the placenta is formed, and leads to the surge of the luteinizing hormone. In a further step this last hormone provokes the secretion of progesterone that allows the full pregnancy of the fertilized egg. Prostaglandins are also involved in this process (Dorrington 1979, Challis 1979, Friday and Ingram 1985).

This rigid and ordered sequence of chemical signals is the basis of the coherent framework of structures and functions that is the placenta. Once an initial threshold of a given hormone is reached a cascade ensues that cannot be stopped and which results in a novel and harmoniously constructed organ.

The Emergence of the Placenta is Not Directly Related to the General Environment

Most fish live in water, and their eggs are released into this medium, where they are fertilized and where they develop independently of their parents. Placental fishes occur in the same habitat.

The same also holds for amphibians. *Pipa americana* has a well developed placenta, whereas *Xenopus laevis* eggs develop freely, but both species live in water. Moreover, the eggs of fish and amphibians possess large quantities of reserve materials and are not in need of a large supply of organic substances from the bodies of their parents (Rabaud 1934).

Although marsupials and most placental mammals are terrestrial, they differ sharply in the possession of this organ.

Marine or freshwater conditions do not decide the appearance of the placenta. The majority of marine bivalves (molluscs) are egg laying but some freshwater species are also egg laying or have viviparity alone.

Hence, the emergence of the placenta is not dictated by the general environmental conditions.

Presence and Absence in Closely Related Species

One result that contributes to elucidate the periodic occurrence of the placenta is its presence and absence in closely related species as well as its absence in large groups of animals.

There are about 600 species of cockroaches that lay eggs but only one is known to be viviparous. Viviparity is known from 11 orders of insects. It is rare in some but quite widespread in others.

The situation is most clear within the fish genus *Poeciliopsis* where there are 6 closely related species, three of which lack placentas and three that have this organ.

The frogs and toads belong to the order Anura. Some families, such as tree frogs (Hylidae) have a placenta, whereas their close relatives, the Ranidae, do not possess this organ. The same is true of fishes and reptiles in which close relatives do not have this organ.

The Placenta of Reptiles "Has Evolved on More than 100 Separate Occasions"

South American lizards of the genus *Mabuya*, exhibit: long gestation periods, ovulation of tiny eggs, and placental supply of the nutrients used in development. Two African skinks *Mabuya ivensii* and *Eumecia anchietae* have the same placental features. These *Mabuya* species contrast with other reptiles which exhibit either oviparity or viviparity (with simpler placentas).

Live-bearing reproduction is widespread among lizards and snakes, it is found in among 20% of the approximately 6,000 squamate species and "has evolved on more than 100 separate occasions" (Flemming and Blackburn 2003).

Ostrovsky (2013) adds: "This highly effective mode of parental care evolved independently in more than half of all metazoan phyla". The same opinion is expressed by Moosbrugger *et al.* (2012) who state that the "widely scattered distribution among the animal kingdom indicates multiple, independent origins of extraembryonic nutrition".

The Periodic Table of the Placenta

Column 1 Presence in Invertebrates

The placenta is an event emerging earlier in evolution among the simplest invertebrates (Onychophorans) and appearing again and again in the most diverse invertebrates such as: Cnidarians, Worms, Insects, Molluscs, Scorpions and Ascidians (lower Chordates). Although it evolved, its basic structure and function is not altered by increased organism complexity.

Column 2 Presence in Vertebrates

The same situation was repeated when the vertebrates emerged. It suddenly resurged in sharks and sea horses which belong to two different groups of fish.

Table 5.1 Periodic Table of the Placenta

Increasing grade of complexity	PRESENCE IN INVERTEBRATES	PRESENCE IN VERTEBRATES	PRESENCE AND ABSENCE IN PLANTS
1	ONYCHOPHORANS ENTOPROCTS BRYOZOANS	CARTILAGINOUS FISH SHARKS RAYS TELEOST FISH SEA HORSE	ALGAE Absent
2	SPONGES CNIDARIANS	AMPHIBIANS SURINAM TOAD BLACK SALAMANDER	MOSSES Present
3	RIBBONWORMS FLATWORMS ROUNDWORMS SEGMENTED WORMS	REPTILES AFRICAN SKINKS LIZARDS SNAKES	FERNS Present
4	SCORPIONS FRESH WATER CLAMS	MARSUPIALS BANDICOOTS	CYCADALES No indication
5	COCKROACHES WASPS FLIES	EARLY PLACENTAL MAMMALS LOWER PRIMATES	CONIFERS Present
6	ASCIDIANS SALPS Lower chordates	LATE PLACENTAL MAMMALS HUMANS	FLOWERING PLANTS Present

In this property the main organ has increased in complexity during evolution as is the case in vertebrates and plants. In invertebrates the placenta has not been studied in detail. As a consequence it is not known whether it also increased in complexity as invertebrates reached a higher stage of differentiation. The periodicity is revealed by the fact that suddenly, in sharks, the placenta attains a development that rivals that of mammals, and that in reptiles it emerged on more than 100 separate occasions. Moreover it has a punctuated appearance in both invertebrates and early vertebrates (Original).

Then, without previous announcement, it developed in certain species of toads and lizards. In some marsupials it took particular forms. Only in the placental mammals did it become established as their major feature.

Column 3 Presence and Absence in Plants

The earliest plants showed no signs of its existence, but it turned out in mosses and ferns. It has been less studied in the cycadales and gymnosperms (conifers). It became a permanent organ in angiosperms (flowering plants) where it has been extensively analysed at the gene level (Table 5.1).

Chapter 6

Penis Evolution from Worms to Humans — A Double Penis Occurs in most Unrelated Species

The Penis May Seem Irrelevant in Regard to Periodicity, but it Turns out to have an Equally Significant Punctuated Appearance

The existence of the penis in humans and mammals is such a general feature that one tends to take it for granted as a natural, and obligatory, evolutionary event.

However, within the vertebrates it has been practically eliminated in whole groups such as the birds, and is absent in fishes. But before the fishes arrived it was highly developed in some invertebrates including insects and molluscs.

The penis has an evolutionary history of its own like any other organ that is part of living organisms.

What Characterizes the Vertebrates is a Skeleton Securing the Whole Body

The main structural feature of the vertebrates is their bony skeleton.

How this main body component has arisen from the structure of invertebrates remains a yet unchartered field of research. No direct relevant genetic information is available so far.

The skeleton was a novelty which emerged in the Cambrian deposits (Carroll 1988). What is still more difficult to explain at present in genetic terms is that once the skeleton appeared it changed its shape and number of components in a dramatic way giving rise to frogs as well as whales. Yet it remained an easily recognizable and well-defined structure in which each vertebra (or other component) finds its homologous part throughout the vertebrates. This is why the main groups of vertebrates can be put side by side without difficulty starting with the ancient fishes and finishing with humans.

As stated by Wellik (2009): "The axial skeleton of all vertebrates is comprised of similar structures that extend from anterior to posterior along the body axis: the occipital skull bones, cervical, thoracic, lumbar, sacral and caudal vertebrae. Despite significant changes in the number and size of these elements during vertebrate evolution, the basic character of these anatomical elements, as well as the order in which they appear, has remained strikingly similar".

Vital Organs in Mammals, Such as the Penis, May have Bones or Lack them

Since vertebrates are characterized by their skeleton, it could be supposed that bones had shaped the totality of their body structure. This is not the case.

During the evolution of mammals several external organs have been formed without an internal bone edifice. It could also be thought that they are of secondary importance. But no, they may be of large dimensions and perform vital functions.

Large ears that can be oriented in different directions have the important function of allowing a better location of sound. They consist of a nearly transparent tissue having no bones, as in the Antelope jackrabbit.

Protruding snouts have no bones that extend to the terminal region of this organ. They are distinct in the Aardvark, Elephant shrew and Proboscis monkey. The name of these last two animals derives from the resemblance of this body part with the elephant's trunk. This large organ is used by the African elephant in many ways (among them courtship) and it can be seen, when compared with the animal's skeleton, to be free of bones.

Mammals derive their name from the mammary glands that are common to all their females. Such an important organ protrudes from the frontal part of the trunk in humans and in other animals is located in the lower ventral region, as in cows. No bones are present in mammary glands.

The penis represents an intermediate situation. In humans it has no bones but in the dog and the otter it contains a bone. The penis bone is a feature of all Carnivores. Man is an exception, as in most primates a skeletal part is found in their penis called *os penis* or *baculum* (Fig. 6.1).

Penis Erection is Created By Fluid flow which Substitutes with Equal Efficiency the Function of Bones

Worms possess a hydraulic skeleton created by the fluid of the body cavity. In velvet worms their many legs are hollow and unjointed. The rigidity of these legs is not maintained by bones but by water pressure. The muscles present in the body wall oppose the hydrostatic pressure that moves the animal (Margulis and Schwartz 1982). It is thus apparent that water alone performs with equal efficiency the function that in vertebrates is accomplished by the mineralized tissues of the bones, and in insects and myriapods by the outer skeleton.

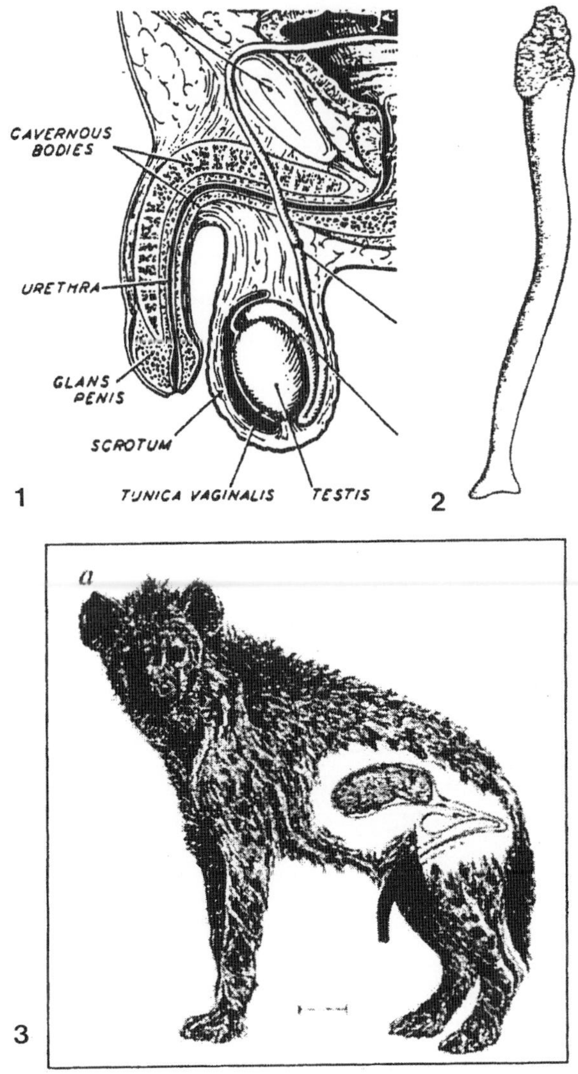

Figure 6.1 Penis in mammals. (**1**) **Primates**. *Homo sapiens*. Penis with its ejaculatory canal and seminal vesicle which is also connected with a prostate gland. (**2**) **Primates**. The bone (os *penis*) develops in the penis of many mammals and nearly all primates man being an exception. (**3**) **Penis in females**. **Spotted Hyaena**, *Crocuta crocuta*. Spotted female hyaenas give birth through the "hypertrophied clitoris, similar in size and structure to the male penis". Females copulate with as many as four males, copulation takes place through the female penis, because they have no external vagina.

Hydraulic pressure is also used in the erection of the penis.

When a pulmonate snail copulates, both the sensory and locomotory organs of the head and foot are flaccid and crumpled. This condition is followed by a change in blood flow within the body. A large proportion of the blood volume in the snail is involved in the dilatation of the penis and other genital structures. The same happens in humans during copulation. A previously flaccid penis becomes erect as the result of a change in blood flow. Also the double penis of lizards is everted hydrostatically by blood and lymph (Arnold 1986).

An Effective Penis with Coherent Accessory Structures Occurs Already in Flatworms

The flatworms of the class Trematoda, known as flukes, are one of the major groups of parasites. Most are no longer than a few centimeters but they may attain a length of 7 meters.

These invertebrates are highly complex. They have a mouth that leads into a muscular pharynx. The pharynx passes into an oesophagus that connects to a long intestine. The reproductive system is elaborate in both males and females: There are usually two testes. From each testis arises a sperm duct which unite anteriorly and then enter the copulatory organ (called penis or cirrus). This has the shape and function of the penis present in other invertebrates and vertebrates. Moreover, it is accompanied by an ejaculatory duct, prostate gland, seminal vesicle and vas deferens (corresponding structures accompany the human penis).

During copulation the penis of one worm is inserted into the vagina of the other worm and sperm are ejaculated. The prostate gland provides semen for sperm survival which travels down the uterus to be stored in the seminal vesicle (Barnes 1980) (Fig. 6.2).

The Sudden Occurrence of the Penis in Simple Phyla of Invertebrates

The animals of the phylum Gnathostomulida have transparent bodies due to the absence of an external cuticle. They also lack a circulatory and a respiratory system as well as a skeleton. These extremely simple invertebrates have already a fully functional penis. *Problognathia* has a penis stiffened by a stylet. The animal injects sperm packets between the skin and gut of its mate. The sperm swims to a storage sac called a bursa, fertilization taking place internally.

A separate phylum is built by the Gastrotricha. They are also transparent worm-like animals without circulatory, respiratory or skeletal organs, their body fluid acting as an hydraulic skeleton. The marine species are hermaphrodites. In *Dactylopodola* packets of sperm are transferred from individuals that behave as

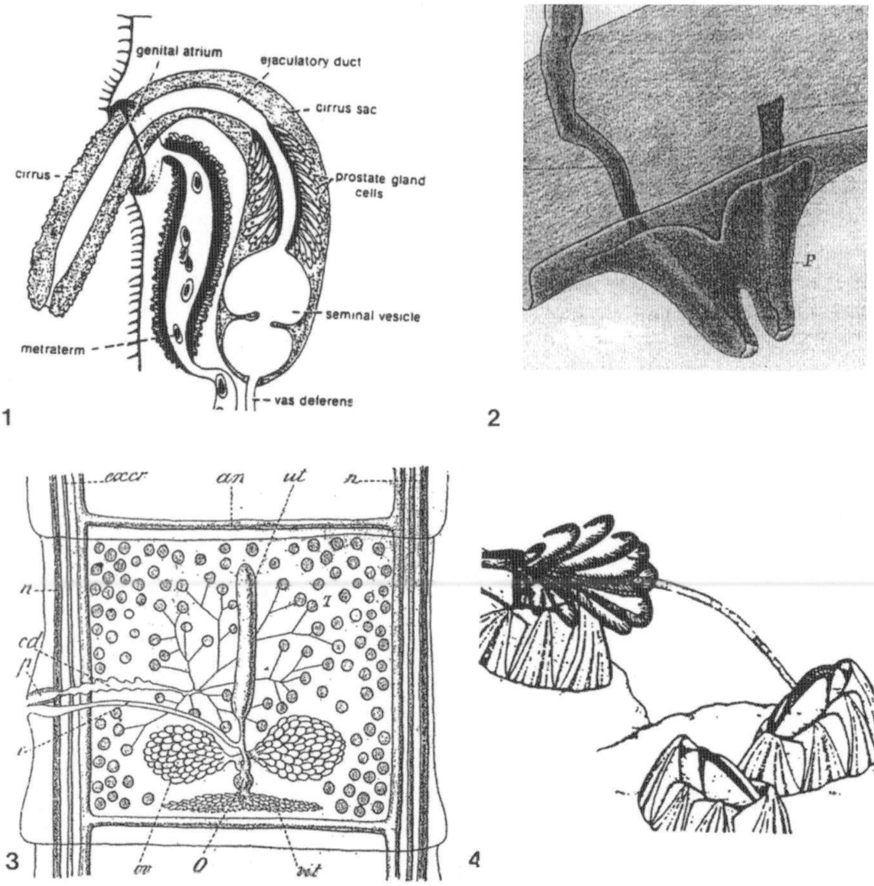

Figure 6.2 Penis in worms and crustaceans. (**1**) **Flatworms**. Flukes, *Fasciola hepatica*. The protruding penis (cirrus). This organ has a seminal vesicle, a prostate gland and an ejaculatory duct. (**2**) **Crustaceans**. Isopoda, *Cymothoa aestroides*. Penis and part of ventral wall. (**3**) **Flatworms**. Cestodes, *Taenia saginata*. Body segment showing ovaries (ov) and penis (p). (**4**) **Crustaceans**. Barnacles. Copulating barnacles. The long penis of the individual on the left is inserted on the individual on the right. (Based on photograph from R. Sisson.)

males to those that behave as females. The end of the sperm duct serves as a penis.

All Acanthocephalans, or spiny-headed worms, are parasites of vertebrates. More than 600 species live in soil and in fresh or salt water, nearly anywhere in the world depending on the habitat of their hosts. Circulatory, digestive and respiratory organs are lacking. The sexes are separate and reproduction is entirely sexual. The males have a pair of testes from which sperm ducts lead to a penis. The male bursa holds the female while sperm are released through the penis into her vagina. Both fertilization and embryonic development take place inside the female (Margulis and Schwartz 1982).

The Longest Penis of any Animal Occurs in Invertebrates

Barnacles are Crustaceans which are exclusively marine attaching usually to rocks. They act both as functional males and functional females, possessing eggs, testes and a penis. When searching for a mate, a male extends the penis and seeks receptive females by sweeping the penis among neighbors (Fig. 6.2).

The penis of barnacles is a long, flexible, muscular structure which can stretch to many times its resting length, in addition it grows rapidly in the weeks before mating. The penis of the barnacle is extended by hydrostatic pressure and is "the longest penis of any animal (up to eight times their body length)" (Neufeld and Palmer 2008). In *Semibalanus balanoides* successful mating takes place in individuals located as far as 35 mm from their partners. This penis has impressive abilities. The penises of *Balanus glandula* from wave exposed shores are shorter, stouter and more massive than those from quiet waters. Barnacles have a dramatic capacity to change the size and shape of their penises according to coastal wave action (Neufeld and Palmer 2008).

Molluscs with a Single Large Penis

Several species of marine gastropods (molluscs) house their body in beautiful shells. What is remarkable is that they exhibit a single and large penis as a normal feature. *Buccinum undatum* has a most distinct penis that is exerted from the back part of the mantle. An equally impressive penis sticks out over the tentacles of the male of *Busycon perversens* (also a marine gastropod).

The same happens among herbivorous land gastropods like the snail *Helix*. Its single penis, which has a long flagellum that looks like a whip, is accompanied by a dart sac and a copulatory bursa (Barnes 1980).

One Penis is not Enough — A Double Penis Occurs in Crustaceans and Molluscs

Many organs in living organisms are found as pairs: lungs, kidneys, testicles, ovaries.

Most vertebrates, including humans, have only one penis but the invertebrates became more advanced in this respect, they build two penises instead of a single one. Among the collected specimens of *Semibalanus balanoides* (Crustaceans) one had two penises (3.1 and 3.4 mm in length) with quite normal appearance. The formation of a double penis was attributed to a genetic mutation but may have resulted from an injury, because if the penis is removed from these animals it can regenerate. Sister groups of barnacles (Crustaceans), as well as their ancestors, have double penises. This animal was considered to represent a reversion to the ancestral type (Hoch and Yuen 2009).

Parasitic Isopoda (Crustacea) also exhibit a double penis, which is identical to that of free living forms such as *Asellus aquaticus* (Bullar 1876) (Fig. 6.2).

The presence of a double penis was reported in *Rapana venosa* (molluscs) inhabiting the Black Sea (Ukraine). The double penis is divided into two parts, similar in shape, with typical "hooks" on their end. The main part of the penis is 20 mm (Fig. 6.3). Two other cases of double penis were reported from Brazil (*Leucozonia nassa*) and Tunisia (*Heraplex trunculus*) (Govorin 2009).

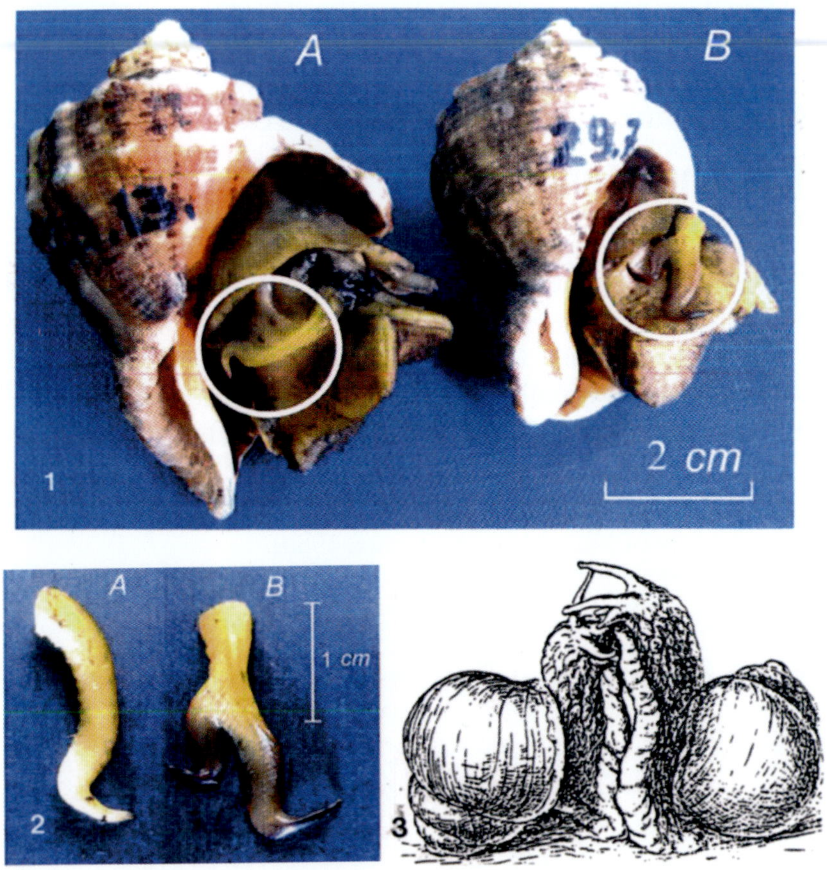

Figure 6.3 Penis in molluscs. (**1**) **Molluscs**. *Rapana venosa*. Males of the veined rapa whelk: **A**, normal single penis; **B**, animal with double penis. (**2**) **Molluscs**. Gastropods, *Rapana venosa*. Isolated single penis (left) and double penis (right). (**3**) **Molluscs**. Pulmonates, Landsnails. Copulation in hermaphrodite snails. One animal may function as male extruding its penis, while the other functions as a female.

A double penis turns up in other molluscs. It has been described in *Helix aspersa* (Gastropods) which has two but may even have three penises (Alaphilippe *et al.* 1960).

Insect Genitalia Display Great Morphological Diversity

The form of the components of the external genitalia of insects are very diverse being of considerable value in species determination. The female genitalia are usually simpler and less varied than those of males.

The penis (or phallus) may be simple or bilobed and is accompanied by a series of appendages. These terminal abdominal segments have excretory and sensory functions in all insects and participate in mating.

Besides the penis there are in insects additional structures that grasp and hold the partner during mating. The penis, and these structures, are so diversified that it is difficult to establish their homologies when species are compared.

Typically there is a median tubular penis which often has an inner tube that is everted during insemination. Lateral to the penis there is a pair of lobes. This three-part complex is well exemplified in many beetles (Gullan *et al.* 2000).

Right-handedness Prevails — When two Penises are Present the Right One is Usually Used, the Left One is a Spare Part

The number of penises varies in earwigs (insects). Some families have two penises (right and left) others a single penis. Following the collection of over 600 specimens, and study of their behavior, Kamimura (2006) discovered that when double penises are present only a single penis is usually used per copulation. Although there was no detectable morphological differentiation between the left and right penises, males of *Labidura riparia* predominantly used the right one for insemination (90% of cases).

This right-handedness developed without any experience — of the wild-caught males only a few were left-handed. The left penis is merely a spare intromittent organ.

Behavioral asymmetries in the use of the organs of vertebrates have been widely recorded, such as the right-handedness of human hands, but in invertebrates these are rare.

Fishes do not have a Penis But Sharks and Rays Use a Modified Fin Instead

Bony fishes (about 23,500 species) usually fertilize their eggs by releasing the sperm in sea water at sites where the eggs have been released or deposited. In a few species there is an internal fertilization.

However, in Cartilaginous fish (sharks and rays) fertilization is internal in all 810 species. The male passes sperm into the female's cloaca via a modified pelvic fin that functions as a rudimentary copulating organ (Burnie 2004).

In most amphibians (frogs and toads) fertilization is external and in newts and salamanders it is internal, the sperm being transferred to the female in gelatinous capsules. However, in Caecilians the male has a penis-like organ which he uses to insert his sperm directly into the female (Burnie 2004).

Double Penises in Lizards and Snakes

The emergence of a double penis, which rarely appeared in invertebrates, bypassed the fishes and many early reptile groups to become suddenly firmly established in Lizards.

Lizards are described as the most successful group of reptiles with about 4,500 species distributed in 19 families. These include Chameleons, Iguanas, Geckos, and different types of Lizards. Among them is the family Lacertidae (wall and sand lizards) known for the presence of a double penis.

Arnold (1986) made an extensive study of this family investigating in detail 740 specimens distributed over 26 genera.

The lacertid double penis is a paired and intromittent organ that is quite unlike that of other squamate reptiles (the group that includes all lizards, snakes and worm-lizards) (Halliday and Adler 2004). The distal lobes of the retracted organ are folded in a complex manner and there is a supporting structure called the *armature*. This incorporates blood vessels and retractor muscles participating in erection. Some lacertids possess a double penis that is unarmatured, which is considered a primitive condition. There are also intermediate cases between the two major types. The penises are covered by minute spines and hooks and are everted hydrostatically by blood and lymph.

Like in insects, very similar species have penises that differ in size, asymmetry and simplification of their components (Arnold 1986).

Differently from insects, males of the lizard *Anolis sagrei* demonstrated a significant tendency to use the left and right hemipenis alternatively when mated daily for 7 days with different females. The two intromittent organs are independent, only one of them being inserted into the female cloaca. The manner in which the male mounts the female determines which of the two penises will be used (Tokarz 1988, 1989) (Fig. 6.4).

Double penises have also been described in snakes of the family Colubridae comprising 60% of all snakes. Two Brazilian species of *Tantilla* have double penises which are naked or covered by spinules (Sawaya and Sazima 2003) (Fig. 6.4).

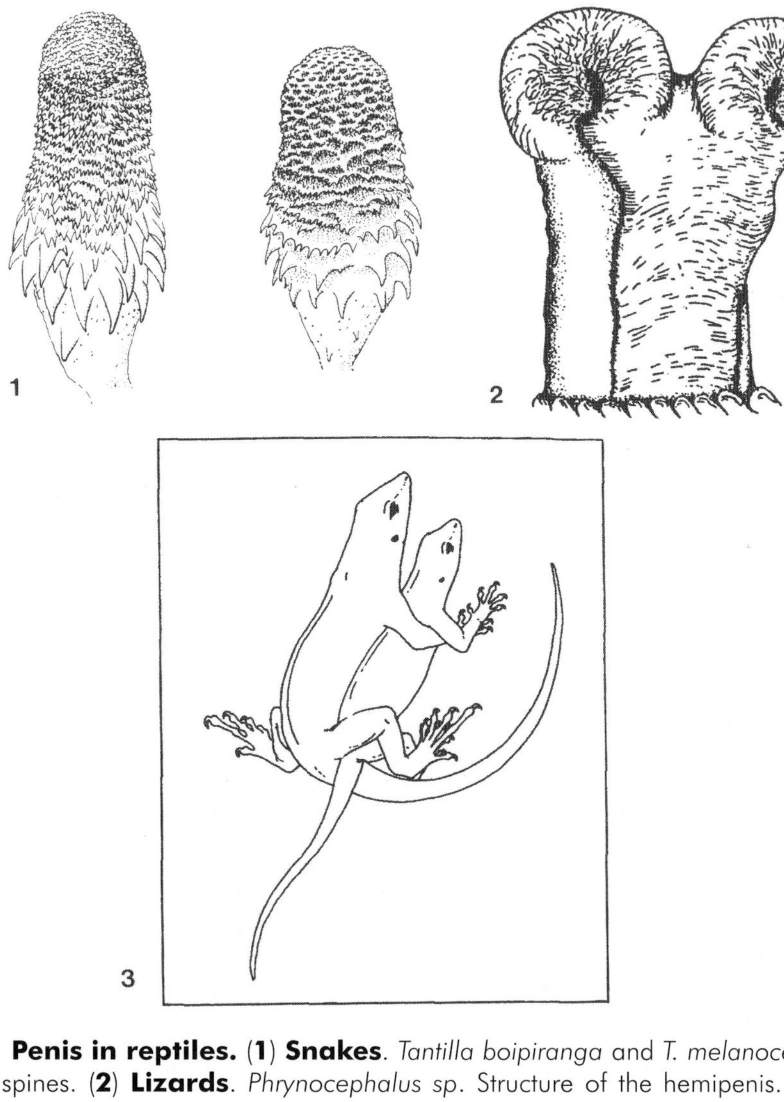

Figure 6.4 Penis in reptiles. (**1**) **Snakes**. *Tantilla boipiranga* and *T. melanocephala*. Penises with spinules and spines. (**2**) **Lizards**. *Phrynocephalus sp*. Structure of the hemipenis. (**3**) **Lizard**. *Anolis sagrei*. Copulatory posture. The male's tail is pointing to the right which means that the right hemipenis is intromitted. Drawing based on photographs of copulating lizards.

The Birds were Neither Allowed to have Penises Nor Teeth — DNA Activation and Inhibition are Responsible in both Cases

Birds comprise over 9,000 species that have been described and illustrated with the utmost detail in the "Handbook of the Birds of the World" edited since 1992 by del Hoyo, Elliot and Sargatal, Lynx Edicions, Barcelona, Spain (17 volumes). What is unique about this work is that it covers all species and subspecies from all over the world.

As Herrera *et al.* (2013) point out "One of the most puzzling events in evolution is the reduction and loss of the phallus in birds". The same could be said of their teeth. Equally unexplained is the fact, that although their reptilian ancestors had teeth, or tooth-like serrations on their bills, none of the extant species of birds have teeth (some exhibit only bill serrations such as the hornbills).

The fishes that preceded the birds, have highly developed teeth. These are even replaced throughout their life time (*e.g.* cartilaginous species). Most mammals, that arrived later in evolution than the birds, have their mouth covered with teeth. Humans, who emerged only a few million years ago, were endowed with the luxury of having two dentitions, their milk teeth are discarded and replaced by a set of new ones.

Hence, the impressive ability to create and recreate teeth in fishes and mammals finds its opposite in the birds where they were eliminated.

The same concerns the penis. Occurring in simple invertebrates and in some families of reptiles, it is a general feature of mammals including humans, but among birds only about 3% have been allowed to get a penis. Exceptions are: the ostrich, geese and ducks (*e.g. Anser anser* and *Anas platyrhynchos* respectively) which have well developed penises. The others are obliged to use an inefficient cloaca. This is one of the reasons why birds copulate several times in succession.

The genetic events leading to the elimination of the penis are now being understood. Landfowl (chick, quail) which lack phallus were compared with waterfowl (duck and geese) which possess an intromittent penis. The lake Duck (*Oxyura vittata*) has a penis 20 cm long extending to circa half of the bird's body length (McCracken 2000). The gene involved in this process is *Bmp4*. Inhibition of the Bone morphogenetic protein signaling (*Bmp* gene) in chick genitalia rescues cells from cell death and prevents phallus regression whereas activation of *Bmp* signaling in duck genitalia induces a chick like pattern of cell death. Distal *Bmp* activity is necessary and sufficient to induce cell death in genitalia.

The duck and geese are Anseriforms and considered to be ancestral but the most basal group of birds are the Palaeognathae (*e.g.* the emu which, like its close relative the ostrich, has a penis). The role of the *Bmp4* gene in the evolution of this organ is depicted in Fig. 6.5. McCracken (2000) concludes that: "the evolutionary

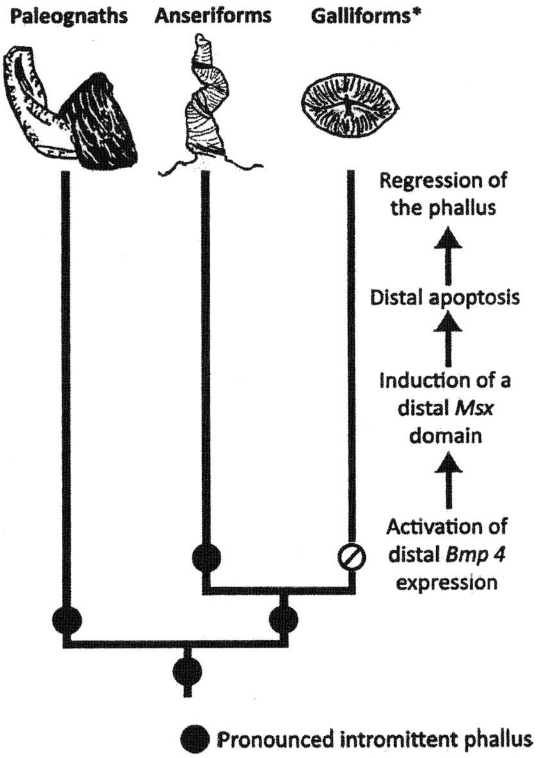

Figure 6.5 Penis in birds. Evolutionary loss of penis in galliforms. Black circles indicate presence of an intromittent phallus, open circle, with diagonal line, indicates loss of intromittent phallus. Asterisk indicates all galliforms with the exception of *Cracids*, which have a fully intromittent phallus. The *Msx* gene is a downstream effector of the *Bmp* signaling. *Bmp4* bone morphogenetic protein gene that induces apoptosis (cell death). Paleognaths (e.g. emu, ostrich). Anseriforms (waterfowl), Galliforms (landfowl).

reduction of the intromittent phallus in galliform birds (chicken) occurred not by disruption of outgrowth signals but by *de novo* activation of cell death by *Bmp4*".

Of particular significance is that *Bmp* genes were also found to participate in the elimination of teeth. Hence, a single group of genes is responsible for the two basic traits, in the birds, that before seemed unrelated (Herrera *et al.* 2013). As these authors point out, "selection pressures" were invoked to "explain" this phenomenon, but the genetic data disclose instead that a DNA event is the immediate cause of the penis elimination.

Marsupials have a Two-lobed Penis

The division of the penis into two separate elements is not only present in invertebrates and reptiles but suddenly emerged when the mammals arrived on the planet. These were the marsupials which exhibit a penis that became divided into two lobes.

Whereas the vagina of placentals is a single structure, marsupial females have two vaginae. This double condition in the female genitalia is often matched in the male by a two-lobed penis (Macdonald 2002) with the exception of the kangaroos in which the penis is undivided (Wilson and Mittermeier 2015).

The Female Clitoris and the Male Penis are Homologous Structures

With the exception of the primitive monotremes all mammals bear their young live and the penis is a general feature of their body. The labia minora flanking the vagina are homologous to the part of the male penis surrounding the urethra, and the labia majora are homologs of the scrotal swellings in the male. Females have a glans clitoridis that is homologous to the male glans penis and the main body of the clitoris contains erectile tissue comparable to the male corpora cavernosa (Wilson and Mittermeier 2009). This strict homology implies that there is a latent bisexuality, which is also patent in the presence of nipples in the human male. Testicular feminization, found in mammals including humans, is due to the *Tfm* gene located in the X chromosome.

Among mammals the size of the penis varies considerably since it is mainly related to body size. A pygmy mouse which is 8 cm long has a penis that only measures a few millimeters, whereas in the Gray Whale (*Eschrichtius robustus*) the penis can be up to 1.7 meters long. Although enormous, it is not very large in comparison to its body length which attains 15 meters (Wilson and Mittermeier 2014).

Female Hyaenas have a Large Penis that Functions as a Birth Canal

Hyaenas are Carnivores known for their high mental ability and an extraordinary social system, although there are only four living species they show a remarkable social behavior which includes aggressiveness and dominant females having priority of access to food.

Brown and Striped hyaenas have long hair which they raise in situations of conflict, but the Spotted hyaena (*Crocuta crocuta*) has short fur. The female genitalia of *C. crocuta* are almost indistinguishable from the male's. The clitoris is large like a penis and the vaginal labiae are fused to form pseudo-testes. Females give birth through the penislike clitoris. In the other hyaena species the sexual organs are normal (Macdonald 2002).

The female clitoris is similar in size and structure to the male penis and is also erectile. Copulation takes place through the large clitoris.

Neonatal mortality is high due to the large foetuses being squeezed through the narrow peniform clitoris, the birth canal being 60 cm, twice the length of that in a typical mammal of similar size (Frank *et al.* 1995) (Fig. 6.1).

How to Reverse Sex by Changing Temperature or the Environment

Sea slugs and oysters, can reverse their sex, from male to female and back again, several times in a season (Margulis and Schwartz 1982). Some bony fish change their sex in the course of their adult life: males that become females are larger and accomodate more eggs. Females that become males assume dominant positions in the social group (Burnie 2004).

The worm *Bonellia* has environmental sex-determination, albeit with genetic involvement. If a larva develops independently it becomes female, if it is influenced by pheromones produced by the adult female's proboscis it develops into a male.

Sex is strongly determined by the temperature at which the embryo develops in turtles and crocodilians (Abercrombie *et al.* 1990). In many turtles low temperatures produce male offspring and high temperatures produce females. In crocodilians males are produced at intermediate temperatures and females at low and high temperatures.

Environmental contaminants such as pesticides and their metabolites (DDT and others) also caused sex reversal in the American alligator (*Alligator mississippiensis*). Besides male juveniles from contaminated lakes in Florida exhibited reduced phallus size (Halliday and Adler 2004).

Effects of Hormones on Vertebrate and Invertebrate Genitalia

Testosterone is the main male sex hormone in vertebrates. It is a steroid produced by the testes, responsible for maintaining testes and for male growth spurt at puberty. It promotes protein synthesis, sexual behavior and maturation of germ cells.

Its action in humans has been widely studied and the effects of testosterone on sexual organs and sexual behavior of lizards, fishes, frogs, birds and rodents has been the object of several studies showing that gonadal steroids play a direct role in regulating the differentiation of sexual traits (Lovern *et al.* 2004).

In invertebrates Köhler *et al.* (2007) have shown that echinoderms respond to androgens (male sex hormones, *e.g.* testosterone and its derivatives) and in *Branchiostoma* an estrogen receptor has been cloned. In aquatic molluscs sex steroid receptor-like proteins are abundant in gastropods, bivalves and cephalopods and estrogen receptor-like proteins are present in Crustacea.

Coherence Associated with Penis Formation — The "Lock-and-Key" Correspondence

The penis is never alone. It is usually accompanied by several additional organs and structures that complete and enhance its function. Some of the organs are the prostate gland, seminal vesicle and vas deferens. Among the structures are "hooks", lobes and other terminal segments that enhance mating.

A more impressive type of coherence is especially evident in insects. Biologists have for years been intrigued by what they called the "lock-and-key" correspondence between the male and female genitalia. This is a phenomenon with the following characteristics: (1) The penis and the vagina of insects are very diverse in structure. (2) They frequently exhibit species specific morphology in otherwise similar species. (3) The genitalia are complex but correspond tightly between the sexes. They are considered to fit only the conspecific female genitalia. (4) The particular structure of the male sensory appendages fails to fit the plate of the "wrong" female. As Gullan *et al.* (2000) point out there have been a series of interpretations proposed in an attempt to explain this coherence, most of them based on hypothetical types of selection. As they point out, none of them elucidate the many well documented examples of "lock-and-key" coherence.

The Emergence of the Penis is not Directly Related to the General Environment or Organism Complexity

A significant aspect of the evolution of the penis is that it occurs in marine species, such as the molluscs and crustaceans, in which the sperm could be easily released into the water, as is the case in many fish. Besides, the existence of this organ in vertebrates, does not represent an evolutionary solution that is related to life on land. A simple cloaca exists in fish and amphibians. Reptiles living on land or in

Table 6.1 Periodic Table of the Penis

Increasing grade of complexity	ABSENCE IN INVERTEBRATES	PRESENCE IN INVERTEBRATES	ABSENCE IN VERTEBRATES	PRESENCE IN VERTEBRATES	PRESENCE OF DOUBLE OR BILOBED PENIS
1	CNIDARIA CTENOPHORA MESOZOA NEMERTINA	FLATWORMS FLUKES	BONY FISH *Most species*	FISH SHARKS RAYS Modified fin	CRUSTACEANS Double
2	ROTIFERA KINORHYNCHA ENTOPROCTA NEMATODA	GNATHOSTOMULIDA *Problognathia*	AMPHIBIANS FROGS	AMPHIBIANS CAECILIANS Penis-like organ	MOLLUSCS Double
3	NEMATOMORPHA ECTOPROCTA PHORONIDA BRACHIOPODA	GASTROTRICHA *Dactylopodola*	AMPHIBIANS TOADS	REPTILES LIZARDS SNAKES	INSECTS Double
4	PRIAPULIDA SIPUNCULA ECHIURA ANNELIDA	ACANTHOCEPHALA Spiny-headed worms	REPTILES TURTLES	BIRDS OSTRICHES DUCKS	REPTILES LIZARDS SNAKES Double
5	PENTASTOMA ONYCHOPHORA POGNOPHORA	MOLLUSCS *Buccinum* CRUSTACEANS *Balanus*	REPTILES CROCODILIANS	MARSUPIALS Most species	MARSUPIALS Bilobed
6	ECHINODERMATA CHAETOGNATHA HEMICHORDATA	INSECTS Many species with additional structures	BIRDS Most species	PLACENTALS Most species	PLACENTALS Single

The main feature from the point of view of periodicity, is that the penis is absent in a large number of invertebrate phyla as well as in most groups of vertebrates. It emerges suddenly in totally unrelated groups of invertebrates and vertebrates to establish itself as a permanent organ in mammals (Original).

water have copulating organs, in addition to the cloaca, that function as penises. In birds the situation is more extreme since most species (though they live on land) do not have a penis.

The emergence of the penis occurred under the following conditions:

1. The organ emerged in totally unrelated groups.
2. It appeared in environments where a need for it did not seem apparent.
3. Its emergence failed to occur in situations in which it could have been useful for the propagation of the organism (birds have had 150 million years of evolution at their disposal).
4. Its degree of differentiation, as well as its number per individual are unrelated to organism complexity.

The Periodicity of the Absence and Occurrence of the Penis in Invertebrates

The absence of penis, in whole groups of invertebrates, is as significant for the understanding of its periodicity as its presence. Thirty-one phyla of invertebrates, described in detail by Margulis and Schwartz (1982), reveal the following: The penis is absent in: Cnidaria, Ctenophora, Mesozoa, Nemertina, Rotifera, Kinorhyncha, Entoprocta, Nematoda, Nematomorpha, Ectoprocta, Phoronida, Brachiopoda, Priapulida, Sipuncula, Echiura, Annelida, Pentastoma, Onychophora, Pognophora, Echinodermata, Chaetognatha and Hemichordata (22 phyla).

A single penis is present in: Platyhelminthes, Gnathostomulida, Gastrotricha, Acanthocephala, Mollusca, Arthropoda (Crustaceans and Insects) (6 phyla).

Placozoa and Porifera, have specialized cells but no true organs, and as such a penis cannot be expected to be formed. Tardigrada are an uncertain case because a cirrus is mentioned but the presence of a penis is not explicit.

The Periodic Table of the Penis

Column 1 Absence in Ivertebrates

Twenty-two phyla are distributed, according to increased complexity.

Column 2 Presence in Invertebrates

Six phyla are included. In them the penis appears without previous announcement.

Column 3 Absence in Vertebrates

The lack of penis is most evident from fishes to birds. As the vertebrates emerge it is inhibited in most groups.

Column 4 Presence in Vertebrates

Unexpectedly the penis emerges in special groups from fishes to birds to become well established only in the marsupials and the placentals.

Column 5 Presence of Double or Bilobed Penis

This is a punctuated event occurring from crustaceans to marsupials.

Chapter 7

Regeneration Starts in Crystals, Expands in Plants, but Slows Down in Higher Vertebrates

The most Significant Aspect of Regeneration is its Ability to Produce, without External Intervention, The Original Pattern — Memory of Cellular Order

By definition regeneration is the "Restoration by regrowth of parts of the body which have been removed, as by injury or autotomy" and "It involves the re-establishment of local tissue differentiation" (Abercrombie *et al.* 1990).

This definition misses several significant properties. The "parts of the body" that are added not only replace those that were removed but rebuild a pattern identical to the original one. Moreover, this pattern agrees with the coherence of the organism as a whole.

Once an injury arises in the form of an amputation, no mutations occur, or there is a sorting out of cell types, as seems to be the case in cancer, where uncontrolled growth and chaotic cell formation ensues. In regeneration the appropriate genes are called to action and the cells only develop a preprogrammed cellular order that does not deviate from the original configuration of the organism. This program is actually a memory which remembers the whole organization of the cell, the organ or the organism depending on the species. Without it no identical pattern could be recreated.

Besides the whole process is essentially independent of the environment, in its direction and development, and is also independent of the organism's advantage or disadvantage, it simply takes place as an obligatory consequence of DNA activation at the proper sequences followed by signaling molecular cascades that, within hours, restore the original pattern, whether the organism likes it or not.

Regeneration is a Pure Atomic Process — Before it Appeared at the Biological Level it Existed in Crystals and Minerals

Of fundamental importance is that regeneration started in crystals before biological evolution arrived.

It was Louis Pasteur, the developer of modern microbiology, who described for the first time that regeneration occurred in crystals. If a crystal of ammonium bimalate gets its end broken, regeneration ensues by the addition of identical atoms, which lead to the restoration of the original pattern (Blaringhem 1923). The same happens in other crystals. Following breakage, the liquid crystals of ammonium oleate regenerate, building the original configuration (Bonner 1952) (Fig. 7.1).

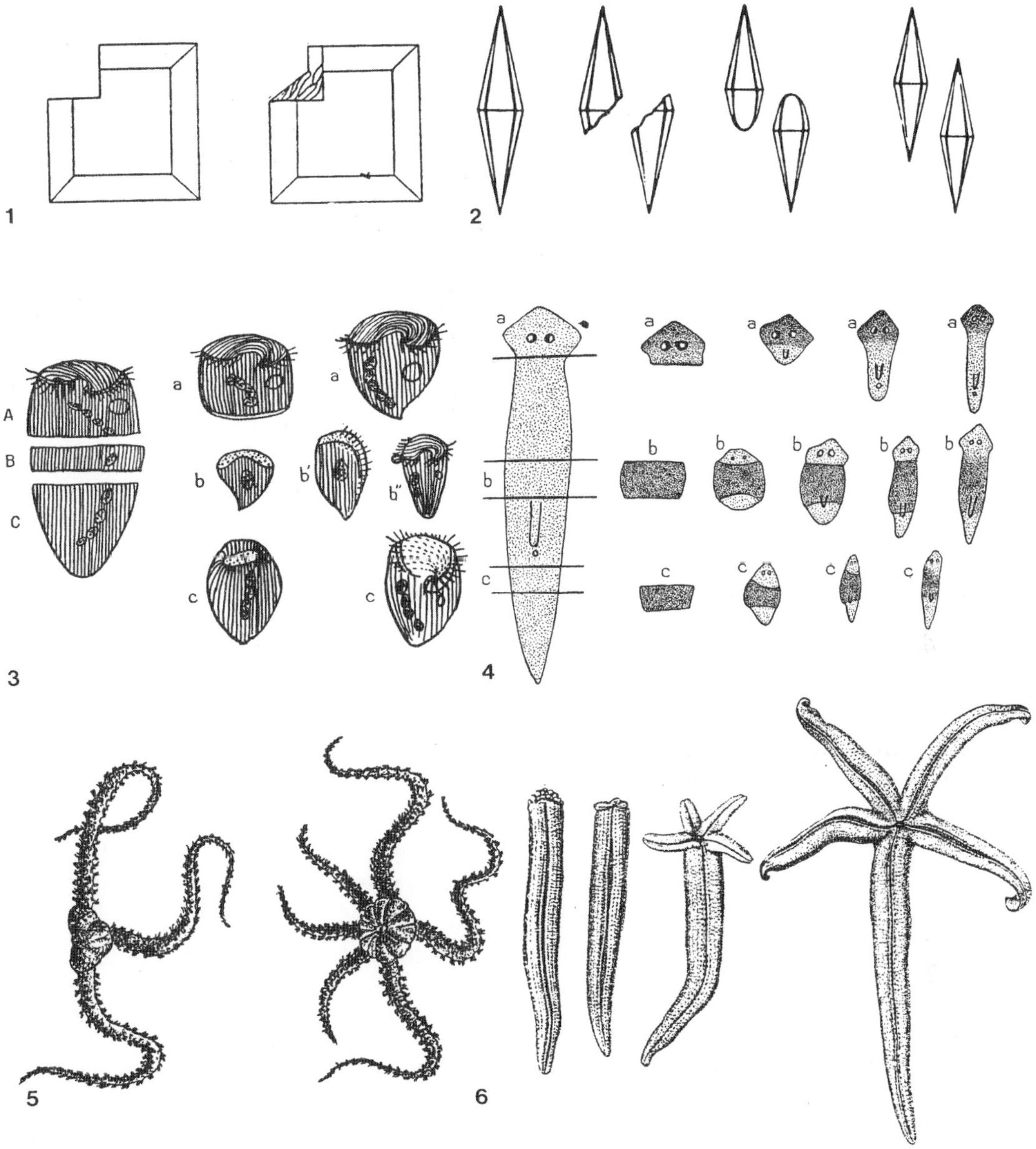

Figure 7.1 Regeneration in crystals, protozoa and invertebrates. (**1**) **Crystals of ammonium bimalate**. Crystal with its end broken (left) and under regeneration which leads to the restoration of the original pattern (right). (**2**) **Liquid crystals of ammonium oleate**. Following breakage the crystals regenerate building the original pattern. (**3**) **Protozoa**. *Stentor sp*. Stages of regeneration from three separate fragments. (**4**) **Flatworms**. *Planaria maculata*. Regeneration of complete individuals from three separate fragments. (**5**) **Echinoderm**. **Ophiuroids**. An ophiuroid after being divided in two parts (left). Shortly after, one half regenerates building the original pattern (right). (**6**) **Echinoderms**. **Starfish**, *Linckia multiflora*. Three successive stages of the regeneration of the whole body from the end of a single arm.

Present research on the atomic configurations taking place during crystal formation show that disorder is transformed into order during crystallization depending on the temperature of the solution. Disordered structures in minerals (such as feldspars) are formed at high temperatures, whereas low temperatures lead to an ordered assembly of novel atoms. The high temperatures do not allow the atoms to easily find their correct places in the whole, while the low temperatures give them the opportunity to move sufficiently slowly to occupy the correct site. Crystalline "defects" become coherent structures in feldspars due to the inherent tendency for ordering in the atoms of aluminum and silicon (Wenk and Bulakh 2004).

Protozoa and Algae Rebuild their Whole Body from Small Isolated Fragments. Released Chemicals Determine the Pattern

As early as 1885 E.G. Balbiani used the unicellular organism *Stentor* to carry out an experiment that became classic. He cut the protozoa into three pieces and each of them was able to rebuild a complete animal (Fig. 7.1).

A similar experiment was later performed by Hämmerling (1931, 1963) in the alga *Acetabularia* which is also unicellular. Its body consists of an elongated stalk terminated by a circular cap. This umbrella structure varies in shape from species to species. Hämmerling sectioned the stalk of the alga into small pieces and found that each was able to regenerate a complete alga with its specific umbrella cap. Two phenomena were established: the control of morphogenesis by the nucleus of the alga and the existence of a morphogenetic gradient along the axis of the stalk. It also turned out that both nucleated and anucleated fragments could regenerate a cap and that the fragments that were further away from the nucleus were those which gave rise to the largest cap. This showed that the chemical messages originating from the nucleus accumulated far away from their origin. These results demonstrated that morphogenetic chemicals, once they were released from the nucleus, were capable of autonomously determining the shape of the organ — they had no need of any further guidance (Brachet 1974).

Hydras have Body Memory and the Ability to Sense Structure Size. These Properties Extend to Higher Organisms

Coelenterates are invertebrates that already form complex organs. Among them, *Hydra*, has been extensively studied as a model organism due to its regeneration power.

Berking (1997) performed a series of basic experiments that elucidated this phenomenon in detail:

1 — Transverse sections of *Hydra, independently of their position* in the original animal, were able to form normal individuals.
2 — The tissue had a *polarity* which determined where the head and the foot were formed since both were missing in the initial fragment.
3 — Fragments as small as 1/10 of the whole body developed into animals with normal body proportions. Hence, the control of body size is due to the cells *sensing the size of the whole fragment.*
4 — Body fragments can be dissociated into single cells. These soon aggregate producing multiheaded animals. These subsequently separate into normally shaped animals, an example of how a disordered cell assembly has the *inherent capacity to create order.*
5 — Aggregates of cells obtained from different body regions were combined. Only cells of the middle region of the body were used, but those of a more apical location led to apical structures and those from a more basal origin led to basal patterns. The cells *have a memory* remembering their origin even following dissociation into amorphous aggregates.
6 — Inhibition of similar structures was also found. A head transplanted to a host prevented head regeneration and a foot prevented foot formation in close vicinity, demonstrating the ability to maintain the original *pattern coherence.*

Since this work was published cell sensing and cell memory have been documented at the molecular level in other organisms.

Bacterial cells use two-component signaling systems to sense and respond to nutrients. The first component is a *sensor protein*. The second is the *response regulator protein*. The two interact by means of a series of phosphorylation and dephosphorylation reactions involving specific histidine and aspartate residues (Taiz and Zeiger 1998).

Cell memory became evident during the cloning of mammals. Highly differentiated lymphocytes (*B* and *T* cells) are able to accomplish a complete reprograming *i.e.* to reiterate the original potency state. Fibroblasts are also capable of reprograming cellular development (Serov and Serova 2004).

Cell memory is also a general process in plants. Epigenetic mechanisms during differentiation allow changed states to persist through cell divisions due to the cell's molecular memory (Baulcombe and Dean 2014).

In humans, as many as 10^{11} cells die in each adult, each day and are replaced by other cells (Gilbert 2000). What an opportunity for total confusion. One of the reasons that this does not happen "is that normal cells can somehow perceive and arrest aberrant cycles of cell division" a form of cell sensing (Venkitaraman 2005).

Rigorous Reestablishment of the Original Body Pattern in Sponges and other Invertebrates

Sponges (Porifera) are among the simplest invertebrates, they have no tissues or organs and reproduce without the benefit of sex. Fragments may break off and be moved away by ocean currents. Such fragments continue growing as individual sponges. They may also produce amoeboid cells that disperse and form new sponges.

Ribbon worms (Nemertina) have an anterior sensitive proboscis. When irritated these animals release their proboscis which they then regenerate. The same happens to the proboscis in echiurans (spoonworms).

Priapulids are marine worms that possess a caudal appendage involved in gas exchange. This can be removed without killing the animal and is soon regenerated.

Annelids, characterized by their many external segments, can regenerate lost parts and can reproduce by budding.

Genes Participating in Flatworm Regeneration have Been Identified

Flatworms (Turbellaria) reproduce asexually by fission as protozoans do. At the same time they have great powers of regeneration. Any piece about one tenth the size of an adult flatworm will regenerate into a complete worm (Fig. 7.1). The reproductive organs are rebuilt in the regenerated fragments. Besides, under starvation conditions, senescence can be prevented — juvenile flatworms are produced that are "potentially immortal" (Russell-Hunter 1979).

The molecular mechanisms involved in the regeneration of planarians are being elucidated. The *Wnt3* genes induce a wave of proliferation in neighboring cells. Low levels of *Wnt* signaling cause head regeneration, whereas high levels of this ligand result in tail regeneration. Also the repression of the function of *β-catenin* (a protein) leads to head induction and conversely the activation of *β-catenin* results in tail induction (Li *et al.* 2015).

Genes have Been Isolated, Homologous to those of Vertebrates, that Determine Regeneration in Crustaceans and Insects — Organ Memory

Animal regeneration takes two forms: (1) The animal regenerates by reorganizing its tissues as in *Hydra* with little or no cell division, or (2) via large cellular proliferation by the formation of a specialized and transient structure the *blastema* (a mass of dedifferentiated cells). Regeneration via *blastema* occurs in planarians, crustaceans and echinoderms, these cells having retained a *memory* of their lineage. Regeneration of limbs has been identified in 35 genera of crustaceans and 38 genera of insects. There are differences but also highly conserved similarities between these two groups.

Regeneration in the crustacean *Uca pugillator* (crab) is controlled by steroid and retinoid hormones. They have the capacity to cast off their limbs at a predetermined point at the base of the leg. Following injury the base is sealed immediately by a membrane and a *blastema* develops. Nerves are required for regeneration of limbs and severed nerves release fibroblast growth factors like those identified in higher vertebrates. RNA interference has been successfully used to knockdown genes allowing the study of gene function by creating knockout mutants. There are at least 9 different growth factor genes and 4 receptor genes in mammals which are homologous to those of invertebrates. These initiate a cascade of organizational and positional signals. Crab retinoid receptor genes in *Uca* have been isolated, cloned and sequenced and found to be similar to their vertebrate counterparts (Hopkins 2001).

In fruitflies, crickets and cockroaches, amputation leads to regeneration by formation of granular cells, clot formation (after 60 minutes) and a compact plug (by 24 hours). A *blastema* forms by day 4 and after 16 days the leg is restored (Das 2015).

Damage to *Drosophila* larval imaginal discs elicits a robust regenerative response from the surviving tissue. In this fly there appears to be a mechanism that *senses tissue damage* and extends larval phase to coordinate it with tissue regeneration. This mechanism includes the inhibition of the transcription of a gene encoding a neuropeptide that in turn promotes the release of the steroid hormone ecdysone. Retinoid biosynthesis is also involved in this sensing process (Halme *et al.* 2010).

As Many as 694 Genes Decide Ordered Regrowth of Organs in Echinoderms — Whole Body Memory and Mineral Determination

The echinoderms are divided into: (1) Sea lilies and feather stars, (2) Starfish, (3) Brittlestars and basket-stars, (4) Sea cucumbers, (5) Sea urchins. For years their regenerative power has been described as "exceptional" but is only in the last decade that the molecular processes involved became elucidated.

Starfish have been the animals of choice. They can regenerate entire arms lost either by self-induced amputation (as in lizard's tails) or by traumatic loss caused by accidents or predators (Fig. 7.1).

The regenerative process comprises three main phases: (1) repair by wound healing, (2) differentiation events and (3) proper arm regrowth leading to the rebuilding of a whole coherent organism. Light and electron microscopy revealed that undifferentiated cells were intermixed with numerous white blood cells which frequently appeared in the wound area during the first two phases. Cell debris and microorganisms were then digested.

Echinoderm regeneration involves a series of growth factors: the *BMP/TGFβ* signaling pathway, the *ependymin* pathway and *hox* regulators. *Hox* genes are known to function as regulatory transcription factors and to pattern development in multicellular organisms.

Using gene amplification, in situ hybridization and other DNA techniques, Khadra *et al.* (2014) cloned a collection of *Hox* genes that are specifically expressed in regenerating arms. They identified 19 DNA sequences, 14 in starfish (*Echinaster sepositus* and *Asterias rubens*) and 5 in the brittle star (*Amphiura filiformis*).

In sea cucumbers the *ependymin* gene, which is known to be involved in vertebrate regeneration, is overexpressed during regeneration of the echinoderm *Holothuria glaberrima* (Suarez-Castilho *et al.* 2004). *Apostichopus japonicus* is another sea cucumber in which the *Wnt6* and *Hox6* genes were found to be expressed in the intestinal regeneration of this animal (Sun *et al.* 2013).

Brittlestar regeneration of arms is directed by a whole series of genes and transcription factors. A total of 694 genes and 194 proteins were found undergoing differential expression during the initiation of their regeneration process (Czarkwiani *et al.* 2013, Purushothaman *et al.* 2015).

Sea urchin spines have unique physical properties. They consist of a large single crystal of magnesium-containing calcite. Spine mineralization is driven by specialized cells located in the dermis and spine regeneration is initiated by a wound-healing process. Calcification then takes place. *Notch* signaling genes participate in spine regeneration and other genes bind to specific microRNAs which act in the silencing of transposable elements (Reinardy *et al.* 2015). This means that the ability of single crystals to regenerate, described above, occurs also at the level of living organisms where the mineralization is driven by molecular processes originating in specific genes.

The Tail Fin of Fishes can be Rebuilt when a Battery of Genes is Activated

The fish tail is a tissue where it is easy to study regeneration. It is a simple and radial structure with accessory functions related to animal survival.

The tail of fish does not contain an extension of the notochord or muscles as is the case in terrestrial vertebrates. Its main constituents are 16 to 18 dermal bones radially arranged forming a fan-like frame. These are accompanied by collagen fibers, arteries, nerves, fibroblasts and mesenchymal cells. Regeneration follows steps: (1) Subsequently to amputation the wound is closed within hours. (2) There is an epithelial wound healing that leads to the creation of a wound epidermis. (3) Within 1 to 2 days a mass of proliferating undifferentiated progenitor cells (the *blastema*) is formed at the distal end of the mesenchyme. (4) This is followed by proliferation of blastema cells, their differentiation and tissue reconstruction. (5) The new tissue remodeling reconstitutes the original tissue architecture.

The blastema is actually a collection of activated cells derived from a population of dormant and widely distributed stem cells. The blastema cells express the gene *msx*, responsible for a family of transcription factors (Fig. 7.3).

Three hours after amputation a battery of genes is activated. The expression of key components of the *Wnt/β-catenin* and *activin-βA* pathways is increased. This is followed by up-regulation of retinoic acid, of the insulin-like growth factor and fibroblast growth factor. All these factors, as well as local cell death, are involved in establishing proper pattern formation (Kawakami 2010, Li *et al.* 2015) (Fig. 7.3).

Fishes can also Regenerate Kidney, Heart and Brain

The kidney plays several functional roles including the removal of waste metabolites, electrolyte and acid-base balance, and blood pressure. Each kidney is comprised of *nephrons*, specialized epithelial tubes, that may number 1.5 million. The zebrafish kidney has amazing regenerative characteristics. It can replace epithelial populations after acute injury and can grow new nephrons (Mccampbell and Wingert 2014).

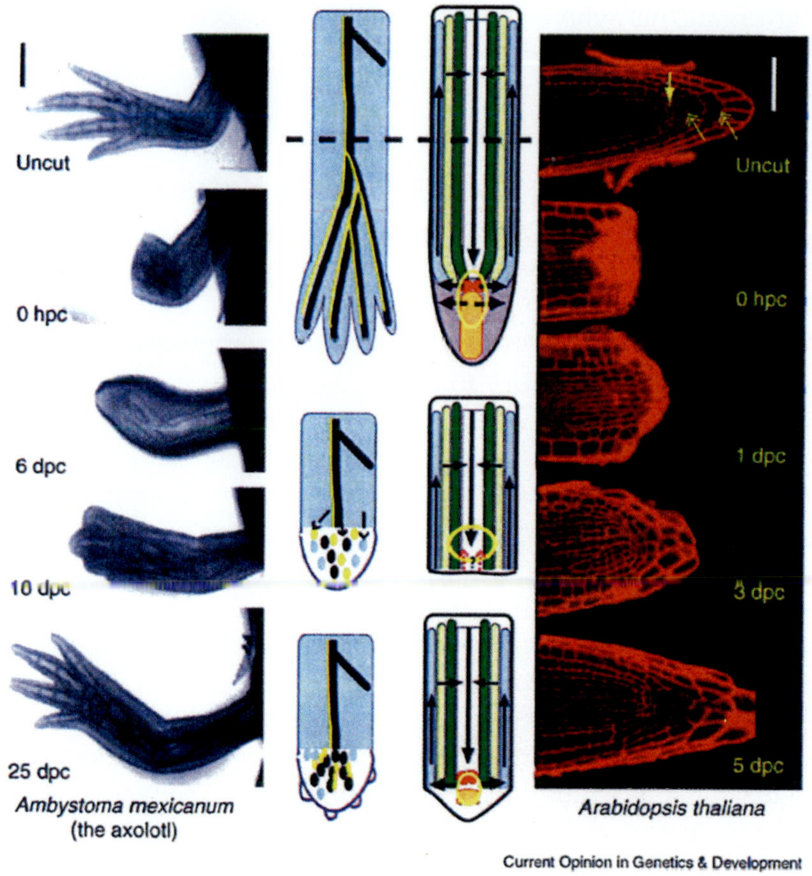

Figure 7.2 Two strategies for regeneration. On outside panels, axolotl limb regeneration through blastema formation and *Arabidopsis* root meristem regeneration without formation of callus (hpc, hours post cut; dpc, days post cut). In the diagram of axolotl, cells migrate from different tissues in the stump to form a heterogeneous blastema. Colors in diagram represent cells from different lineages that retain a memory of the tissue of origin. The blastema regenerates the limb over 25 days with new tissues populated by cells descendent from the same lineage in many cases. In the confocal image of *Arabidopsis*, the arrow shows the position of the QC cells and the double arrow shows the position of the columella cells. In the diagram of *Arabidopsis*, distal (toward tip) cell identities, PIN domains for lateral auxin distribution, and the auxin maximum are removed by the excision (red = QC; orange = columella, yellow circle = auxin maximum). In a potential model, auxin accumulates at the tip and induces PINs localized to redistribute auxin laterally. The auxin maximum determines the position of QC and then induces further patterning through a yet unknown mechanism. Red-dashed arrows indicate the potential for other signals that induce cell identities or mediate auxin flux. Scale bars; axolotl = 1 mm; *Arabidopsis* = 50 μm. QC = Quiescent center cells required for maintaining surrounding stem cells in an undifferentiated state.

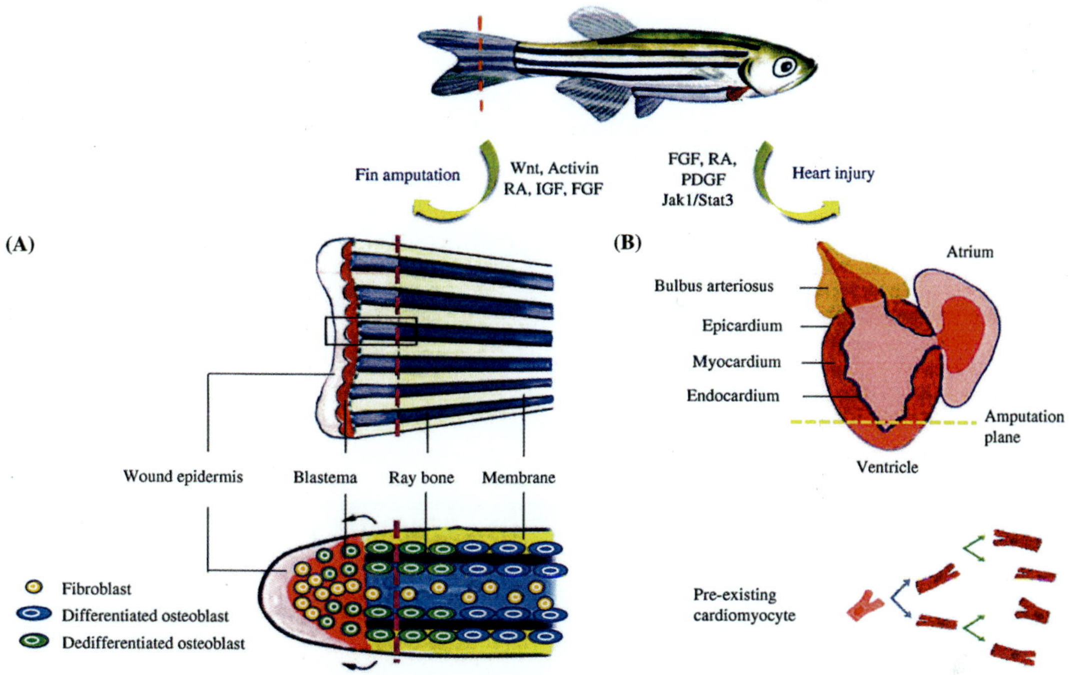

Figure 7.3 Regeneration of fin and heart in zebrafish. (A) Zebrafish fin ray bone regenerates through spared osteoblasts and fibroblasts. The wound heals and a specialized wound epidermis forms after fin amputation. Then osteoblasts dedifferentiate, proliferate and migrate to form the blastema together with fibroblasts. Expression of Wnt/β-catenin and Activin-βA signaling components is increased, followed by up-regulation of retinoic acid (RA), insulin-like growth factor (IGF), and fibroblast-like growth factor (FGF) signaling pathways. A proliferative compartment is maintained at the distal end, while new tissues are generated at the proximal end of the blastema. **(B)** Zebrafish myocardium regenerates mainly through proliferation of existing cardiomyocyte. Endocardial activation is essential for cardiomyocyte proliferation. Activation of the epicardium and neovascularization of the regenerating myocardium also play important roles. Many important signaling pathways, including FGF, RA, PDGF and Jak1/Stat3, regulate heart regeneration.

Heart regeneration is a classic example of regeneration in zebrafish. The adult heart of this species regenerates after ventricle apex amputation surgery. Cell growth was controlled by the incorporation of radioisotopes. Two weeks later the number of novel cardiac cells had increased 10-fold and the pattern was re-established after 60 days, containing cardiac muscles and normal electrical coupling with adjacent cells. Signaling pathways involved in zebrafish heart regeneration include the *platelet-derived growth factor* and *Notch* pathways (Li *et al.* 2015) (Fig. 7.3).

All vertebrates so far studied have *progenitor zones* in their adult brain with various degrees of abundance and regeneration. Teleost fishes have numerous proliferation zones and widespread capacity to produce new neurons. They have the most pronounced and widespread adult neurogenesis of any vertebrate. The brain contains neurogenic niches that harbor and regulate neural stem cells. Several techniques were used in zebrafish to study brain regeneration which included transgenic animals with fluorophore genes. *Notch* genes regulate cell fate decisions (such as glia and neurons) in the central nervous system. In the zebrafish telencephalon high levels of *Notch* signaling have been found, and other genes targeted by *Notch*, are also expressed during neurogenesis in embryonic and adult zebrafish (Kizil *et al.* 2011).

Amphibians Produce New Whole Limbs. Animals with two Heads Found From Worms to Humans

As the vertebrates emerged, regeneration became mainly restricted to single organs, as is the case in fishes. However, in amphibians, this capacity was improved. In frogs (*Rana esculenta*) two additional whole legs may be formed at the side of the original anterior leg (Fig. 7.4) (Aron and Grassé 1939). The legs are complete and functional consisting of nerves, muscles, bones and circulatory system. The axolotl also regenerates its limbs (Fig. 7.2).

During the regeneration of planarians and *Hydra*, animals with two heads were obtained. A comparable situation is found in *Triturus*. When eggs of this amphibian, which had reached the blastula stage, were submitted to prolonged and intense shaking, the blastomeres separated partly and formed a newt with two heads (Bohn 1934). Siamese twins with two heads have also been described in humans (Morgan 1934). This means that this type of reprograming is present in invertebrates, amphibians and humans.

The Tail of a Lizard is Expendable

Confronted by predators, lizards will voluntarily shed their tails. The shed tail of the Texas banded gecko wriggles convulsively for several minutes distracting the predator. The lizard slowly grows a new tail but this is never quite the same as the original. It does not have bony vertebrae but has all the functions of the original one assisting in climbing, swimming and courtship.

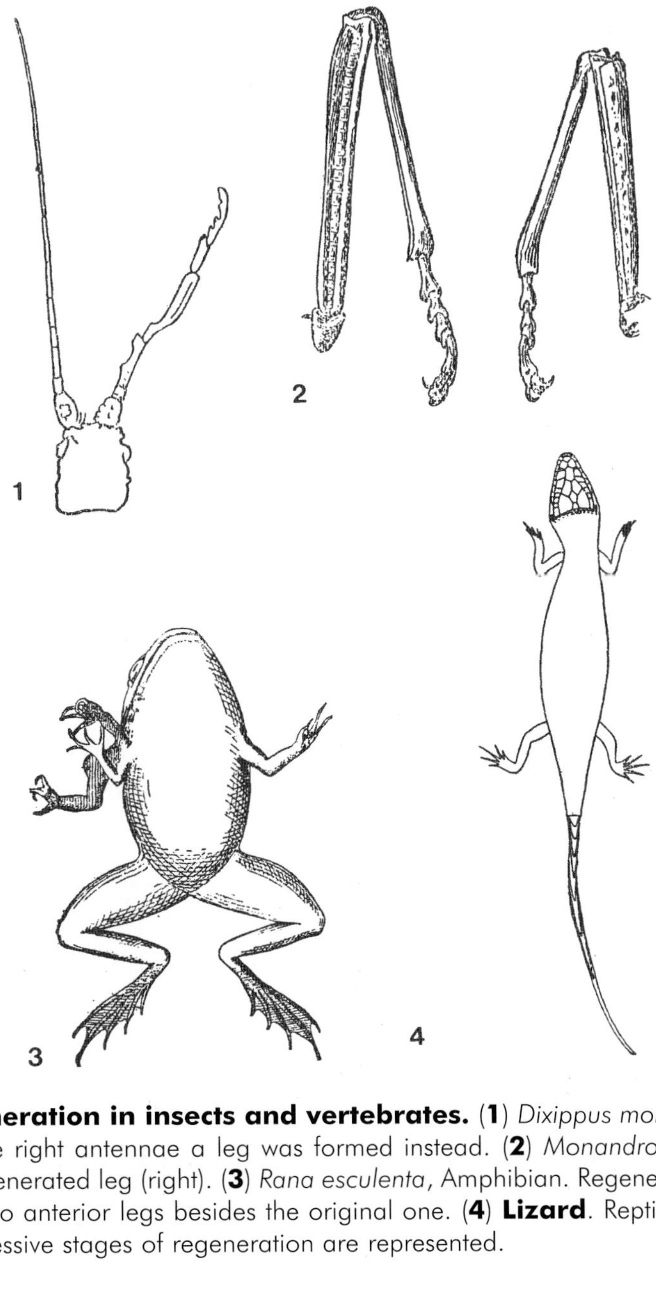

Figure 7.4 Regeneration in insects and vertebrates. (**1**) *Dixippus morosus*, Insect. Following the amputation of the right antennae a leg was formed instead. (**2**) *Monandroptera inuncans*, Insect. Normal leg (left). Regenerated leg (right). (**3**) *Rana esculenta*, Amphibian. Regeneration in frogs resulting in the formation of two anterior legs besides the original one. (**4**) **Lizard**. Reptile. Regeneration of the tail in which the successive stages of regeneration are represented.

Species capable of this regeneration have usually a fragile tail with fracture planes in their vertebrae making a weak point where muscles and blood vessels are modified. The fractures do not occur between vertebrae but across them and the muscles are arranged in such a way that they will come apart neatly (Fig. 7.4).

Another form of regeneration is found in the skin of some geckos. They can escape predators by causing their own skin to reap away. These lizards have preformed zones of weakness in the skin like the weak points in the tail (Halliday and Adler 2004).

Regeneration in the Mammalian Heart, Pancreas and other Organs

Mammals are not able to regenerate complex structures, yet they harbor the potential to regenerate a wide range of injured tissues such as skin, blood, bone, cornea and many epithelial cell types. This is made possible due to multipotent stem cells that are located in the renewing tissues.

Regeneration, although more slow, occurs also in liver, kidneys, spleen, heart and pancreas. In the liver compensatory regeneration, upon injury or amputation, involves limited differentiation and proliferation. Studies in rats, mice and humans disclosed the existence of progenitor cells in the adult pancreas that give rise to novel endocrine cells.

One day-old mice can regenerate their hearts after surgical intervention, due to the proliferation of cardiac cells. This regenerative capacity is lost by the 7th day after birth.

In humans 45% adult cardiomyocytes are renewed through a slow turnover while the other 55% remain after birth. This turnover declines with age as disclosed by incorporation of radioisotopes into cardiac cells. The adult human heart regeneration capacity is considered to be guided by microRNAs, since, in mice and rats, 40 microRNAs were found to robustly boost cardiomyocytes, DNA synthesis and cell division in neonatal animals (Li *et al.* 2015) (Fig. 7.5).

Human Teeth and Hairs are Created Anew and Regularly

Teeth are modified parts of the dermal skeletal materials which are unknown in lower chordates and jawless vertebrates. They are not only confined to the jaws as is the general situation but may even be formed in the palate and the pharynx in some animals.

The first embryologic indication of teeth is an infolding of the epidermis (a *tooth germ*). Subsequently ectodermal *tooth buds* are formed which gradually move to the surface and secrete the *enamel*. Soon cells of the mesenchyme deposit the *dentine* of the tooth which associates with blood vessels and nerves. The tooth works toward the surface and finally erupts. Meanwhile one or more successional

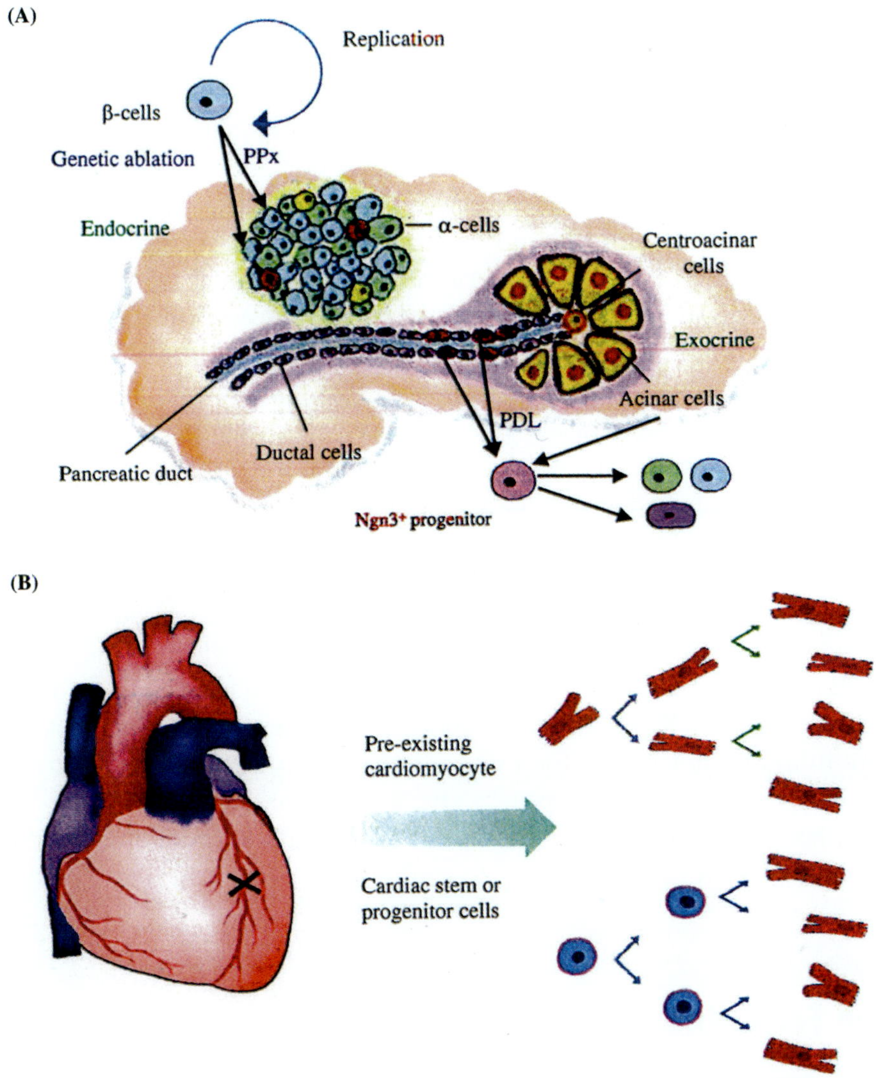

Figure 7.5 Schematic diagram of cellular mechanisms used for regeneration in mammalian pancreas and heart. (A) The β-cell loss by partial pancreatectomy (PPx) or targeted genetic ablation is replaced by replication of pre-existing β-cells. Following partial duct ligation (PDL), pancreatic cells of the ductal lineage (which include acinar and centroacinar cells) are proposed to adopt an $Ngn3^{+}$ progenitor identity and give rise to new ductal and endocrine cells. **(B)** Adult mammalian heart has a modest regenerative capacity after injury (e.g., myocardial infarction, indicated by black cross). Regenerated myocardium may come from proliferation of dedifferentiated cardiomyocytes, differentiation and proliferation of cardiac stem or progenitor cells.

buds may have formed from the germinal material and pushed outward for replacement (Romer and Parsons 1978).

In fishes, amphibians and reptiles, tooth replacement continues throughout life. Through resorption of their bases, teeth are shed, and replaced by a new generation of teeth.

Mammals, including humans, have only one replacement of their teeth but this is, like in the other vertebrates, an independent regeneration process which is dictated solely by developmental events. At birth humans have no teeth. In early years a "milk dentition" is formed that is later replaced by a "permanent dentition". The teeth are pushed outwards and new ones grow from dental buds, replacing the old ones.

This situation has not been described before as an example of regeneration, but it is the same as the seasonal replacement of antlers by deer, which before were not either mentioned as a clear cut case of regeneration but are now included in this phenomenon (Kawakami 2010, Gama-Carvalho *et al.* 2014).

Hairs in humans, are usually regenerated throughout life. They are permanently shed and new ones are pushed forward that replace those that were discarded. Hair is keratinized epidermis. A typical hair includes the projecting shaft and the root sunk in the dermis termed the hair follicle. Initially there is a bulb from which are budded off the cells that build the hair root and shaft (Romer and Parsons 1978).

Like the feathers of birds, human hairs are regenerated during the whole life cycle. Though hair and fish scales are distributed over the body surface in an orderly way, it has been thought that they were morphologically and evolutionally different. Recent studies have shown that the formation of these skin components is governed by the same signaling pathway including a transmembrane protein, ectodysplasin and a receptor. In addition several signaling molecules, such as *shh*, are shared in mammalian hair, bird feather and fish scale (Kawakami 2010).

Mosses and Ferns Know their Body Plan

Heald (1898) made a most careful study of regeneration in mosses carrying detailed experiments in 12 species: (1) The majority of moss leaves "showed a remarkable power of regeneration" producing either rhizoids or protonemata which differentiate later into the production of new shoots. (2) Segments of leaves led to the emergence of protonemata (and buds) after five days. (3) The point of origin of the new cell growth was dependent on exposure to light but was independent of gravity. (4) Regeneration occurred in cells collected from both the ventral side of the leaf and from its margins. (5) The majority of moss stems showed regeneration in the species investigated.

Recently Xu and Huang (2014) have reviewed plant regeneration from the genetic point of view and described the results obtained in the moss *Physcomitrella patens*. It is considered an ideal model for studying plant regeneration because many different types of tissues or cells are easily triggered to undergo cell-fate transition and regeneration. A single cell from this moss species can regenerate into the protonema apical cell, being the result of a reprograming process from differentiated cells into stem cells with expressional changes of thousands of genes (Xiao *et al.* 2012). An interesting discovery was that the expression of the gene *cyclin-dependent kinase A* was induced prior to cell-fate transition and cell division (Ishikawa *et al.* 2011). *Sphagnum* species have shown the highest regeneration power among 9 mosses growing in North America (Graf and Rochefort 2010).

Spores from the leaves of the Chinese fern *Pteris vittata* were sterilized and placed in culture medium where they germinated and formed *prothalli* after 4 weeks. They were transferred to a *callus* inducing medium supplemented with gibberellic acid (GA_3, a plant hormone) accompanied by ethylene. Scanning electron microscopy revealed that prothalli began to swell after 6 weeks of culture in the callus inducing medium and sex organs (antheridia and archegonia) appeared on the surface of the prothalli (Joyce *et al.* 2014).

Conifers Produce Plantlets from Shoot Apices

Conifers comprise large and very old trees such as *Sequoia sempervirens* and *Juniperus excelsa*. Other well known species are *Picea alba* (Norway spruce) and *Pinus roxburghii*.

Cells in tissue culture, on media containing plant hormones, developed into shoots and roots. Immature embryos were dissected and transferred to a basal medium where *callus* developed, followed by the formation of cotyledons and roots (Hakman *et al.* 1985) (Fig. 7.6). Propagation of conifers using embryonic and seedling explants has been achieved successfully in numerous species. Adventitious bud induction was obtained in *Pinus roxburghii* from shoot apices and led to the regeneration of plantlets (Kalia *et al.* 2007).

Flowering Plants are Totipotent

Regeneration in plants is not only related to tissue and organ repair but also to the formation of complete plants. With the exception of stem cells, animal cells generally lose the ability to produce other cell types upon differentiation. Plants, have differentiated cells which are able to regenerate into the full array of tissues (Fig. 7.6).

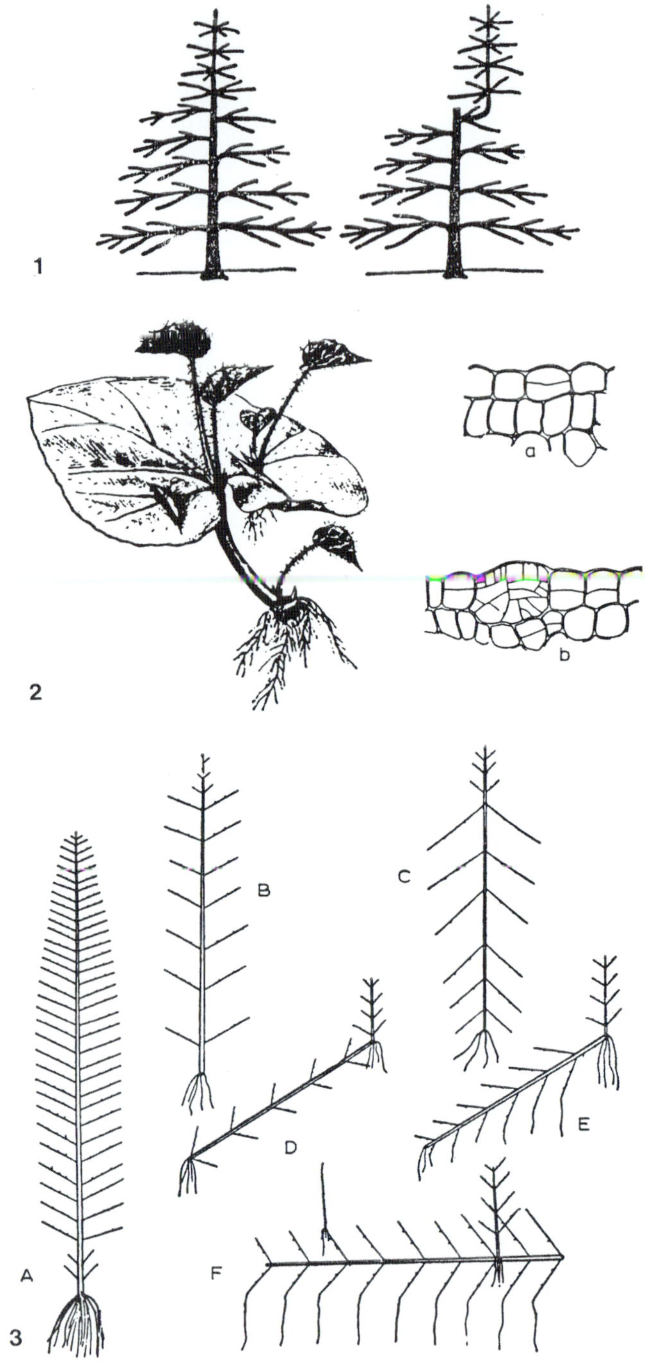

Figure 7.6 Regeneration in plants and Hydrozoa. (**1**) **Conifer tree**. Following decapitation soon the uppermost branch grew vertically and became the main stem. This is a general phenomenon in conifers. (**2**) *Begonia sp*. Leaf-cutting showing regeneration into roots and leaves resulting in a whole plant (left). Transverse section through the epidermis of a leaf of *Begonia* with two stages (a and b) of the formation of a bud from an epidermal cell (right). (**3**) *Antennularia antennina*, colonial hydroid (hydrozoa). (**A**), Normal stalk. (**B**), segment regenerating in the normal vertical position. (**C-F**), pieces regenerating in various positions.

The pluripotency and totipotency of plant cells was demonstrated when a single somatic cell gave rise to a whole plant (Haberlandt 1902).

Higher plants show three main types of regeneration: (1) tissue regeneration, (2) *de novo* organogenesis, and (3) somatic embryogenesis (Fig. 7.7). Plant hormones trigger regeneration and several transcription factors respond to the hormones determining cell fate. This last process requires changes in gene expression which are regulated via epigenetic pathways. Besides, there is a permanent interaction between the three different components: hormones, transcription factors and epigenetic events. Epigenetic regulation mechanisms include modifications in the histone chromosome proteins, chromatin remodeling and DNA methylation.

1 — *Tissue regeneration* (also called tissue repair) is comparable to limb regeneration in amphibians and head regeneration in planarians in which movement of stem cells is obligatory, but plant cells are unable to move, creating a particular situation. Root tip regeneration is an example. The root stem cell niche is at the tip of the root apical meristem. Following excision of the root meristem in maize, the root quickly regenerates a new quiescent center leading to root tip regeneration. The molecular basis of this event was revealed in *Arabidopsis* plants (Fig. 7.2). There were changes in the transport of the *auxin* hormone and the levels of key transcription factors. The high auxin level led to upregulation of the *plethora* genes which promoted *shortroot* proteins to activate the *scarecrow* gene. Regeneration in cucumber and tomato revealed that another plant hormone, *gibberellin*, is involved in wound healing.
2 — *De novo* organogenesis allows detached or wounded parts to develop into new plants. In tissue culture these parts form a pluripotent cell mass, called the *callus*, from which roots and shoots may grow. Several gene families play a role in this process and in rice DNA methylation changes during *callus* formation. Sugimoto *et al.* (2011) have emphasized that the *callus* is organized from a specialized population of adult stem cells.
3 — *Somatic embryogenesis* is a process that returns a differentiated somatic cell back to the stem cell state (dedifferentiation) a confirmation of plant totipotency. In this case a single somatic cell builds an embryo that results into a whole plant. The balance between the hormones *gibberellin* and *abscisic acid* is pivotal in controlling the embryonic state. Auxin and ethylene also participate. Two types of transcription factors studied in *Arabidopsis* (the *leafy* cotyledon genes and the *agamous-like 15* gene), play an essential role in somatic embryo formation. They determine cell fate by cross talk among hormone pathways. Several other transcription factors and epigenetic regulation also participate in somatic embryogenesis (Xu and Huang 2014) (Figs. 7.7 and 7.2).

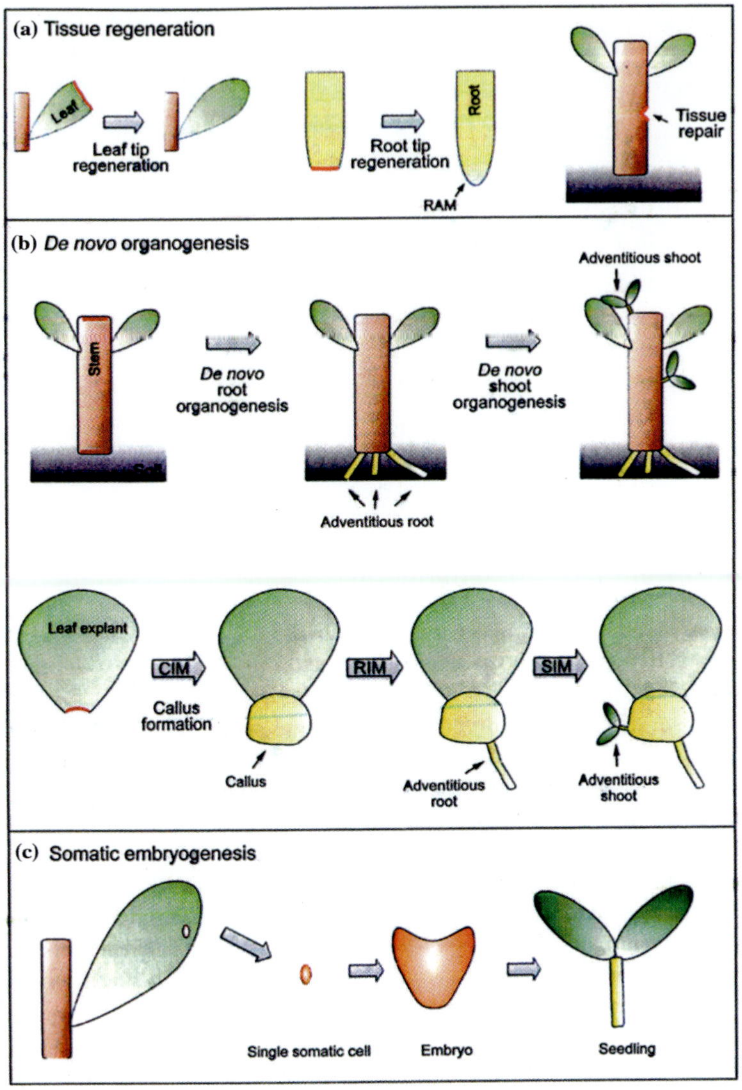

Figure 7.7 Different types of plant regeneration. (**A**) Tissue regeneration. Excised tips of very young leaves (left) and roots (middle) can re-grow via tissue regeneration. Stem tissues can also be repaired after damage (right). (**B**) *De novo* organogenesis. Upper panels show *de novo* root and shoot organogeneses during the propagation of plant materials via cuttings. Lower panels show a plant regenerated from leaf explants induced to form callus, adventitious roots, and adventitious shoots on CIM, RIM, and SIM, respectively, in tissue culture. (**C**) Somatic embryogenesis. A single somatic cell can be induced by auxin to form an embryo, which develops into a whole plant, reflecting the totipotency of plant cells. Red lines in (**A**) and (**B**) indicate wounding sites. *CIM* = callus inducing medium. *RIM* = root inducing medium. *SIM* = shoot inducing medium.

30,000-Year-Old Fruit Tissue Buried in Siberian Permafrost Regenerates into Whole Fertile Plants

Whole, fertile plants of *Silene stenophylla* (Caryophyllaceae) were regenerated from maternal immature fruit tissue of the Late Pleistocene age using in vitro culture and clonal propagation.

The fruits were excavated from never thawed permafrost dated with radiocarbon to be 31,800 years old. Regenerated plants were brought to flowering and fruiting and they set viable seeds.

Earlier investigations led to the finding of viable seeds of *Phoenix dactylifera* (date palm) from the Dead Sea and of *Nelumbo nucifera* (lotus) from China but these were dated to the first and eighth centuries of the Common Era. The *Stenophylla* seeds are from the age of the mammoths.

The placental tissue of seeds was grown by *in vitro* tissue culture method. Organogenesis of adventitious shoots was induced directly from fragments of the placental tissue by the nutrient medium. Rooted plants were transplanted to pots where they developed flowers, fruits and set seeds. All the 36 ancient plants were identical to 29 control extant plants (Yashina *et al.* 2012).

Salamander Regeneration Leads to Reversion of Cancer

The regeneration of the salamander limb is divided into three different stages: (1) Following limb amputation, epithelial cells migrate closing the wound. This wound epithelium then thickens forming an *apical epidermal cap*. (2) A population of epidermal cells migrates under the wound epidermis leading to the formation of a mass of undifferentiated proliferating cells (blastema). (3) These grow in a proximodistal direction regenerating the missing limb.

The emergence of this correct pattern is now understood: (1) A blastema only regenerates structures which are distal to the amputation level. (2) Retinoic acid treatment affects positional identity. (3) *Hox* genes participate in positional identity. (4) Fibroblasts have turned out to be essential for salamander regeneration since they are the earliest blastema cells.

Of importance to the development of anticancer therapies is the finding that the role of salamander fibroblasts in regeneration compares with that of human fibroblasts in cancer. Frog renal tumor tissue was transplanted into a salamander forelimb and began to grow as a tumor. The limb was then amputated and a normal limb without cancer was formed. Another study showed that salamanders, in which a cancer has been induced with carcinogens, spontaneously reverted to normal tissue formation. This evidence has been taken to mean that the regeneration process can reverse tumorigenicity (Fior 2014).

Coherence in the Rebuilding of Body Pattern in Animals and Plants

The coherence is evidenced by the following findings:

1. The production of tumors in cancer is due to a rapid and chaotic increase of cell populations with uncontrolled growth. The tumor has no pattern that is directly related to that of the other organs of the organism. Cancer ignores order and coherence — that is its fundamental difference from regeneration.
2. In regeneration there is a well established program of gene expression that obliges the production of structures and functions identical to the missing ones. Moreover, this cell memory ensures that the original whole pattern, and no other one, is rebuilt.
3. The coherence is most evident by the fact that the different organs are not produced at random but appear in a succession related to their functional importance. Roots that allow the nutrition of a plant by using minerals and water appear first. Secondly emerge the leaves that synthesize sugars through photosynthesis. Last appear the flowers and fruits which enable reproduction.
4. The program is so rigid and extreme that a single cell is sufficient to lead to the formation of a flowering plant, and fragments as small as 1/10 of the whole body of *Hydra* develop into animals with normal body proportions.
5. The establishment of the coherence is accompanied by sensing mechanisms and the intervention of hormones.
6. The strategies for regeneration in amphibians and plants are compared in Fig. 7.2. Axolotl (*Ambystoma mexicanum*) limb regeneration is placed side by side with meristema regeneration in *Arabidopsis*.

Closely Related Species show Great Difference in Regeneration

The large variation of regenerative capacity in related species, has been thoroughly investigated by Li *et al.* (2015) who approached this problem at the molecular level.

Dugesia japonica and *Schmidtea mediterranea* can regrow all missing body parts after transection, regardless of how many times or where they are cut along the axis of the head to the tail. Other species of planarians cannot regenerate as easily. These include: *Phagocata kawakatsui*, *Procotyla fluviatilis* and *Dendrocoelum lacteum* which can only regenerate a head from a tail fragment. This last species revealed that two fate specification defects were present: (1) failure to set up anterior/head identity and (2) inappropriate retained tail identity. These were attributed to aberrant blastema fate-specification. Subsequently RNA interference and the *Wnt* gene signal transducer were found to rescue the regeneration of fully functional heads on tail pieces, indicating the *Wnt/β-catenin* signaling pathway in the inhibition of head regeneration in *D. lacteum*.

Hence, the differences between closely related species are due to differences in specific gene expression.

Lizards, or Lacertilia, comprise about 4,500 species divided into four superfamilies: Iguanas and relatives, Geckos and snake-lizards, skinks and relatives and anguimorph lizards.

Tail regeneration is common in most lizard families. What is significant, for the elucidation of periodicity, is that closely related species do not have the ability to shed the tail voluntarily.

Among the Iguanas and relatives are the chameleons in which this ability is absent.

Chisel-teeth lizards are mainly tropical and the strangest of them are the flying dragons, which belong to the Iguana superfamily. They do not shed their tails.

The anguimorph lizards include the largest of all lizards, the monitors. The Bornean earless monitor is also not able to shed its tail.

Beaded lizards are also part of the Anguimorphs with very large species, e.g. the Komodo dragon. Two species are the world's only venomous lizards: the Gila monster and the Mexican beaded lizard. They do not loose their tails.

Xenosaurs comprise only five American and Chinese species. Their tails have a crest of bony plates, when threatened by predators they gape widely and will bite instead of regenerating their tails.

Amphibians show also great variation of regeneration in closely related species. *Ambystoma mexicanum* (Fig. 7.2) has good whereas *A. tigrinum* has poor ability. The same is true of crustaceans, insects and other groups which were extensively investigated by Maginnis (2006).

Echinoderms are advanced invertebrates with an enormous ability to regenerate. Yet, as pointed out by Khadra *et al.* (2014) "there is a remarkable variability in terms of the degree of morphological and functional recovery, not only between unrelated groups but also between closely related species, and even between organs and parts of the same individual".

The Absence of Regeneration in Birds

In birds there is no indication of body part or organ regeneration except for the feathers that are replaced continuously.

Just as the teeth and the penis were eliminated in most species of this avian group, regeneration of the body main organs was also precluded. A major process of gene repression seems to have dominated when the birds emerged in evolution.

The most probable event is gene repression because the genome of birds and that of humans, when they were DNA sequenced, were found to be most similar. Chicken-human aligned DNA segments tend to occur in long blocks where gene order was conserved (International Chicken Genome Sequencing Consortium 2004).

Absence and Presence of Regeneration Across the 30 Invertebrate Phyla

Margulis and Schwartz (1982) distribute the invertebrate animals into 30 phyla. All these phyla contain thousands of species that have never been studied from the point of view of regeneration. Usually only those that show the phenomenon are mentioned.

1 Phyla Where There is No Indication of Regeneration

Ctenophores reproduce only sexually. Mesozoans, some groups reproduce asexually releasing vermiform larvae. Gnathostomulids are hermaphrodites with no larval stages. Gastrotrichs are hermaphrodites. Rotifers, many populations reproduce by virgin birth (parthenogenesis). Kinorhynchs, have sexual reproduction with separate sexes. Acanthocephalans, reproduction is entirely sexual and sexes are separate. Entoprocts reproduce both asexually and sexually. Nematodes always reproduce sexually. There is no regeneration (Li *et al.* 2015). Nematomorpha have sexual reproduction and no regeneration. Ectoprocts can reproduce both sexually and asexually (involving budding). Phoronids are hermaphrodites. Brachiopods or lamp shells, their reproduction is always sexual with separate sexes. In Molluscs the sexes are separate. Sipunculans, there are more than 300 species with separate sexes. Tardigrades known for their exceptional survival capacity reproduce sexually and by parthenogenesis. In Pentastoma, Onychophorans and Pogonophorans the sexes are separate. Chaetognaths are hermaphrodites.

2 Phyla Where Regeneration is Limited

Nemertines or ribbon worms, release their proboscis which they then regenerate. The caudal appendage of *Priapulus caudatus* (Priapulids), can be removed without killing the animal, which regenerates the appendage. Echiurans, the proboscis can be cast off, but is regenerated later. Annelids, except for the leeches, annelids can regenerate lost parts and can reproduce by budding as well as by sexual organs. They comprise 12,000 species divided into Bristle worms and Clitellates (6,000 species). To these last ones belong the leeches (*e.g. Hirudo medicinalis* which has been used as a bloodletting agent). This is a clear-cut case of a group of species (Leeches) that does not regenerate whereas the remaining ones are able to regenerate. Arthropods do not seem to have special powers of regeneration (insects, centipedes, millipedes, crustaceans, sea spiders, horseshoe crabs and arachnids) but some species of insects and crustaceans can regenerate single organs.

3 Phyla Where Regeneration is Well Established

In Placozoa regeneration is evident. When *Trychoplax adhaerens* is cut it heals rapidly and a complete animal is formed from any cutaway portion. The phylum is represented by this single species. Some Sponges release fragments from their bodies that continue to grow as individual sponges. Many freshwater and a few marine species reproduce by building gemmules, composed of amoeboid cells, that grow to form new sponges. Cnidarians may reproduce by asexual budding (Hydrozoans and Scyphozoans). Anemones may split into two animals and Hydras are the classical example of regeneration ability. *Hydra viridis* is a well known species but their variation in regenerability remains to be investigated. Platyhelminthes or flatworms are capable of extensive regeneration as demonstrated in planarians. Again, there is a tendency to mention only the species that

Table 7.1 Periodic Table of Regeneration

Increasing grade of complexity	ABSENCE OF MEMORY	CRYSTAL AND CELL MEMORY	ORGAN MEMORY	WHOLE BODY MEMORY
1	CTENOPHORES MESOZOANS GNATHOSTO-MULIDS GASTROTRICHS	CRYSTALS MOSSES *Sphagnum* FERNS *Pteris* CONIFERS *Pinus*	RIBBON WORMS ECHIURANS Proboscis PRIAPULIDS Appendages ANNELIDS Lost parts INSECTS Antennae	PROTOZOA *Stentor* PLACOZOA *Trycholax* ALGAE *Acetabularia*
2	ROTIFERS KINORHYNCHS ACANTHOCE-PHALANS ENTOPROCTS	FISH Scales	CRUSTACEANS Limbs FISH Tail Teeth Brain Heart	SPONGES Ameboid cells
3	NEMATODES NEMATOMORPHA ECTOPROCTS PHORONIDS	AMPHIBIANS Skin	AMPHIBIANS Limbs	CNIDARIANS *Hydra*
4	BRACHIOPODS MOLLUSCS SIPUNCULANS TARDIGRADES	REPTILES Scales	REPTILES Tail	FLATWORMS *Planaria*
5	PENTASTOMA ONYCHOPHORANS POGONOPHORANS CHAETOGNATHS	BIRDS Feathers	BIRDS No indication of re-generation capacity	ECHINODERMS Starfish and others
6	BIRDS Absence of memory (except feathers)	MAMMALS Skin Hair Fibroblasts	MAMMALS Teeth Antlers Heart	FLOWERING PLANTS Many species

A phenomenon coupled to the emergence of life is reproduction. This process cannot be easily distinguished from memory. Reproduction is characterized by the copying of a preexisting pattern, but memory is as well a copying of a preexisting pattern. Regeneration is anchored to the crystal world and expands into cells, organs and whole bodies. As organism complexity increased organ memory tended to decrease. This decline became even more drastic in the case of whole body memory which is totally absent in vertebrates. The periodicity is shown by the absence of regeneration in many invertebrate phyla which is accompanied, at the same time, by a high ability of whole body memory among the most simple invertebrates and flowering plants (Original).

regenerate. Obviously the regeneration ability of the 13,000 species of flatworms has never been investigated. Most species of Echinoderms are known to regenerate lost body parts but reproduction is sexual. 6,000 species are found among sea lilies, feather stars, starfish, brittlestars, basketstars, sea cucumbers and sea urchins. The starfishes (1,500 species) are the classical case of regeneration.

The Periodic Table of Regeneration

Four vertical columns were established (Table 7.1).

Column 1 Absence of Memory

Here are included the invertebrate phyla in which there is no indication of regeneration and the sole group of vertebrates, the birds, in which no regeneration has been described (except for the feathers).

Column 2 Crystal and Cell Memory

This leads mainly to the formation of peripheric structures and tissues. It is present in early plant groups and in all the vertebrates.

Column 3 Organ Memory

It is revealed following injury but may also occur naturally. It is present from invertebrates to mammals with the exception of the birds.

Column 4 Whole Body Memory

It extends from protozoa to flowering plants but all the vertebrates lack this program.

Chapter 8

The Eye, the Main Organ of Vision, has had an Ordered Evolution Guided by Self-assembly

The Ability to Direct Light to Specific Sites is Already Present in Crystals and Pure Minerals are Used as Lenses in Animal Eyes

Crystals consist of an ordered three-dimensional atomic structure with characteristic periodicities (Klein and Hurlbut 1985). When irradiated with X-rays, every crystalline substance produces its own pattern, which is dependent on the internal structure. The image obtained is called a "fingerprint" of the mineral.

When visible light is used all transparent substances can be divided into two groups: the first (isotropic) includes such noncrystalline substances as gases, liquids and glass, but it also comprises crystals of the cubic system. In them light moves in all directions with equal velocity. The second group (anisotropic) includes all crystals (except those of the cubic system). When light passes through an anisotropic crystal it is divided into two polarized rays. As concluded by Wenk and Bulakh (2004) "The interaction of light with crystals is directional". Classical examples of this phenomenon are the minerals calcite and tourmaline.

Double refraction, or *birefringence*, is seen when near objects are viewed through calcite, they appear doubled. The light is split into two parts: an *ordinary ray* and an *extraordinary ray*. The light in the ordinary ray is polarized at right angles to the light of the extraordinary ray (Fig. 8.1).

The cornea and the lens of the animal eye are the two structures that bend light rays to focus images on the retina. The refractive index of the lenses (the ratio of the velocities of light in two different media) follows a gradient in which the light rays are bent continuously within the body of the lens (Land 2012).

The animal eye turns out: (1) not only to use assemblies of specific proteins to build its lenses but (2) to use different pure minerals as lenses. In the fossil trilobites (marine arthropods) and the living brittle stars (echinoderms) the lens is made of calcite (Björn 2015) and in the marine mollusc chiton (*Acanthopleura granulata*) the lens is made of the mineral aragonite (Cronin *et al*. 2014).

Calcite and aragonite are produced in the eye following the same atomic self-assembly mechanism that built them outside the cell.

Thus, three processes are involved: (1) In most species, lenses consist of proteins, called crystallins, which are the product of gene action (Charlton-Perkins *et al*. 2011). (2) The lenses deviate light rays, due to their birefringence, like other crystal structures. (3) The eye can even substitute the lens proteins by employing pure minerals in their place with comparable efficiency. Vision is anchored in the mineral world.

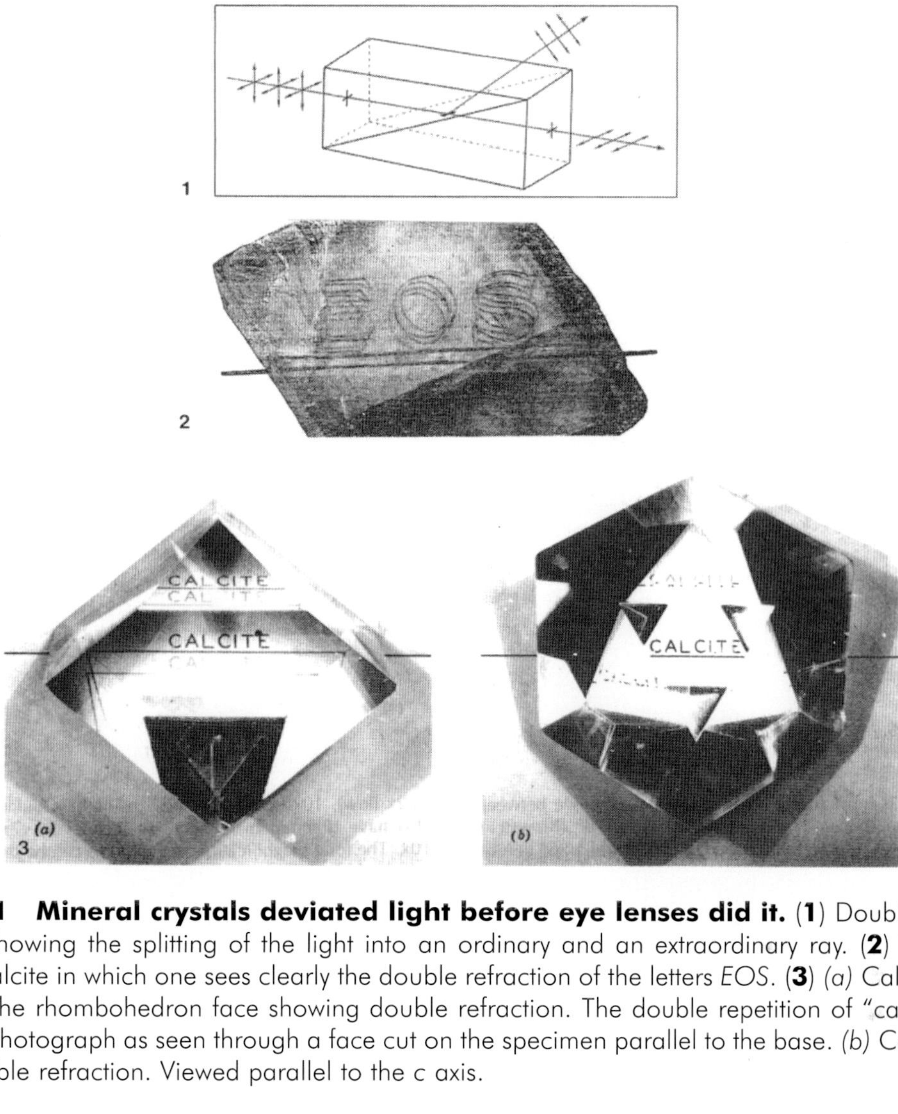

Figure 8.1 Mineral crystals deviated light before eye lenses did it. (**1**) Double refraction in calcite showing the splitting of the light into an ordinary and an extraordinary ray. (**2**) Transparent crystal of calcite in which one sees clearly the double refraction of the letters *EOS*. (**3**) *(a)* Calcite, viewed normal to the rhombohedron face showing double refraction. The double repetition of "calcite" at the top of the photograph as seen through a face cut on the specimen parallel to the base. *(b)* Calcite showing no double refraction. Viewed parallel to the c axis.

Bacteria Act as Spherical Lenses — Their Vision is Considered Similar to that of the Human Eye

The simplest type of cell — the bacterium — already is able to direct light in a well defined way. The cell also changes its shape. The result is that the whole single cell is transformed into a lens.

Cyanobacteria use micro-optics to sense light direction. Bacterial ability to sense light was recognized a century ago but the phenomenon could not be properly analysed. Schuergers *et al.* (2016) have found that the unicellular cyanobacterium

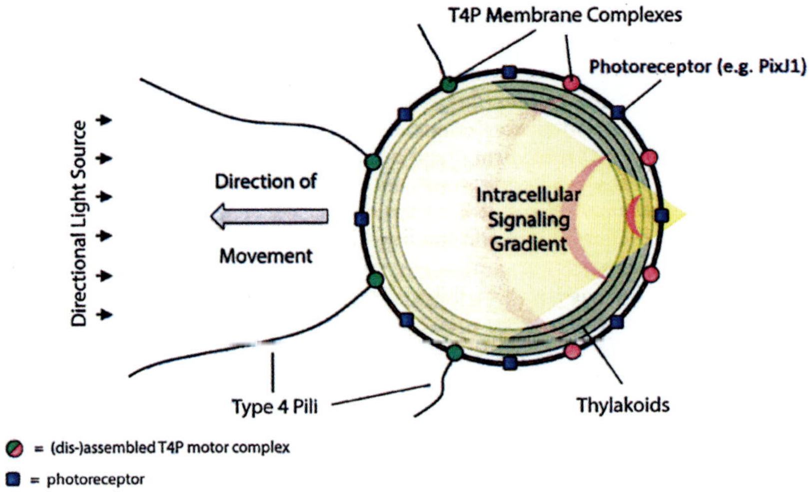

Figure 8.2 Bacteria use micro-optics to sense light direction. Model for control of positive light perception in *Synechocystis*. Directional illumination of the cell produces a sharply focused and intense spot of light at the cell periphery on the opposite side from the light source. The focused spot is perceived by photoreceptors in the cytoplasmic membrane (for example PixJ1) triggering signal transduction via CheY-like response regulators that locally inactivates the T4P motility apparatus, dispersing T4P components including the extension motor PilB1. Consequently, patches of the motor proteins can only form on the side of the cell facing the light source. Pili are extended and retracted at this side of the cell, which therefore moves towards the light. T4P, Type IV pili.

Synechocystis sp. measures light intensity and color with a range of photoreceptors. The single cells sense the position of a light source directly and accurately. This is possible because the cells act as spherical microlenses allowing the cell to see light and to move towards it by means of its pili (protruding proteinaceous filaments).

The bacterial cell functions as a camera eye since a high resolution image of the light source is focused on the edge of the cell opposite to the source, triggering movement away from the focused spot. Light-detecting molecules called photoreceptors respond to the focused image of the light source.

The cell of *Synechocystis* is in terms of volume about 500 billion times smaller than a human eyeball but the authors conclude that vision in this bacterium "actually works by principles similar to vision in humans" (Fig. 8.2).

Plant Cells Function as Eyeballs Condensing Light and Moving Towards it

Plant leaves are actually mosaics of microlenses. With the exception of their roots, plants can be aptly described as living carpets of light-sensitive cells. Most parts of the plant function as light antennae.

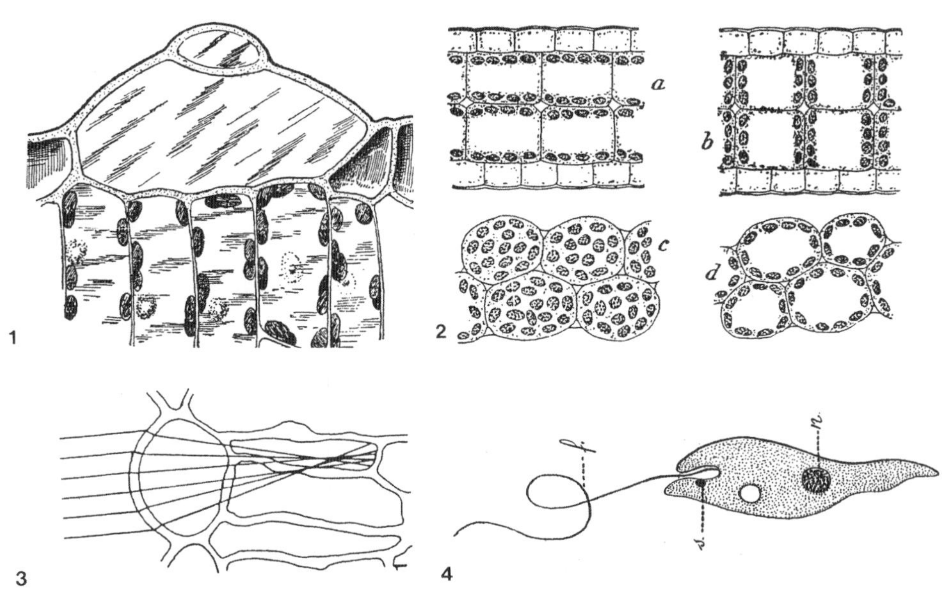

Figure 8.3 Vision in plants and protozoa. (**1**) *Fittonia sp*. The "eye" of the plant *Fittonia*. The epidermal cells of the leaf function partly as a lens by concentrating the light reaching the chloroplasts situated below. (**2**) *Lemna trisulca*. Light orientation of chloroplasts in leaves. The chloroplasts, which contain the pigment sensitive to light, change position in the cell according to the intensity of the light that falls on them. Transversal (upper part) and frontal (lower part) leaf sections under weak (left) and strong light (right). (**3**) Vision in plants. Image formation in the epidermal cells of the leaf of *Medicago sativa*. The lines represent rays of light converging on lower cells with chloroplasts in which chlorophyll is located; the epidermis of this plant functions as a mosaic of micro-lenses. (**4**) *Euglena viridis*. This protozoa already has a simple eye in the form of an eye spot (s). Nucleus (*n*).

The phenomenon is not without special features. The cells of the leaf show several differentiations that do not appreciably differ from the light-sensitive cells of bacteria, protozoa or invertebrates (Fig. 8.3).

It has been known for a long time that in *Fittonia*, and other species, the cells of the leaf epidermis are convex with a biconvex lens at their tip (Niklitschek 1943). The cytoplasm of these cells is transparent and lacks pigment. This allows the transmission of light to the lower palisade tissue, where the chloroplasts, with their chlorophyll pigment, receive the light and transform it into chemical energy. *Fittonia* plants have broad and thin leaves; these form a ground cover in tropical forests where light is rather dim.

Vogelmann *et al.* (1989) and Martin *et al.* (1989) have discovered that, in *Medicago sativa* (alfalfa) which occurs in the fields of Europe, the same efficient light traps are present in the leaves. The epidermal cells in *Medicago* act as lenses that concentrate light, which is focused onto the chloroplasts of the palisade tissue. These cells are plano-convex, as shown by electron microscopy, and they concentrate an incident light beam up to 20-fold. The whole leaf epidermis is "a mosaic of micro-lenses that focuses light". Actually, the photographs of the surface of the leaf epidermis, with its multitude of convex lenses, are not very different from photographs of the surface of the compound eyes of insects.

A comparison between the compound eyes of insects and of the light-sensitive cells of leaves may seem at first sight far-fetched because we tend to regard leaves as having nothing to do with eyes or vision. But there are basic similarities.

1 — The leaf consists of several thousand light-sensitive cell units maintained together. The compound insect eye also consists of several thousand light-sensitive cell units which are tightly coupled.
2 — In *Medicago* there is both a lens and a receptor component. In insects the chitin is transparent and each unit consists of a lens (that is also convex) and a receptor where the light absorbing pigment (retinene) is located.
3 — A phenomenon established long ago is that chloroplasts, which contain the light-sensitive pigments, are able to migrate in cells. The chloroplasts of moss cells, and of other plant species, occupy a central position in diffuse light but move to the cell wall when strong light is applied (Fig. 8.3) (Denffer *et al.* 1971). The cells of the compound eyes of insects (flies) contain pigments that also migrate and do so when light intensity changes. They are in the middle of the cell or move towards one of its ends, thus attenuating the amount of light and changing its spectral composition (Stavenga and Kuiper 1977, Stavenga 1989).
4 — It has recently been found that chlorophyll derivatives are also involved in animal vision and that carotenoids can function as antenna chromophores (Isayama *et al.* 2006). This molecular information brings plant vision closer to that of animals.

Sea Urchins have a Large Number of Genes Responsible for Eye Formation Yet they have no Eyes at All. Like Plants, They See with their Entire Body

The genome sequencing of the purple sea urchin *Strongylocentrotus purpuratus* (Burke *et al.* 2006) led to the surprising discovery of a large number of genes responsible for eye formation: *pax6, atonal, neuroD* and *barh1*, which, in vertebrates, pattern early retina development. In addition, six different opsins and other essential components of the signal transduction cascade of photoreceptor cells were identified (Raible *et al.* 2006). Significant was that these genes were expressed in the tube feet of the adult sea urchin. Using electron microscopy and molecular technology Ullrich-Lüter *et al.* (2011) localized the expression of

Sp-opsin4 and *Sp-pax6* (two genes essential for photoreceptor function) in two distinct groups of photoreceptor cells located in the animal's numerous tube feet. Experiments revealed that sea urchins react to illumination moving towards or away from light sources in a specific way. Sea urchin photoreceptor cells lack screening pigment, indicating that there is a direct relation to the radial nervous system. The photoreceptors are found all over the sea urchin body. Like plants they are a carpet of light detectors.

The Protozoan Eye Already Moves

Protozoa are a heterogeneous assemblage of some 50,000 single-cell organisms that display extreme diversity. They combine the simplicity of being unicellular with the complexity of a large range of organelles and skeletons. Later in evolution, multicellular animals, mainly distributed among tissues and organs, the properties and structures that the protozoa harbored. To exemplify: The digestive system of protozoans consists of an oral region, a pharyngeal basket, oesophageal fibres, food vacuoles and an anal pore. A neuromotor system is also present. This is so developed that it forms a "central motorium" from which radiate nerve fibrils (Kudo 1971).

The same is the case with the eye. *Euglena* (order Euglenida) has an eye spot which is differentiated into a short rod, a cortical layer and an amyloid lens. In an other order (Dinoflagellida) (*e.g. Protopsis*) the eye consists of a layer of light-sensitive bodies containing carotenoid pigments covered by a clear zone that functions as a primitive lens. Other dinoflagellates, such as *Erythropsidinium*, have already a relatively complex eye. The ocellus consists of: (1) a lens, (2) a fluid filled chamber, and (3) a cup covered by light sensitive pigment. It is most remarkable that although one is dealing with a single cell organism the following can happen: (1) the lens can change its shape, (2) the pigment cup can move freely in relation to the lens, and (3) the whole ocellus can protrude from the cell, thereby allowing it to point in different directions. It is thus not surprising that the protozoans use their eye to detect the presence of prey just as later invertebrates and vertebrates are able to do (Fig. 8.3).

What is of significance, from the point of view of periodicity, is that such eye spots are completely absent in other protozoa (Friday and Ingram 1985). The phylum *Mastigophora* consists of 19 orders, but in only three of them does the cell contain an eye spot: Euglenida (*Euglena*), Dinoflagellida (*Protopsis*) and Volvocida (*Volvox*). The same is true of the many thousands of protozoan species that lack eyes and which comprise the other phyla: sarcodina (*Amoeba*), sporozoa (*Gregarina*), cnidospora (*Ceratomyxa*) and ciliophora (*Paramecium*) (Barnes 1980).

Flatworms have Eyes of the Most Primitive Type

Eyes are common in the class Turbellaria (planarians) and are of the pigment-cup type. Two is the usual number, but 4 or 6 are not uncommon. Land planarians may even have many eyes.

Other species of this group have the most primitive type of eye found in animals. These are patches of photoreceptors and pigment cells in the epidermis distributed down the sides of the body from which pigment cups evolved.

Another group of flatworms, known as flukes (Trematoda), are parasitic but some have four distinct eyes whereas the Cestoda, which are also parasitic, have no eyes or very rudimentary ones.

Highly Complex Eyes, Simple Eyes and no Eyes Occur within the Cnidarians. The same Species has Four Different Types of Eyes

Within this Phylum the class Hydrozoa comprises *Hydras* and hydroid medusae. Hydras lack eyes but these medusae have small ocelli. Remarkable is that the tissues of *Hydra* have differentiated into some of the most complex cell types — the specialized stinging cells called nematocysts, yet vision has not developed although they have a nervous system that consists of neurons.

Another Cnidarian class, the Scyphozoa or jellyfish, comprise some 200 species including the cubomedusae. *Aurelia* has ocelli consisting of simple pits containing pigment and photoreceptor cells, but all cubomedusans have complex eyes containing a lens and a retina-like arrangement of sensory cells and can orient to small points of light. This highly complex eye is not far from the vertebrate one in its organization, form and function (Koyanagi *et al.* 2008). Light sensing starts with phototransduction (in photoreceptor cells) and is followed by a molecular cascade that involves an opsin-based pigment. This triggers the G_5-type G protein-mediated phototransduction in the ciliary-type visual cells of the jellyfish lens eyes. The box jellyfish (*Tripedalia cystophora*) has 24 eyes of four different types: lower lens eye, upper lens eye, pit eyes and slit eyes. These different eye types are close to those of vertebrates (Petie 2012). The box jellyfish has no brain yet it uses its vision to move properly (Harland *et al.* 2012).

A third class of Cnidarians, the Anthozoa (corals and anemones) although they are close relatives do not even have specialized sense organs (Barnes 1980).

DNA sequences responsible for opsin building were recently found in hydras, sea anemones and hydroid medusae. This means that the potency for vision is present but cannot be fully realized.

Closely related to the Cnidarians are the comb jellies (Phylum Ctenophora) containing 50 species. They have no eyes and their only sense organ contains a statolith.

Hence, closely related species, closely related classes, as well as closely related phyla, may have eyes or lack them.

The Segmented Worms Display Absence and Presence of Simple and Complex Eyes

The phylum annelids comprises the segmented worms and includes the earthworms and leeches.

The class polychaeta (marine worms) has species without eyes, with very simple eyes and highly developed ones.

The eye of *Vanadis formosa* consists of: (1) Iris. (2) Cornea. (3) Lens. (4) Accessory retina. (5) Optic nerve. (6) Retina. (7) Pigmented layer. (8) Nuclear layer. (9) Plexiform layer and (10) Basal lamina. The small lateral retinas are sensitive to different wave lengths of light as are the principal retinas. Similar accessory retinas are found in some fish and cephalopods. The bulging eyes of *Vanadis tagensis* possess means of accommodation.

The class oligochaetes, which contains the earthworms, lack eyes, except for a few aquatic forms that have simple pigment-cup ocelli.

The general *epidermis* in leeches (class hirudinea) contains dispersed *sensory* cells of a number of types which include *photoreceptor cells*. Eyes in leeches consist of a cluster of photoreceptor cells surrounded by a pigment cup. Their eyes vary in number from 2 to 10 according to the species.

Spiders have no Compound Eyes

Although closely related to the insects, known for their compound eyes, no spiders have compound eyes (Harland *et al*. 2012). But the eyes of some spiders surpass those of all other arachnids in degree of development. There are eight eyes arranged in two rows of four, but other arrangements occur. In a few families the eye number is reduced to six, four or even two, and some completely lack eyes.

The *indirect eye* of a spider consists of: lens, rhabdome row, tapetum, vitreous body and retinal cells. The *direct eye* is simpler being composed of: lens, photoreceptors, and vitreous body. The eyes of most spiders are unable to form an image owing to an insufficient number of receptors, but in jumping spiders the number of receptors is extremely large. Each of a jumping spider's eight eyes is a separate camera eye, the retina consisting of a mosaic of photoreceptors, each photoreceptor covering different parts of the image.

The Transition to Compound Eyes in Myriapods

Myriapods comprise centipedes, millipedes and other groups. The head bears sometimes ocelli, but true compound eyes are not present.

Among the centipedes many species (and a few cave-dwellers) lack eyes. Other centipedes possess few to many ocelli. In the scutigeromorphs, the ocelli are so clustered and organized that they form structures close to compound eyes.

Eyes may also be totally lacking, as in the flatbacked millipedes, or there may be 2 to 80 ocelli.

Millipedes without eyes have integumental photoreceptors.

Horseshoe Crabs are Living Fossils with Rudimentary Compound Eyes

Horseshoe crabs are living fossils. *Limulus polyphemus*, common in the USA and Mexico, looks like its ancestors that flourished over 300 million years ago. Although they are called crabs they are not crustaceans but are close relatives of the spiders.

The *lateral eyes* of horseshoe crabs are peculiar in a number of respects: (1) They are compound eyes made up of units of 8 to 14 cells around a rhabdome. (2) Each unit has a lens and a cornea and is called an ommatidium. (3) In contrast to the compound eyes of insects their ommatidia are not compactly arranged. (4) Pigment is present but does not form a shifting screen between ommatidia. (5) The cornea and the lens are structurally different.

Crustaceans Show Large Variation in Number of Units in Compound Eyes

Crustacean eyes are of two types: median and compound eyes.

1 — The *median eye* is composed of three or four pigment-cup ocelli of the *inverse type*. Each ocellus consists of a few photoreceptor cells, which possess rhabdome microvilli on their opposing surfaces. In most ocelli no lens is present. The median eye is probably largely an organ of orientation in larvae.
2 — Two *compound lateral eyes* are found in the adults of most species. (a) Eyes may be located at the end of a stalk or may be sessile. (b) The corneal surface is often greatly *convex* resulting in a wide visual field. (c) There is a relationship to the environment. Crustaceans living in well-lighted habitats possess eyes with developed screening pigment. Nocturnal crustaceans have eyes in which the screening pigment is reduced or absent. (d) The number of ommatidia in crustacean compound eyes varies enormously. A single eye is composed of 25 ommatidia in *Armadillidium* or of 14,000 ommatidia in the lobster *Homarus*.

Among the crabs, lobsters and shrimp there are blind species particularly among deep-sea forms and cave-dwelling crayfish, but compound eyes when present are usually highly developed.

The Compound Eyes of Insects are an Extreme Case of Ordered Eye Association

Ocelli are absent in many adult insects, but when present there are usually three. Ocelli can detect changes in light intensity and may be very sensitive to low intensities. Compound eyes are laterally situated on the head. The number of facets is greatest in flying insects. Extreme reduction of facets is found in certain parasitic and cave-dwelling insects. The facets are larger in nocturnal than in diurnal insects. In some crepuscular flying neuropterans and some water beetles the compound eyes are divided into upper and lower halves.

The compound eye of an insect is composed of hundreds of individual ommatidia. These consist of a corneal lens followed by a crystalline cone which focus incoming light rays onto the rhabdom. Light passes from this clear structure into eight retinal cells that surround it. Each retinal cell contributes a nerve fiber to the optic nerve. Of importance is that the whole ommatidium is surrounded by pigment cells which prevent the leakage of light from one ommatidium to another.

These compound eyes form *erect images*. In a camera-type eye the image is optically inverted when it falls on the retina and has to be inverted again by the brain. No such optical inversion is involved in the mechanism of the compound eye (Simpson and Beck 1965). Camera eyes are found in molluscs such as *Helix* and *Octopus* as well as in mammals (Friday and Ingram 1985).

The origin of compound eyes is elucidated by the finding that a Cambrian fossil *Anomalocaris* (half a billion years old) has well preserved visual surfaces composed at least of 16,000 hexagonally packed ommatidial lenses in a single eye, rivalling the most acute compound eyes of modern insects (Paterson *et al.* 2011).

The Molluscs — Eye Complexity Becomes Close to that of the Human Eye

In molluscs, not all species have eyes. Primitive eyes that contain photoreceptors and pigment cells in a simple pit are found in *Patella* (a Gastropod). In others (*e.g. Murex*) the pit has become closed over and differentiated into a cornea and a lens (Fig. 8.4). The most highly developed eyes in this group are considered to be superior to those of fish.

Another class of molluscs (Bivalvia) comprises the clams, oysters and mussels. The scallop *Pecten* has many small blue eyes located along the mantle edge. In *Spondylus* the ocelli consist of a cornea, a lens and a retina in which the photoreceptor cells are directed away from the source of light.

The squids, cuttlefish and octopods build the class cephalopoda. Their eyes are strikingly similar in organization to those of vertebrates.

The eye in cephalopods consists of: (1) A spherical housing containing cartilaginous plates. (2) The lens is a rigid sphere suspended by a ciliary muscle (it has a fixed focal length). (3) An iris diaphragm which controls the amount of light entering the eye. (4) The pupil is slit-shaped the slit being kept horizontal. (5) The retina contains closely packed photoreceptors that are directed towards the source of the light.

The eye is thus of the direct type, instead of indirect as in vertebrates. The cephalopod eye forms an image but the animal's perception is considered to be different from that of the human eye.

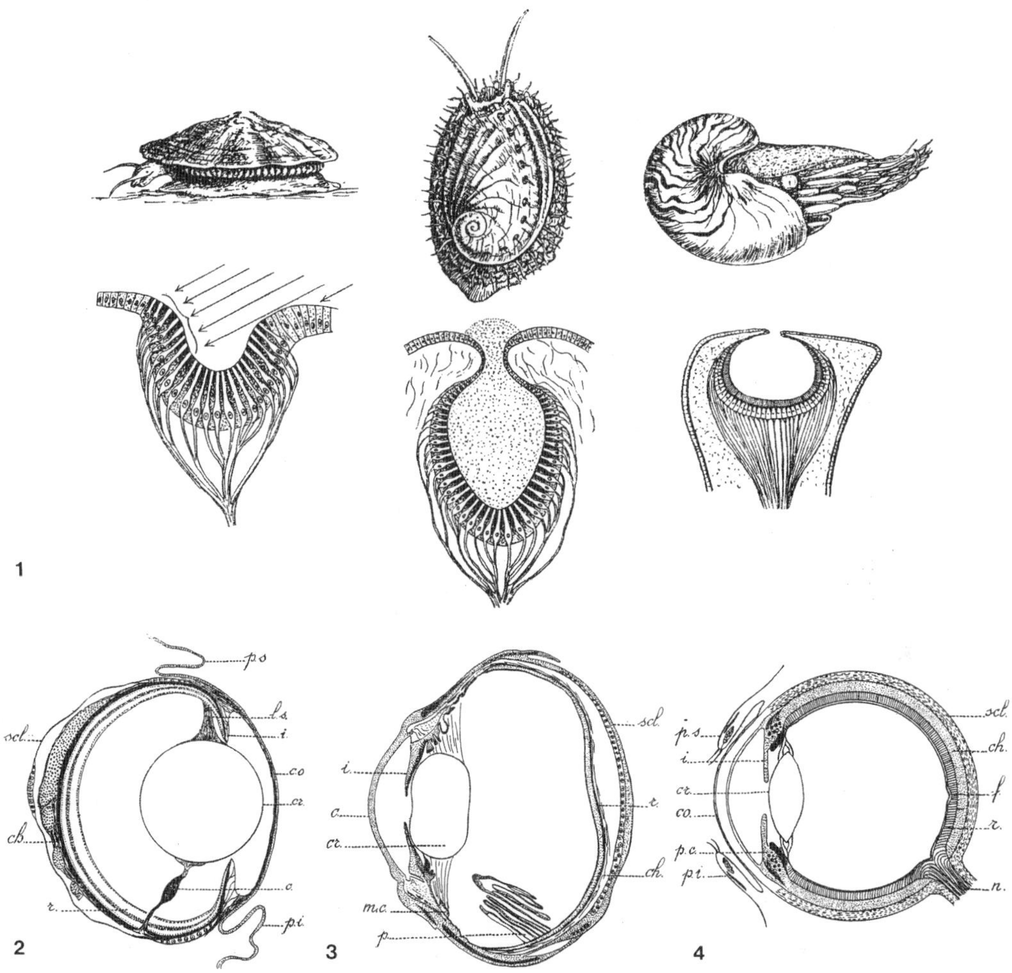

Figure 8.4 Structure of eyes in invertebrates and vertebrates. (**1**) **From simple to complex eyes in molluscs**. Transformation of an open eye to a closed one. Common limpet *Patella vulgata* and its eye (left). Ear shell *Haliotis lamellosa* with partly closed eye (middle). The mollusc nautilus *Nautilus pompilius* has an eye with a lens (right). (**2**) *Oncorhynchus mykiss*, Rainbow trout. Transversal section of the eye of a fish (co, cornea; cr, crystallin; *i*, iris; *r*, retina). (**3**) *Phalacrocorax carbo*, Great cormorant. Transversal section of the eye of a bird (*r*, retina; *i*, iris; *cr*, crystallin; *c*, cornea; *p*, pecten). (**4**) *Homo sapiens*. Diagram of the eye (*r*, retina with fovea, *f*).

The squid has eyes with different functions. The *large left eye* is directed upward and responds to faint light from the surface, and the *little right eye* is directed forward and responds to bioluminescent light.

The eyes of *Nautilus* are large and are carried at the end of short stalks, but they are much less complex than those of other species of cephalopods. The eye lacks a lens and is open to the external sea water through a small aperture, the pupil (Fig. 8.4).

Ascidians Make the Transition Between Invertebrates and Vertebrates, Yet they have Some of the Simplest Eyes

Sea squirts or ascidians are tunicates that belong to the Urochordata. They are also called invertebrate chordates (Burnie 2004).

Although they build the evolutionary bridge between the invertebrates and the vertebrates, they have no complex eyes. Instead their vision is among the simplest known.

Their sensory apparatus consists of two pigmented sensory organs: the anterior otolith and the posterior photoreceptive ocellus.

The ocellus is a multicellular structure composed of a pigment cell, three lens cells and about 30 photoreceptor cells. Like vertebrate retinal photoreceptors, the photoreceptor cells of ocellus are ciliary and hyperpolarized (Esposito 2014).

Fish have Good Vision Yet their Eyes Differ Sharply in Organization from those of Octopuses

The fish eye consists of: cornea, iris, lens, suspensor muscle, choroid and retina. The structure is remarkably similar in functional performance and adaptive significance to that of the mollusc *Octopus*, but differs dramatically in details of how its parts are assembled and work.

1 — In the eye of an octopus or squid the photoreceptors *point toward* the direction of incoming light. In the eye of a fish the photoreceptors *point away* from the direction of incoming light, so the light must pass through the tissue of the retina to stimulate the photoreceptors. This means that the fish retina has an inverted position, whereas the retina of octopus and squid is noninverted.
2 — Nerve cells in the fish retina, unlike those in octopuses, form networks that extensively process visual information before signals go to the brain.
3 — In these molluscs the nerve cells that convey visual signals from the retinal receptors to the brain leave the eye directly in multiple optic nerves. In contrast, the nerve cells leaving the fish eye retina gather into a single optic nerve. Although differing sharply in organization excellent vision is obtained in both cases (Hill *et al.* 2004).

The fish lens allows for good spatial resolution and good focus. This is due to the ability that fishes have to move the eye lens along the optical axis or even change its shape, which permits accommodation. In teleost fishes multifocal lenses allow for correct focusing of multicolored images (Gagnon 2010) (Fig. 8.4).

Repression and Expression of the Same Genes Appear as the Source of Eye Absence and Transformation — Larvae and Adults have the Same Genes Yet Larvae have Eyes Whereas Adults Lack them

As molecular and genetic information accumulates it becomes evident that the same basic molecular events and the same genes are present across most living organisms, indicating that it is the autonomous atomic evolution of DNA, combined with the expression or repression of specific genes, that leads to different solutions during evolution.

Larval development supports this interpretation, because although the larva and the adult of an animal have the same genes, they may have different vision. The situation is extreme in some annelids, brachiopods and bryozoans in which the adults have no eyes whereas the larvae have simple eyes. Hence, eye absence appears as the result of the repression of genes that were active in the larva.

Another gene transformation occurs in gastropod larvae which undergo a transition from ciliary to rhabdomeric photoreceptor cells during embryonic development. These are two cell types that usually appear separated in different animal groups (Randel and Jekely 2016).

A third example is the larval development of decapod shrimps in which the larvae have compound eyes with hexagonal facets which change to square facets as they turn into adults (Land and Fernald 1992).

Hydras and sea anemones have no eyes but DNA sequences responsible for opsin building were found in these two groups which means that the potency for vision is present but cannot be fully realized (Koyanagi *et al.* 2008). In the coelenterates closely related species, closely related classes as well as clearly related phyla may have eyes or lack them.

Reptiles have a Third Eye Located on the Top of the Head which Functions as a Sky Compass and Senses Ultraviolet Light

Two surviving species of the ancient order Rhynchocephalia (*Sphenodon punctatus* and *Sphenodon guntheri*) have in addition to large lateral eyes a third eye located on the top of the head. They are living fossils, which at present inhabit New Zealand. Their heyday was over 200 million years ago but they look like most other reptiles (Halliday and Adler 2004).

This third eye (Fig. 8.5) was lost in later orders such as crocodiles, birds and mammals, but remnants of this organ can be found in most of these groups. It "represents evolution's earlier approach to photoreception" (Schwab and O'Connor 2005). The third eye is a complex structure consisting of: a cornea, a lens (that is surprisingly similar to that of lateral eyes), vitreous cavity, pigment epithelial layer, ciliated photoreceptors and ganglion cell layers.

The function of the third eye has been recently investigated in detail. It is a sky polarization compass in *Podarcis sicula*, *Sceloporus jarrovi*, *Tiliqua rugosa* and *Uma notata*. Following a series of experiments the lizards were trained to swim from the center of a water maze onto a hidden platform (the goal) positioned at the periphery of the maze in a single compass direction. Lizards learned to swim to the hidden goal and the intact third eye was found to be required to perform the orientation. When the eyes on top of the head were painted black the animals were completely disoriented (Beltrami *et al.* 2010).

It turns out that the third eye is physiologically and anatomically related to the pineal gland present in the lower part of the brain of most vertebrates. Pineal and pineal-related organs are present in the early vertebrates such as lamprey, teleosts and frogs. These organs can discriminate between different wavelengths of light. The same applies to the reptilian parietal eye in *Iguana iguana* and *Anolis carolinensis* which was found to sense UV light (Wada *et al.* 2012).

Bird Eyes are Known for their Excellent Vision

The fovea is a depression in the retina containing no rod cells but numerous cone cells. It is a region that allows acute diurnal vision. It is found in diurnal birds, lizards and humans. The eyes of birds have a lateral and a central fovea (Rabaud 1932) (Fig. 8.4).

The eyes of owls, which are nocturnal birds, are extremely sensitive being larger than those of mammals and containing a bigger lens and pupil. The retina has a high proportion of rods (which only detect black and white) and relatively few cones (which enable color vision) permitting the eye to function at very low light intensities. The owl has an intense tunnel vision that allows to judge the distance to prey with lethal precision (Perrins 2003).

The eye of a bird of prey, like a hawk, is relatively larger than that of a human. The lens is held in position by a ring of bony plates a feature characteristic of bird eyes. Focusing is achieved by adjustment of the shape of the lens by the ciliary muscles and the amount of light entering the eye is controlled by the iris. The group of cells, located on the back of the eye, called the pecten is especially well developed in birds of prey but its exact function is not known (Perrins 1976).

There are 10 Generally Recognized Optical Eye Types

Cronin *et al.* (2014) described in detail 10 types of eyes that have evolved in the various branches of the animal kingdom. Whereas vertebrates possess only one of them, invertebrates possess all 10. These types are: (1) *Simplest pigment-pit eyes*. They are nothing more than crude photoreceptor-lined invaginations in the epidermis. They are considered to be an ancestral type. (2) *The pinhole eye of Nautilus*. Although lacking a lens the eyes of *Nautilus* are large, the pupil is mobile and varies in size. Pinhole eyes are thought to work like a pinhole camera. An inverted and blurry image is created on the retina. (3) *Concave mirror eyes*. In this eye type a concave mirror rather than a lens is used to focus the image on the retina. It is found in bivalve molluscs and three classes of crustaceans (amphipods, ostracods and copepods). They also appeared in a deep-sea fish. Like lenses concave mirrors focus images by bending rays of light but not by refraction (as in a lens) but by reflection. The mirror's image can be upright or inverted. (4) *Camera eyes* are the principal eye type of all verbebrates but this type is found as well in molluscs, annelids, crustaceans, cnidarians, spiders, scorpions, ticks, mites and in insect ocelli. All camera eyes are a single optical unit and the image is inverted on the retina. (5) *Compound eyes* are the most common type in animals. The ommatidia create in insects an erect view of the world. Compound eyes fall into two main subtypes: (6) apposition eyes and (7) superposition eyes, the difference lying in the number of units that provide light to a single visual unit (rhabdom).

Apposition eyes come in three main forms: (8) *Focal apposition eye* (insects and crustaceans), (9) *afocal apposition eye* (butterflies), (10) "*neural superposition eye*" that despite its name is an apposition eye (higher flies).

Eye Evolution is Characterized By the Introduction of Novel Solutions and a Type of Lens has Evolved at Least 8 Times

Land and Fernald (1992) also divided eyes into ten types but their classification is different in several ways.

The simplest eyes of invertebrates are the *pit eyes*. They contain from 1 to about 100 receptors. They are found in all but five of the 33 groups of multicellular animals.

The *single-chambered eyes* of aquatic animals have a spherical lens. This type of lens has evolved at least 8 times in: (1) fish, (2) cephalopod molluscs (excluding *Nautilus*), (3 to 6) at least four times in gastropod molluscs, (7) the annelids and (8) once in the copepod crustaceans. The eye is not such a perfect structure as we tend to think. These lenses have a defect which is chromatic aberration.

Multiple lens eyes are also found in aquatic animals. Copepod crustaceans (e.g. *Pontella*) have a triplet lens in the male and a doublet in the female. *Copilia* eyes move to and fro in the longitudinal plane, scanning the water in front of the animal.

Compound eyes have evolved a number of times. The classical eye (diurnal insects and crustaceans) is built up from the elementary contributions of all the separate ommatidia (Fig. 8.6).

A recently discovered novel type of eye by Nilsson (1988) appears in the crab *Portunus*. This eye uses lenses and mirrors in both apposition and superposition configurations.

The Number of Eyes Varies from One To 30,000

Fish, amphibians, reptiles and mammals have usually only two eyes, but this is not the case among the invertebrates. They may have no eyes or thousands of them.

Eyes in leeches vary in number from 2 to 10 according to the species. Spiders have eight eyes arranged in two rows of four, but other arrangements occur, and the number can be reduced to complete lack of eyes.

Horseshoe crabs compound eyes consist of only 8 to 14 units. Other evolutionary solutions led to an extreme variation of eye number in crustaceans, from 25 units to 14,000.

One single compound eye of the fly *Drosophila* contains 750 individual units but in a dragonfly the number is 30,000. However, some ant species have only 20 units and the tropical ant *Eciton* has compound eyes consisting of a single giant ommatidium.

Hence, eyes may combine, building optically efficient units, by starting with one unit and putting together: 8, 14, 25, 750, 14,000 and 30,000 single eyes.

Eyes that remain individualized can also attain large numbers. Nemerteans (which are close to flatworms) have two to six eyes but they may have on their heads several hundred isolated eyes. In some millipedes the number of ocelli varies from 2 to 80. The scallop *Pecten* has along its mantle three rows of tentacles with 30 to 40 eyes.

Eyes can Occupy the most Unexpected Positions Occurring in Legs and Wings

Eyes tend to be located on the frontal region of the head in living organisms but this position is far from being general. By manipulation of genes in *Drosophila* it has been possible to obtain flies with eyes on the legs, on antennae and even on wings (Halder *et al.* 1995) (Fig. 8.7).

Ancestral vertebrates had a third eye, besides the two lateral eyes, medially situated on the forehead and directed upward. It was universally present in the oldest ostrachoderms.

In other organisms eyes may also be located differently. The larval form of the widespread black dragonfish *Idiacanthus fasciola* has elongated stalks arising from the head, the eyes being situated at the extremities of these hornlike appendages. Among the sharks the location is still more extreme. The adult smooth hammerhead *Sphyrna zygaena* has eyes at the end of wing-like flaps that protrude from both sides of the head.

Instead of two a fish may have four eyes. This is why *Anableps anableps* is called the "four eyes" fish. Its eyes have divided pupils and retinas enabling it to see above and below the water's surface simultaneously. Consequently it can keep a look-out for predators as it searches for food. A variant of this solution is found in the spookfish *Dolichopteryx binocularis*. As the name indicates its eyes have a most unusual anatomy being tubular in shape and pointing obliquely upwards like a pair of binoculars. This fish is considered to be the truly 4-eyed fish, because on the side of each tubular eye there is a second eye complete with retina and lens (Burnie 2004).

A phenomenon described since the early days of biology is the migration of eyes in flatfish during the transition from juvenile to adult. The hatching young of *Solea vulgaris* differs little from other fishes in the lateral position of its eyes,

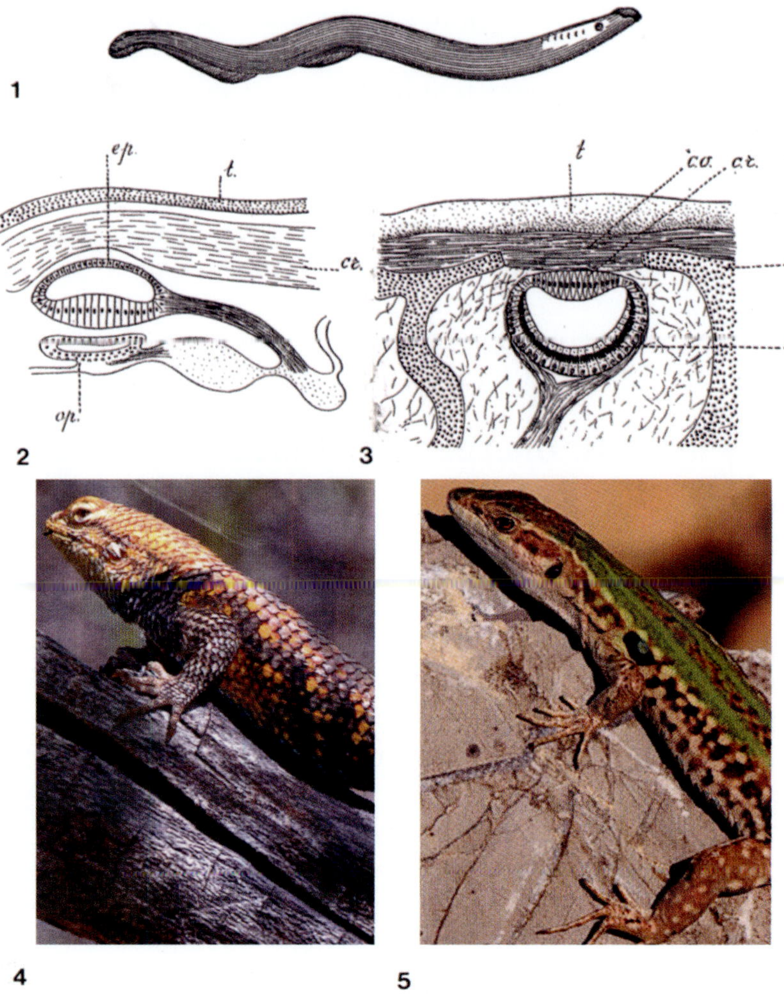

Figure 8.5 Third eye on top of head. (**1**) *Lampreta fluviatilis*, Brook lamprey. This ancestral group of fish appeared 500 million years ago. Lampreys have pineal-related organs. (**2**) *Lampreta sp.*, Lamprey. The parietal eye and pineal organ. *op*, parietal eye; *ep*, pineal organ; *cr*, cranium. (**3**) Lizard. Parietal eye: *r*, retina; *cr*, crystallin; *co*, cornea; *vc*, cranium. (**4**) *Sceloporus magister*, Desert spiny lizard. A lizard species with a third eye located on top of head. (**5**) *Podarcis sicula*, Lizard. Although this species looks like other common lizards it retains an ancestral third eye on the top of its head.

but at an early age the animal begins to lean to one side, at the same time that one eye migrates across the top of the head to take up a position beside the other eye. As the flatfish reaches the adult stage both eyes lie on the upper side of the body. The rotation may be to the left or to the right. In dextral fishes the left eye migrates, in the sinistral type it is the right eye that moves (Paxton and Eschmeyer 1998).

The Source of Coherence — Self-organization of Eye Components During its Development

Self-assembly is a process that starts even before the atomic level. Elementary particles: the leptons and quarks self-assemble following their own rules combining into protons, neutrons and other larger particles as a result of their energy states.

Self-assembly was first discovered by biochemists who showed that the components of certain viruses: their RNA and proteins could be isolated. Subsequently when they were allowed to combine they self-assembled producing infectious viruses, without external intervention. Later self-assembly was demonstrated at the organism level when whole organisms were self-assembled from single cells such as sponges and even human cells self-assemble combining into functional organs. The coherence that has so much impressed biologists can be seen at present to derive mainly from the process of self-assembly that is inherent to matter and only allows combinations taking place according to the initial energetic properties of elementary particles, atoms, molecules and macromolecules.

The extreme coherence of the eye has now been shown to have this cause. Eiraku *et al.* (2011) have demonstrated the self-organization of the optic cup in tissue culture experiments by following, in vitro and in vivo, the formation of the eye in the mouse. The retinal primordium first appears as the optic vesicle which is an epithelial evagination from the diencephalon. Subsequently its distal portion invaginates to form the optic cup, which develops into the outer and inner layers of the retina. Using a three-dimensional embryonic stem cell culture system they found that the initially homogeneous cell aggregates formed hollowed spheres which spontaneously subdivided and built the optic cup. Important was that the formation of the cup occurred in the absence of a lens or ectodermal tissues indicating that the invagination was not caused by external structures but "occurred in a self-directed fashion".

Their study revealed that the initial retinal component possesses a "latent intrinsic order involving dynamic self-patterning and self-formation" driven by a sequential combination of local rules within the epithelium.

Ubiquitous Genes and Identical Regulatory Molecular Cascades Shaped the Eye Throughout its Evolution

A series of experiments has shown that highly divergent structures of vertebrates and invertebrates are directed by the same genes and identical molecular cascades.

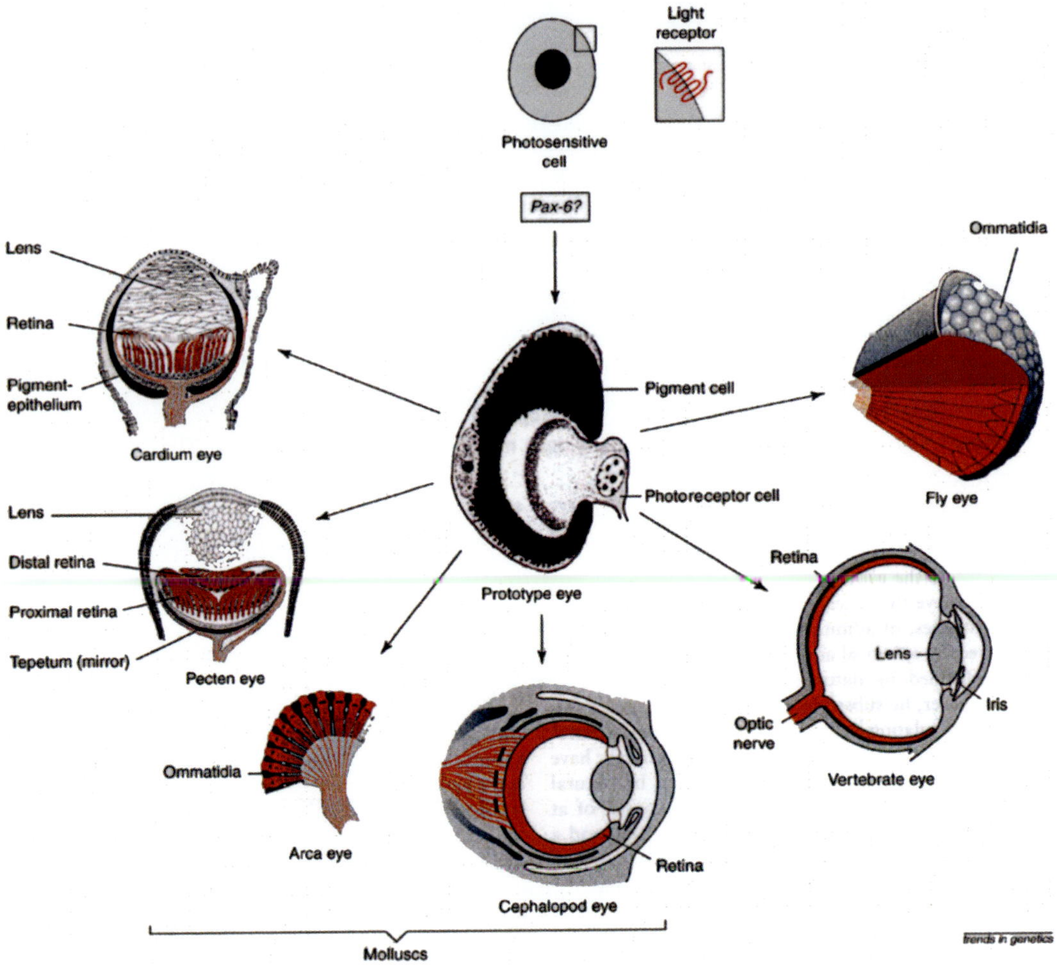

Figure 8.6 General scheme of eye evolution. Evolution of various eye-types from a common ancestral prototype. As a first step, photosensitive cells with a light receptor (opsin) have evolved. Under the control of the *Pax 6* gene, the photosensitive cell assembles with a pigment cell to form an organ, the prototype eye. By evolution, the various eye-types are generated from the prototype: the compound eye of insects; the camera-type eye of vertebrates; and the large spectrum of eye-types in molluscs ranging from the primitive camera-type eye in *Cardium*, the mirror-plus-lens eye of *Pecten*, the compound eye of *Arca* to the highly evolved cephalopod eye, that greatly resembles the vertebrate camera-type eye. *Cardium*, *Pecten* and *Arca* are bivalve molluscs.

The *Drosophila* retina is patterned by a developing wave driven by the *Hedgehog* signaling protein. A *hedgehog* homologous gene, called *Sonic hedgehog*, is present in zebrafish and is expressed in the first retinal neurons. This gene also drives a wave of neurogenesis across the retina, strikingly similar to the wave in *Drosophila*. To explore these events in the fish retina the eyes were stained with an antibody against the activated form of a protein kinase which revealed that the same domain of expression and type of wave spreading occurred as in *Drosophila* (Neumann and Nüsslein-Volhard 2000).

The results from the hedgehog gene led to the conclusion that this gene was already used to pattern a primordial eye structure before vertebrate and invertebrate lineages diverged and thus supports a common origin of the animal eye.

Opsins are the protein components of animal pigments responsible for catching photons. The light signals received by the opsins are amplified by G-proteins. Both the opsins and the G-proteins are of general occurrence. It also turned out that the molecular cascades are essentially the same throughout all animals.

As more information accumulated, the presence of these molecules was found to extend to the algae and the bacteria. Rhodopsin was tentatively identified in cyanobacteria and the flagellated alga *Chlamydomonas* (Gehring 2014). In addition the whole genome sequencing of other cyanobacteria identified one sensory rhodopsin in *Anabaena.* Sensory rhodopsins were also present in the ancient lineage of bacteria, the Archea (*Halobacterium salinarum*) (Spudich 2006).

Lenses are formed by crystallin proteins and *α-A crystallin* is ubiquitous missing only in bony fish.

The genetic similarity is even more extreme. Gehring and collaborators, as a result of their extensive experimental work in *Drosophila* and other organisms, found that "All bilaterian animals share the same master control gene, *Pax6* and the same retinal and pigment cell determination genes" which led them to conclude that "the different eye-types originated monophyletically and subsequently diversified by divergent, parallel or convergent evolution" (Gehring and Ikeo 1999, Gehring 2014). The term "bilaterian" refers to the property, of most multicellular organisms, to have just one symmetry plane that separates them into two halves. The coelenterates and the echinoderms, with their radial symmetry are excluded. However, opsin-based pigments have also been described in the box jellyfish (a coelenterate) and they follow the same molecular cascade in which G-proteins participate (Koyanagi *et al.* 2008). The same situation takes place in echinoderms. Sea urchins contain six different opsins which participate in the same molecular process (Raible *et al.* 2006). It also turned out that the *Pax6* gene is present in sea urchins (Ullrich-Lüter *et al.* 2011). So far opsin genes were not found in sponges but a single *Pax6* gene is present (Gehring 2014) (Fig. 8.6).

The loss of *Pax6* function led to animals without eyes in both mammals and insects. Also the transfer of the mouse *Pax6* gene to *Drosophila* resulted in the the induction of extra compound eyes in the fly (Onuma *et al.* 2002).

Additionally they demonstrated that *Notch* signaling regulated the expression of *eyeless* and *twin of eyeless* in *Drosophila* and that activation of *Notch* signaling in *Xenopus* caused eye duplications which indicated that the gene regulatory cascade is similar in vertebrates and invertebrates.

1 2 3

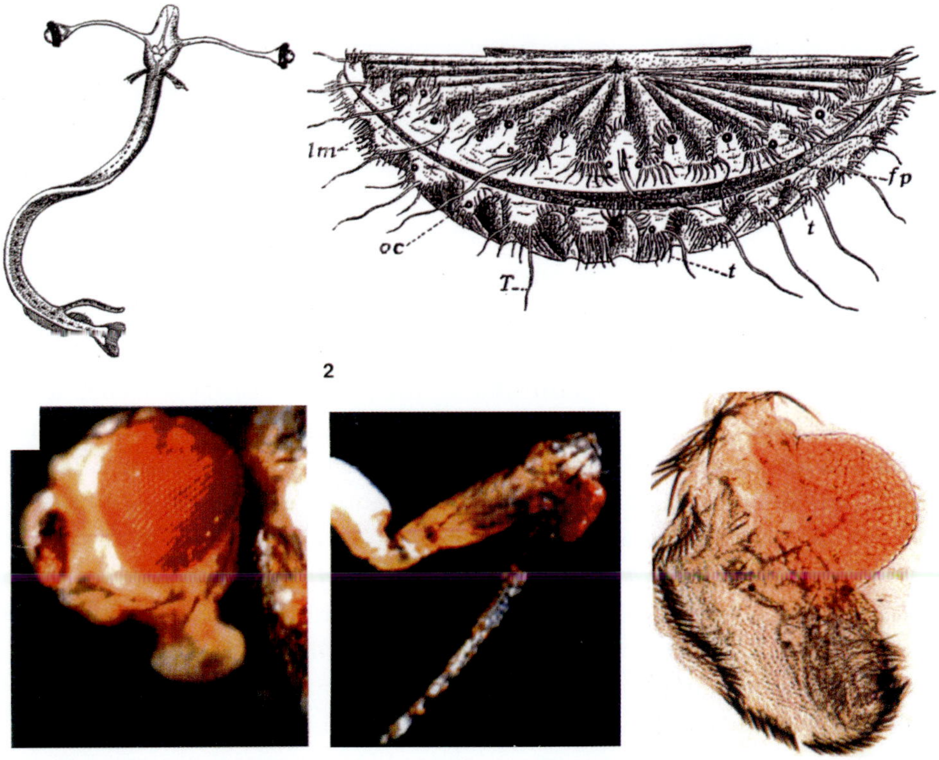

Figure 8.7 Extreme variation in eye location occurring naturally and obtained by genetic intervention. (**1**) *Stylophtalmus paradoxus*, Fish. Larval form of fish with eyes at end of stalks that lives at a depth of 2,000 meters. (**2**) *Pecten maximus*, Mollusc. Not less than 13 dark blue eyes are located along the mantle of this scallop. These are advanced eyes with a retina and a crystallin. (**3**) *Drosophila melanogaster*. Genetic manipulation in the fly *Drosophila* has led to the production of extra eyes at the most unexpected locations. Large normal eye accompanied by two small eyes on antenna and foreleg (left). This is a transgenic fly obtained by expression of the normal *eyeless* gene. Location of an eye on a leg (middle). Large eye on a wing. Additional eyes on the wing were also obtained using the *eyeless* gene (right).

Presence and Absence of Eyes Across Living Organisms — A Key to Periodicity

The many thousands of highly diversified species of Protozoa have evolved for no less than 1.6 billion years. Yet among the many orders in which they are subdivided, eye spots emerged in only three of them, being represented by only a few species (Barnes 1980). Although they had the ability to initiate vision this was strictly limited.

The same situation occurs in the Bacteria which appeared in evolution earlier than the Protozoa. More than 5,000 species have been described but many more are still unidentified. They build 16 different groups and for many years they have been known to be photosensitive. Recently some cyanobacteria (*Synechocystis*) were found to use micro-optics to sense light direction. So far only a few species are known to have acquired vision.

As the invertebrates were reached (1.0 billion years ago) although they became multicellular, and developed complex cell types, many of the groups that they built lack eyes or have only rudimentary ones.

Phyla lacking eyes or where eyes have not been described are: Placozoa, Porifera, Eumetazoa, Mesozoa, Gastrotricha, Acanthocephala, Gnathostomulida, Pogonophora, Echiura, Priapulida, Pentastomida, Phoronida, Bryozoa, Entoprocta and Brachiopoda. This means that these animals reproduced and survived efficiently without eyes for millions of years.

Other invertebrate phyla have eyes but only of a rudimentary type: Rhynchocoela, Rotifera, Kinorhyncha, Nematoda, Sipuncula and Tardigrada. This situation changes dramatically in other invertebrate phyla. In them the eye may be highly developed attaining the complexity of the human eye. These are: Cnidarians, Molluscs and Annelids. But, other invertebrate groups that have had the same time to evolve, never went further than the simplest type of eye: planarians (flatworms), bivalvia (molluscs), earthworms (annelids), centipedes (myriapods) and starfish (echinoderms). Ascidians, that make the transition between invertebrates and vertebrates also belong to the category of simple eyes.

Significant is that within most groups that exhibit advanced eyes there are species in which eyes are totally absent: Hydras, corals, anemones (coelenterates), some species (molluscs) and earthworms (annelids).

The same is the case concerning compound eyes. Among the crabs, lobsters and shrimps there are blind species.

Simple eye-spots have evolved independently at least 40 times and probably as many as 65 times (Land and Fernald 1992).

The Periodic Table of Vision

Table 8.1 consists of three vertical columns in which living organisms are distributed according to an increasing grade of complexity.

Column 1 Absence of Eyes

It starts with the simplest eubacteria and finishes with the echinoderms.

Table 8.1 Periodic Table of Vision

Increasing grade of complexity	**ABSENCE OF EYES**	**RUDIMENTARY EYES**	**COMPLEX EYES**
1	MINERALS EUBACTERIA PROTOZOA Most species CNIDARIANS *Hydra*	CYANOBACTERIA PLANT CELLS PROTOZOA *Euglena* CNIDARIANS *Aurelia*	CNIDARIANS *Tripedalia* ANNELIDS *Vanadis*
2	PLACOZOA PORIFERA EUMETAZOA MESOZOA GASTROTRICHA ACANTHOCEPHALA	RHYNCHOCOELA ROTIFERA KINORHYNCHA NEMATODA	CRUSTACEANS *Homarus* Most species INSECTS Most species
3	GNATHOSTOMULIDA POGONOPHORA ECHIURA PRIAPULIDA PENTASTOMIDA	SIPUNCULA TARDIGRADA	MOLLUSCS SQUIDS CUTTLEFISH OCTOPODS
4	PHORONIDA BRYOZOA ENTOPROCTA BRACHIOPODA	FLATWORMS *Planaria* ANNELIDS MARINE WORMS	FISHES AMPHIBIANS
5	MYRIAPODS Many species ANNELIDS EARTHWORMS Most species	CRUSTACEANS Some species MYRIAPODS Some species SPIDERS	REPTILES BIRDS
6	FLATWORMS CESTODA Some species ECHINODERMS Sea urchins	MOLLUSCS *Patella* ASCIDIANS	MARSUPIALS PLACENTALS

As is the case with other properties, in vision the eye attains a high degree of complexity even in the simplest invertebrates. Contrasting to this feature is that thousands of species of invertebrates have existed for millions of years and survived efficiently without eyes. However, there is an increase in eye complexity that tends to follow the organism's phylogenetic position. Insects and molluscs do not fit easily into this progressive development due to particular solutions. The eye became a permanent feature of vertebrates after being a punctuated event in invertebrates. Important is that the same gene, Pax6, is present in all species that have eyes and even in those that lack eyes. This indicates that gene expression and repression are responsible for its periodic emergence (Original).

Column 2 Rudimentary Eyes

These occur already in the cyanobacteria, include plant cells, and appear as well in ascidians.

Column 3 Complex Eyes

Most advanced eyes emerge already among the simplest invertebrates (cnidarians) and recur throughout the invertebrates, leading to quite different solutions, to become a general feature of vertebrates.

Chapter 9

Flight in Air, an Explosive Event throughout Invertebrates and Vertebrates

Flight is an Evasion from Gravity. It Occurs in Water, in the Water-air Interface, and in Air

Flight is a liberation from the force that attracts all bodies to the center of the earth. It is an evasion from gravity.

This process is usually considered to occur as a movement from the ground into the air, but it takes place equally in the water and in the interface that separates water from air.

When flight occurs as a movement from the ground (birds, insects, bats) the animal is obliged to return to the earth either to feed or to reproduce. Bats and birds may capture their food in the air but are obliged to build their nests on the ground or on living structures that are part of the ground (e.g. trees). In this case flight implies a movement which not only encounters the resistance of the air, but demands a combination with another medium to sustain life.

Water, on the other hand, although it offers a higher degree of resistance, has the advantage of allowing feeding and reproduction to take place without changing medium. Swimming in water allows to counteract gravity by moving upwards without great effort. Moving from the bottom of the ocean to its surface is a relatively easy task made possible by using the propelling power of fins and flippers (fish and marine mammals respectively). Sperm whales in search of giant squids descend to depths of 1,000 meters and can remain under water for 90 minutes. Subsequently they ascend to the surface by moving upwards, as much as one kilometer, without any difficulty.

Not only marine mammals, but most fish (about 24,500 species) "fly" in water, as they regularly move from the bottom of the ocean, river or lake, to their surfaces, counteracting gravity as spontaneously as birds. Most fish move in water primarily by using their tail fin, but the dorsal pelvic and pectoral fins also participate actively in the ascending and descending movements. Active flying is particularly evident in manta rays (*Manta birostris*). "Measuring up to 7 meters across, the manta is the world's largest ray, swimming through the open ocean like a large bird on slowly flapping wings" (Burnie 2004). Zoologists have not hesitated to call the swimming of these fish as true flight. The "underwater flight" of the manta, as they also call it, is achieved by its triangular pectoral fins which beat up and down about once every 4 or 5 seconds. The manta can even accelerate rapidly somersaulting out of the water with a massive impact. It breaks through the water-air interface moving into a different medium. Another species, the eagle ray, can leap clear of the water's surface like flying fish.

Usually it is this movement, out of the water and into the air, that has been described as flight in fishes and in octopuses. But this is only a particular case of the flying process, although a remarkable one, because it shows that in given species it has become a regular process to change from flight in water to flight in air.

Flying fish use either 2 or 4 specialized fins that function as wings in their complex flight out of the water and octopuses move into air "like rockets" by expelling water out of their mantles.

Hence, the evasion from gravity called flight is a regular event both in air and in water. Its occurrence on the interface between these two media only reinforces the notion that it is a process not necessarily confined to a single medium.

Flight Starts in the Brain

No wing, be it of a bird or an insect, can move by itself. It only changes position when the animal's brain gives the signal. Flight starts in the brain and not in the wing.

The function of the brain in flight was studied in detail in dragonflies, because they are considered a "superb example of an animal that captures rapidly moving prey". The dragonfly compound eye is among the largest of insect eyes. Neurons were found to be sensitive to object motion. These are of two types: small target motion detectors and target-selective descending neurons, which transmit target motion to the protocerebrum down the ventral nerve cord to the thoracic ganglia. These neurons elicit the movements of the wings (Olberg 2012).

The avian cerebellum is considered a key site to determining and responding to wing motion. Two main types of neurons are responsible for this process. These neurons, project to hindbrain areas, including the medial column, and parts of the cerebellum are important for the coordination of motor output to the flight muscles (Altshuler *et al.* 2015).

Many insects have wings that are expanded during flight and which are moved close to the abdomen when the animal is at rest. In Coleoptera (beetles) there are two pairs of wings of quite different dimensions: the small compact elytra and the large transparent hindwings. When beetles flight all the four wings are expanded but when at rest the large hindwings, are packed under the smaller elytra, by means of a series of folds. Such an impressive process of ordered wing movements demands neural control since the whole event follows a well defined sequence of foldings.

The brain of an insect weighs between 1 milligram and 1 microgram. The whole body of the common ant *Lasius niger* weighs 0.6 mg and the number of cells in its brain is thought to be about 10,000, whereas in the human brain this number is 10 billion cells and the average weight is 1,400 g (Changeux 1985).

Being a dragonfly, a beetle or an ant makes no difference. Although, minute in each case, their brain is able to make decisions that control and coordinate the highly complex flight of insects (Fig. 9.1).

Figure 9.1 Flight in insects with 6, 4 and 2 wings. (**1**) *Stenodictya lobata*. Drawing of a fossil of an extinct species with 6 wings (4 large and 2 winglets). (**2**) *Orthetrum brunneum*. Dragonfly with four transparent wings. (**3**) *Aglais urticae*. Butterfly with four wings which are large and covered with opaque microscopic scales. (**4**) *Microdon myrmicae*. Fly with two transparent wings.

The Brain is Informed by Sensors on the Wings, Ears and Beak

Following the decision to fly, the brain turns then to the information that it receives from other body parts to achieve optimal performance.

Aerodynamic forces are influenced by wing shape as well as motion and the wing's neuromuscular system is informed by a series of key sensory systems: the eyes, inner ear, and innervated feathers.

The lifting surface of a bird's wing is composed of feathers that can be spread and folded to modulate force generation. Primary flight feathers are asymmetrically shaped, with a stiff leading edge and a long flexible trailing edge, which helps to withstand the force of the oncoming wind. Owls have evolved a serrated leading edge structure that contributes to silent flight.

Wing shape is modified according to circumstances: (1) *Passive*, involves transient shape changes during flight. Bent feathers function as winglets to reduce drag. (2) *Active*, leads to: wing folding or expanding, twisting and wrist flexing.

The ability to sense pressure and magnetic fields is used by birds to control altitude and to direct navigation. The sensing of pressure variation is considered to be localized in the ear. At least three different avian structures are able to transduce magnetic signals: (1) the photoreceptor cells of the eye, (2) the inner ear and (3) the magnetic structures found at the base of the upper bill in birds. All three relay information about the intensity of magnetic fields (Altshuler *et al.* 2015).

Hormones Integrate the Energy Metabolism Used in Flight

The movement of the wing in insects depends on the wing base and the flight muscles. Its skeleton consists of a number of individual sclerites. The muscles deliver the power for active flight and adjust the wings during flight. The sclerites mediate the power transfer from the thoracic muscles to the wing. They enable adjustments to different flight modes and allow the wings to be moved into a resting position over the abdomen (Hörnschemeyer and Willkommen 2007).

Insect flight involves mobilization, transport and utilization of energy reserves at extremely high rates. Hormones, synthesized and stored in neuroendocrine cells, integrate flight energy metabolism. The control of hormone release in the locust includes both stimulatory and inhibitory factors. The hormones induce lipid release, mobilization and transport during flight (Horst 2003).

Birds also rely on fat to fuel migratory flights. Dietary lipid fat composition affects motion in birds and has been shown to affect flight performance (Price 2010).

The Origin of Birds and their Flight is a Source of Debate and Controversy

As several authors acknowledge, the origin and evolutionary history of birds is poorly known in comparison with that of reptiles and mammals. Fossils are rare, most probably because the skeletons of birds are fragile.

The well-known five fossils of *Archaeopteryx*, from the limestone deposits in Bavaria, are about 150 million years old (Upper Jurassic). But there is a gap in the fossil record until the Cretaceous (130 million years ago). *Hesperornis regalis* from this period, still retained primitive teeth, had small wings and a weak sternum. By the end of the Cretaceous the toothed birds disappeared. All the fossil species of the present era are toothless animals.

Archaeopteryx was characterized by grasping forelimbs, long tail and a skeleton resembling that of reptiles. It was capable of gliding or primitive flight since it had feathers of modern design. By late Cretaceous *Ichthyornis* had a keeled sternum allowing the insertion of bulky flying muscles and maneuverability of the tail.

This crucial transformation was little understood until new fossils were found in China with intermediate characters. A new bird species, *Sinornis santensis*, was discovered in lake sediments that belong to the lower Cretaceous (135 million years old). The species has a toothed snout, a pelvis and many other primitive features that are present in *Archaeopteryx* and *Ichthyornis* but it also exhibits a broad sternum, has a wing-folding mechanism and other modern flight functions (Sereno and Chenggang 1992).

Controversy and debate have dominated in the last decade due to the discovery of feathered dinosaurs and the re-examination of the earliest fossils. The results from phylogenetic trees have also added to these disputes (Zhou 2004). The most controversial issue has been the finding of feathered dinosaurs and their close relatives, such as oviraptorosaurids and dromaeosaurids, which have been proposed to be flightless birds. Other fossils have added to the confusion. *Protoavis texensis*, a late Triassic bird is considered 75 million years older than *Archaeopteryx*. Some authors have argued that it is a chimera belonging to two different species. *Rahonavis*, a long-tailed bird from the late Cretaceous of Madagascar, is also considered a composite of avian and theropod traits. Other fossils have been added to this list of controversial birds (Fig. 9.2).

Arboreal Monkeys were Not able to Develop Wings or Gliding

There are two separate hypotheses on the origin of avian flight: (1) *Terrestrial or "ground up"* coupled with the dinosaurian ancestry of birds, implying that it

Figure 9.2 Flight in reptiles. (**1**) *Pterodactylus kochi*, **Flying reptile**. Life restoration of the fossil *Pterodactylus*. It was an efficient flier; it soared out to sea, with its two large wings, feeding on fish. (**2**) *Microraptor gui*, **Feathered dinosaur**. Although not a bird this tiny dinosaur had feathers on its arms, on its legs and tail and it may have flown. The long feathers show how complex they were from the start. A series of Chinese fossils disclosed that highly evolved feathers appeared in reptiles before birds emerged. These reptiles are from the early Cretaceous (145 to 100 million years ago).

began as a flapping process to avoid predators. (2) *Arboreal or "trees-down"* which is linked with the thecodont origin of birds in which the wing was acquired by gliding efforts (Heers and Dial 2012, Provini 2013). Against the first hypothesis is that the large-sized terrestrial dinosaurs could not be the ancestors of birds. The second hypothesis favours the assumption that avian flight originated in the trees taking advantage of gravity (Ostrom 1986, Feduccia 1999).

One serious difficulty encountered by these assumptions resides in the fact that the monkeys have had an arboreal life, jumping down from trees, for 45 million years, without having acquired wings or wing rudiments. As Napier and Napier (1985) have pointed out: "The arboreal monkeys are the most generalized of all primates, both structurally and behaviorally. They are also the most successful in terms of wide geographic range and high population numbers". Other monkeys, like the macaques, are ground-living, and face numerous predators that they could have avoided if they had been able to fly. Both groups have diversified into not less than 242 species (Burnie 2004). Yet there are no signs that they developed wings as they continuously jump from trees or run on the ground.

At least gliding could have developed in the monkeys since it occurred in different mammalian orders: the marsupial greater gliders, the rodent gliding squirrels and the dermopterans or gliding "lemurs". However, not a single monkey species is known to have acquired gliding power. The conclusion is inescapable: although random mutations and selection, had millions and millions of years at their disposal, they were not able to produce flight or gliding ability in monkeys.

Hypotheses on the Origin of Bird Flight Do Not Apply to the Origin of Flight in Insects and Bats — Authors Agree that Selection is Impotent to Furnish an Explanation

As McFarland (1981) points out the precise course of evolution of insects is uncertain, because of gaps in the fossil record. These gaps are attributed to the absence of intermediate forms of insect wing formation. But these probably never existed. The earliest insects were wingless as demonstrated by the fossil remains and the same geological studies show that the emergence of insects with wings (in the Carboniferous era about 310 million years ago) was an explosive process. The insects covered the skies with their transparent wings as attested by the numerous fossils of this period.

Two hypotheses, quite different from those used for birds, have been proposed for the origin of insect flight: (1) the *gill theory* in which wings evolved from gills and (2) the *paranotal theory* assumes that lobes developed from the side of the thorax and abdomen becoming gliding planes.

Contrary to those used for birds, these hypotheses resort to embryonic features because it would be most inappropriate to assume that insects started flying by jumping from trees or as a result of avoiding predators on the ground.

The same applies to bats. Mammals had evolved since the Triassic 230 million years ago. They were able to diversify into small rodents and large whales, yet none of their species was able to acquire wings. By the Early Eocene (58 million years ago), bats not only appeared suddenly but they rapidly expanded into the major land masses. As acknowledged by most authors, explaining the origin of bat flight by selection posed insurmountable difficulties. These became even worse with the discovery of ecolocation which added to the complexity of this problem (Fig. 9.3).

Figure 9.3 Flight in birds and bats. (**1**) *Anhinga anhinga*, Anhinga (Darters). Like cormorants this bird species has a plumage, which due to the microscopic structure of the feathers, is permeable. In this figure the bird dries out its plumage. (**2**) *Larus hemprichii*, Sooty gull. All gulls are heavy-bodied and long-winged, and have fully webbed feet. (**3**) *Rousettus aegyptiacus*, Common fruit bat. The wings of bats are nearly as transparent as those of many insects.

The result has been the creation of a series of hypotheses (so many that they have been summarized in a table) that reflect the inability to approach this phenomenon on a scientific basis by resorting to selection (Giannini 2012).

Flight in Air Arose Only Once Among the Invertebrates Which had at their Disposition 600 Million Years to Diversify into the most Different Animals

The Chordates, which include the vertebrates, build only one phylum. But the invertebrates have diversified into many phyla yet in only one of them, the Arthropoda, did flight emerge, and within this phylum only the class insecta developed this ability. This was not due to lack of time, as at least 600 million years have elapsed, since the invertebrates emerged.

Such a punctuated appearance of this function is one of the characteristic features of biological periodicity. No other ancestral groups exhibited wings and no other invertebrate phyla that were later formed were able to fly. An additional characteristic is that insect flight at its origin (350–270 million years ago) attained full capacity. The fossils of the giant dragon fly *Meganeura*, which inhabited the forests of the Carboniferous period, had wings that measured 73 cm from tip to tip. Many bird species have smaller wings and the insects that live at present have wings of smaller size.

Some insect species (e.g. *Dytiscus marginalis*) are well adapted to aquatic life. The adults carry air supplies beneath their wing cases and have hairy hindlegs adapted to propulsion. Like in fishes, their swimming in water is a form of flight.

As the Vertebrates Emerged no Fish, Amphibian or Reptile Could Fly in Air Until Suddenly the Pterosaurs Populated the Sky

The second time flight in air appeared it showed no direct connection with that of insects. As the vertebrates emerged, the fishes, the amphibians and most reptiles could not fly. The flying reptiles, the pterosaurs (230–195 million years ago) had wings built by a different type of skeleton, but which were equally efficient in allowing them to fly.

The pterosaurs became extinct during the Cretaceous period leaving no survivors. Later, the situation repeated itself. The birds, which emerged 150–135 million years ago, are supposed to have as ancestors flightless reptiles, such as the

Thecodonts, the main reptile group during the Triassic period (Walker 1974). They had no direct relation to the pterosaurs (Fig. 9.2).

The Same Scenario was Repeated in the Mammals — After 160 Million Years of Evolution Only One Order, Among 21, was able to Develop Flight in Air

The mammals started their evolution about 160 million years ago, the marsupials appearing first around 63 million years ago and the placentals about 58 million years ago. During their long evolution the mammals diversified into 21 orders. Within a large number of flightless families and species, appeared a single order that had no direct connection with the birds or the flying reptiles, but which got highly effective wings. These were the bats which emerged about 58 to 45 million years ago. These dates fluctuate. As research progresses most geologic events tend to be earlier than previously considered.

Fossils that represent intermediate stages have so far not been found, and those that exist are so well preserved that they disclose the stomach contents of the first bats (Colbert 1980).

In bats the large membrane that allows flight starts on the back of the head or on the neck, extends along four fingers (the first finger, the thumb being free). It also extends to the legs and embraces the tail (Macdonald 2002) (Figs. 9.3 and 9.4).

Characteristics of Air Flight in Fishes — Fishes Travel Faster in Air than in Water

Equally showing no signs of direct relationship to closely related species was the emergence of flight in air among fishes. These vertebrates arose about 425 million years ago and are represented today by a large number of species.

The teleosts appeared 190 million years ago. Among them a few species fly, and do so in an environment that is adverse to their respiration, namely in air. They are obliged to return within minutes to water otherwise they would die, but their flight shows a degree of complexity considered not inferior to that of insects (Hanström and Johnels 1962, Beazley 1980).

Fishes do not glide passively like some squirrel species, but use powerful muscular contractions that oblige the body and the wings to vibrate like those of

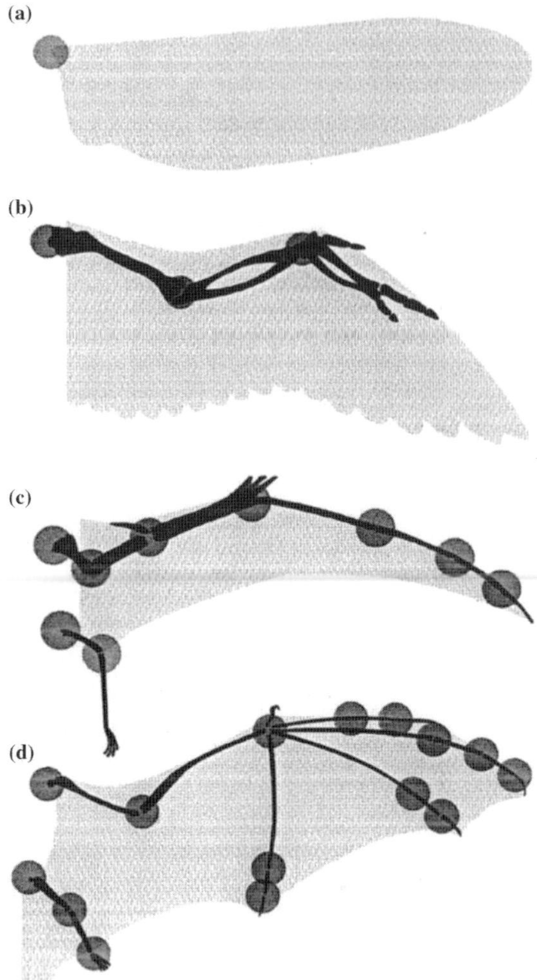

Figure 9.4 Surface of the wing, with increasing number of joints involved in movement, from insects to bats. Aerodynamic surface of the wing in light shading in (a) insects, (b) birds, (c) pterosaurs, and (d) bats; for vertebrates, the internal bony skeleton is depicted in dark shading. Circles indicate joints that may potentially be under active neuromuscular control. The numerous wing joints in bats can aid in dynamic control of wing motion three-dimensional geometry. The single joint of the insect wing, depicted by these authors, is found in most insect orders. However, in Coleoptera the upper pair of wings, the elytra, have a single joint but the larger hindwings have several joints that allow them to be folded under the smaller elytra. In this case the wing motion of the insect is not a simple event but demands as well a complex neural intervention.

insects (as much as 70 contractions per second). Fishes can circle around a boat, can stop in air, and modify their flying direction at ease, like birds (Hanström and Johnels 1962).

The marine forms of flying fishes populating the tropical seas, such as *Exocoetus*, are part of a family (with 48 species) which have all well-developed pectoral fins. These are used in flight as gliding wings. An enlarged lower part of the caudal fin is employed in thrust while the fish is taking off from the waves. As Rayner (1986) puts it: "The flight performance can often be impressive, with speeds up to 13 m/s and glides up to 50 m being recorded". Flying fish are also quite common in the Mediterranean Sea (*e.g. Cheilopogon heterurus*, *C. furcatus* and *C. exiliens*) (Louisy 2015).

Later studies revealed that the fresh water fish of the African swamps *Pantodon buchholzi* and the flying hatchet fish *Gasteropelecus sp.* "undoubtedly leave the water on short flights that involve flapping of the pectoral fins". This last flying fish has extremely well-developed pectoral muscles that make up 25% of the body weight, which is more than most birds (Davenport 1994).

Hence, there are two groups of flying fishes: (1) those in which the caudal and pectoral fins vibrate and (2) those which beat the fins both during taxing and in flight. The gurnards are fish which also have greatly enlarged pectoral fins and seem to represent an intermediate stage (Greenwood and Thomson 1960, Davenport 2003).

Earlier interpretations of the origin of flight in fishes considered it to be the result of an escape from predators. More recent work has led to the conclusion that fishes travel considerably faster in air than in water. This allows a reduction in energy costs of transport, a solution used by aquatic birds and mammals, such as dolphins (Rayner 1986) (Fig. 9.5).

Squid Launch into Air Faster than they Swim

Squid can exhibit brief periods of flight above water. They not only swim in water but also fly in air, or better stated, they fly both in water and air.

The animals propel themselves through the air by forcing water out of their mantles like a rocket. Four species of squid had velocities in air that could reach 37 body lengths per second whereas maximum seen in water was a mere 11 body lengths. This information was combined with photographs of flying *Stenoteuthis pteropus* of the coast of Brazil, showing that this phenomenon was more common than currently thought (O'Dor *et al.* 2012).

The movement in air reduces the cost of long-distance migration (Fig. 9.5).

Figure 9.5 Flight in squids and fishes. (**1**) *Stenoteuthis pteropus*. **Flying squid**. Squid launching into air. Squids not only swim but can also fly like rockets by expelling water out of their mantles. (**2**) *Cheilopogon melanurus*, **Flying fish** using two fins as wings. (**3**) *Cypselurus heterurus*, **Flying fish** with four fins used in flight. (**4**) *Myliobatis aquila*, **Eagle ray**. A ray moves through the water using the wing-beat motions of birds, and for this reason is considered to fly in a different medium than air. Rays use vertical movements of their large pectoral fins to generate forward movement. The eagle ray can even leap clear of the water's surface like flying fish.

Gliding, Like Flight, Re-emerges Following Periods of Latency—Gliding is Not an Accidental Event: It Recurs from Insects to Placental Mammals

Gliding takes two forms: it is *optional* as in certain insects and birds, or is a *regular* event appearing in amphibians and recurring in reptiles, marsupials and placentals. This form of movement is rare in invertebrates but is used, as a secondary adaptation, by flapping insects (Rayner 1986). In birds gliding is used by species with very large wings such as the Albatross with a wingspan of over 3 m.

Amphibians are impressive gliders. Tree frogs (e.g. *Rhacophorus nigropalmatus*, Wallace's flying frog) inhabit Madagascar, southern Africa, southern Asia and South and Central America. They are terrestrial as well as arboreal and belong to two families Rhacophoridae and Hylidae. Many of these species parachute relying on spreading their webbed feet to achieve a gliding surface, but some can flatten their bodies so that the entire animal contributes to gliding generation. They glide from tree to tree steering themselves in mid-air (Norberg 1990). There are not less than 50 species of *Rhacophorus* frogs.

Among the living gliding reptiles is *Draco volans*. About 20 species of this animal inhabit South East Asia. They glide by expanding their ribs which can be folded down against the animal sides like an umbrella. Other reptiles, the geckos (e.g. *Ptychozoon kuhlii*), have lateral body membranes that allow them to glide by parachuting (Norberg 1990) (Fig. 9.6).

Among the mammals gliding emerges in three well separated groups. Marsupials are represented by the Squirrel glider (*Petaurus norfolcensis*). The furry gliding membrane of this possum extends from the front toes to the back foot. The long furred tail acts as a glider when parachuting as far as 50 meters.

Dermoptera mammals glide by extending a strong and large membrane that surrounds their body. This kite-shaped structure is attached to the neck and to the tips of the fingers, toes and tail. The Malayan flying lemur, as it is usually called (*Cynocephalus variegatus*), is able to glide more than 100 meters between trees (Burnie 2004).

Rodent mammals are represented by the Giant flying squirrel *Petaurista elegans*. It is usually called a flier but it is actually a glider that can travel more than 400 meters, being able to change its glide angle with its front legs. To achieve gliding it stretches out its limbs extending large patches of furred skin between front and rear legs in a parachute-like shape.

There is a great diversity among mammalian gliders. There are 66 extant species in 22 genera and 6 independent clades. Of special significance is that "Among these taxa, there are cases of extremely similar non-gliding species that are sister to gliders (e.g. *Gimnobelideus* and *Petaurus*), and of initial stages of gliding in living forms that are closely related to gliders (*Hemibelideus* and *Petauroides*) (Johnsson-Murray 1987)" (Giannini 2012) (Fig. 9.6).

Figure 9.6 Gliding. (**1**) *Draco volans*. Glides by expanding its ribs that can be folded down. (**2**) *Petaurista elegans*. Giant flying squirrel (Rodentia, placental mammals). (**3**) *Petaurus australis* (Marsupials). (**4**) *Cynocephalus variegatus*. Called Malayan flying lemur as well as colugo (Dermoptera, placental mammals).

Wings are Formed as Rapidly as they are Eliminated — Genetic Significance of the Presence and Absence of Wings in the Same Insect Species

According to the fossil record the first insects were wingless (425 million years ago). Silverfish, bristletails and springtails and other insects without wings were initially the representatives of this fauna (Fig. 9.7).

An explosion of species diversification was accompanied by the emergence of the wing (280 million years ago). Mayflies and giant dragonflies covered the skies

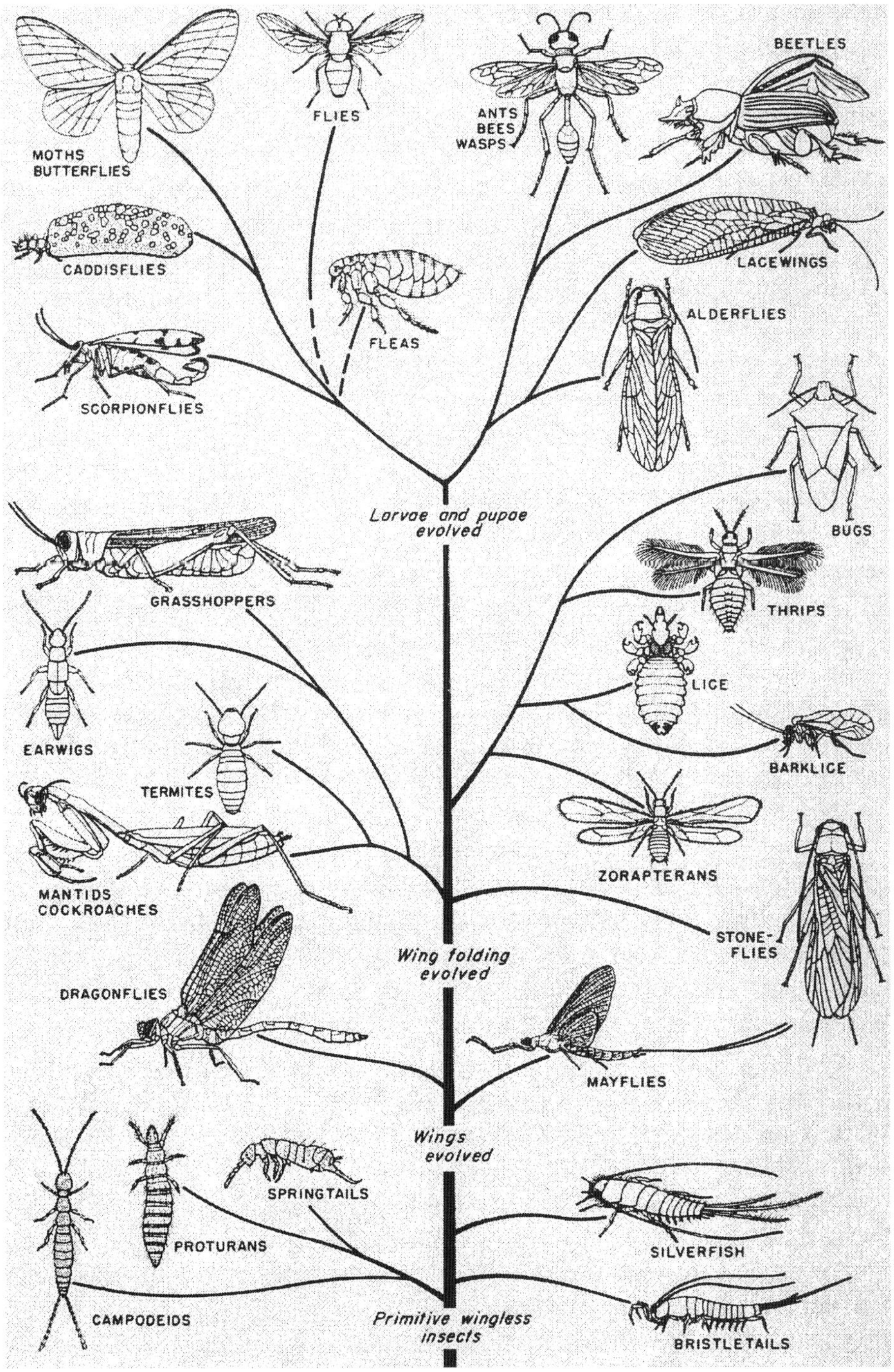

Figure 9.7 The evolution of insect orders. The primitive orders are at the bottom and the later more specialized ones at the top of the tree.

above the forests of ferns and gymnosperms. The wing became an obligatory structure until, much later, some insect species lost their wings (e.g. fleas). Hence, the wing was produced or eliminated independently of the evolutionary stage, indicating that it was formed as rapidly as it disappeared.

This phenomenon is elucidated by the fact that in several insect species some individuals have wings whereas others lack them. In the common North American termite *Termes flavipes*, the queen, workers and soldiers have no wings whereas the males possess four large ones (Fig. 9.8). The same occurs in another order in which *Oligotoma saundersii* males have four wings and the females are wingless. In a third order this event is even more patent. Females may have wings or no wings and the males are also alated or apterous. Closely related species may also have wings or lack them as in the Psocoptera (Romoser 1973).

Thus, the wing comes and goes in a most unexpected way in individuals with the same genetic constitution (be they females or males).

Hox Genes Decide the Formation of Wings as well as their Suppression

Several findings lead to a still better genetic explanation of wing formation and elimination.

Firstly, the well known *Hox* genes, that are responsible for body pattern formation in insects, birds and most other organisms, have been found to be responsible for the switch of the wing-promoting genes: (1) mayfly nymphs with four small wings have sites for *Hox5*, (2) insects with four large wings have sites for *Hox5, 9 and 10* and (3) modern butterflies with four extra large wings contain sites for *Hox5, 8, 9 and 10* (Carroll 2005).

Secondly, the gene *Sex combs reduced* (*Scr*) mediates the inhibition of wing formation in insects (Carroll *et al.* 1995, Chesebro *et al.* 2009).

This information explains the occurrence of individuals with identical genetic constitutions having their genes responsible for wing formation being expressed or repressed. Taken together the data reveal: (1) That a wing is built, or not, by decisions occurring at the DNA level (wing-promoting and inhibition genes). (2) Its formation is immediate without intermediate forms. (3) The structure that is formed is fully coherent since it is accompanied by the appropriate structures. (4) The wing is functional allowing a perfect flight at once.

Genetic manipulation in *Drosophila* has corroborated this interpretation.

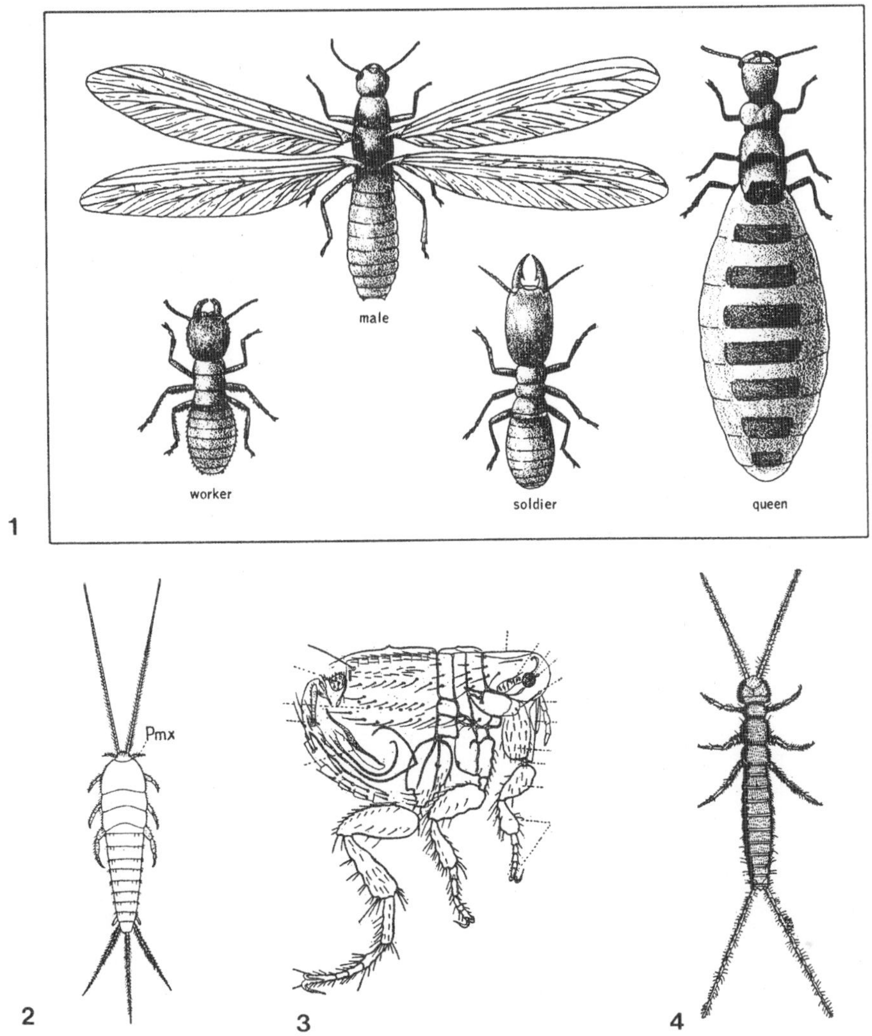

Figure 9.8 Presence and absence of wings in the same species and wingless insects. (**1**) *Termes flavipes*, North American Termite. The males of this insect have wings but the workers, soldiers and queen are wingless. (**2**) *Lepisma sp.*, Silverfish. A primitive insect having no wings. (**3**) *Xenopsylla chaeopis*, Flea. An insect with no wings that appeared late in evolution. (**4**) *Campodea staphylinus*. Another ancient insect (Campodeids) without wings.

How to Produce a Fly with Four Wings by Directed Genetic Intervention without Selection or Successive Random Mutations

Confirmation that combination of genes may suddenly produce a structure and function that has not been available for millions of years, is furnished by the results in *Drosophila*. A normal fly has only two wings, but molecular biologists have been able to produce flies with four wings. What is remarkable is that these flies are not monsters but are quite well-formed and their new wings are equally normal. These accessory wings are also partly functional. Moreover, contrary to the generally accepted idea on the mechanism of evolution, these flies were neither obtained by selection of individuals that originally displayed rudimentary wings, nor by random gene mutations that led to successive progenies with bigger and better wings. The flies with four wings were obtained directly, without any progressive transformation.

The studies of Lewis (1978), Bender *et al.* (1983), Peifer *et al.* (1987), Duncan (1987) and Lawrence (1992), led to the production of four wing flies by a simple molecular manipulation combining mutations occurring in the genes *bithorax* and *postbithorax* (in another set of experiments the inactivation of the gene *ultrabithorax* was involved). These genes were somatically recombined following irradiation of larvae the result being the immediate appearance in the progeny of normal flies with four perfect wings (Fig. 9.9).

Insects with four wings are known to exist in nature and to be quite common, the butterflies are an example. Moreover, the position of the wings in the butterflies is the same as in the four-winged *Drosophila*. Among the vertebrates, fishes may display four wings as well.

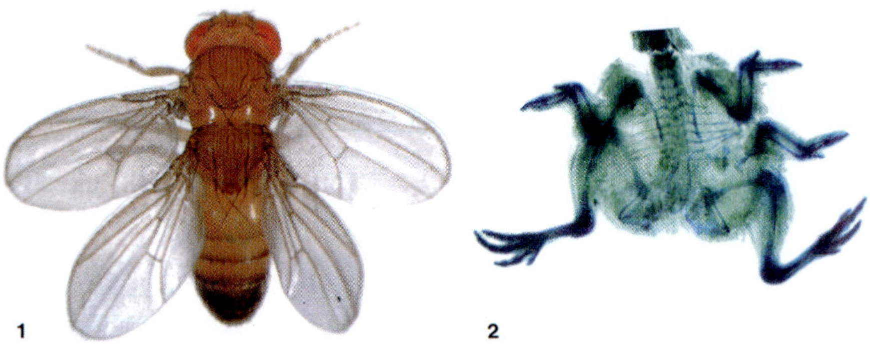

1 2

Figure 9.9 Production in the laboratory of flies with four wings and of birds with four wings due to the gene similarity between invertebrates and vertebrates. (**1**) *Drosophila melanogaster*, Fruit fly. Four-winged fly produced by combining the gene mutations *bithorax* and *postbithorax*. (**2**) *Gallus gallus*, Chicken. Genetic intervention leads to the production of a bird with four wings. This was achieved by modifying the location of a growth factor in the bird embryo.

How these Molecular Transformations Elucidate the Mechanism Behind Biological Periodicity — Flies with 8 Instead of 6 Legs

The implications of this molecular transformation at the chromosome level are significant for the understanding of evolution and of biological periodicity. They reveal the following. (1) The structures appear as "ready-made". (2) They are accompanied by blood flow, control by the nervous system, access to large amounts of energy. (3) The position of the wings is not accidental but occupies the dorsal part of the body. (4) The genetic transformation apparently re-enacts in days what has occurred millions of years ago. (5) The whole event took place in the DNA sequences of the fly.

Lewis (1992) was also able to produce in the laboratory *Drosophila* flies with 8 legs instead of the normal 6. Spiders which are phylogenetically located close to the insects are characterized by having 8 legs. Again, one is recreating, by genetic techniques, animals with the structures that are found in nature on other groups. The extra legs, like the extra wings, are identical in structure and function to the previously existing ones and appeared also "ready made" without intermediate random mutations or selection.

The Same Genes Determine the Formation of Insect and Bird Wings — Periodicity Could Hardly be Conceived if the Wings of Insects and Birds had Different Origins

The wing of an insect has been considered by zoologists to be analogous to that of a bird, but not homologous. The reason was that the two structures are so different that they seemed to have no relationship. Any comparison seemed impossible and unreasonable.

The recent work on embryonic development in the fly *Drosophila* and vertebrates (including birds and mammals) has disclosed that the same type of genes are responsible for the formation of corresponding regions of the body in invertebrates and vertebrates. These genes (called homeotic genes) not only control the formation of similar structures but have also maintained the same order in the chromosomes of the distantly related species. They exhibit an extreme conservation of organization and of internal order.

The wings of birds are mainly formed by the bones and muscles of the anterior limbs. It has been found, by molecular analysis, that the genes responsible for limb formation in the chicken are homologous to the genes that lead to the production of the wings in *Drosophila*. The results show that the wing of an insect is determined by the same type of genes that decide the formation of the wing of a bird. The analogy has become a homology since it is the product of the ordered expression of the same DNA sequences (Affolter *et al.* 1990, De Robertis *et al.* 1990, Holland 1992, Lawrence 1992, McGinnis and Kuziora 1994).

How to Produce a Bird with Four Wings — Genetic Manipulation Led to Formation of Extra Wings in Birds Which Appear as a "Surprise" to the Animal

Since the genes are the same the direct production of a normal wing or of extra wings can be expected to occur in birds, as it does in *Drosophila*.

This type of reasoning was soon confirmed by experiments carried out in different laboratories. The application of the protein *fibroblast growth factor* to the chick embryo changed the expression of *Hox* genes inducing the formation of either an additional pair of wings or an extra pair of legs (Cohn *et al.* 1997). These results demonstrated that the wings of birds can be produced at will in the laboratory (Fig. 9.9). The structure combined with the function emerged whether the organism liked it or not, they came as "a surprise" to the animal, that was obliged to use it.

What was obtained by genetic intervention has at times occurred spontaneously under domestication, but was at that time considered a curiosity. Ducks and cocks with four legs and birds with four wings have been depicted and photographed before in captivity (Shepherd 1939, Hough 2010).

The Wing can Change its Body Position — Wings can be Formed on Eyes

Among homeobox genes the most interesting is *eyeless*, which corresponds to the mutation *Small eye* in mice and to *Aniridia* in humans (Quiring *et al.* 1994). Following a series of experiments transgenic flies were obtained in which eyes were formed on the wings, legs and antennae (Halder *et al.* 1995). The eyes are real eyes because they are identical in structure and contain functional photoreceptors.

This means that a single master control gene can start the cascade of eye development which involves not less than 2,500 other genes needed for eye formation (Gehring 1998).

It turns out from these experiments that:

1 — A single gene *eyeless* is the switch for a developmental molecular cascade.
2 — Hundreds of other target genes are under its hierarchical command which implies an impressive order in gene action.
3 — This fly gene has its functional homologue in mice and humans, a proof of the conservation of its structure and function from invertebrates to vertebrates.
4 — The gene *Antennapedia* is also expressed in the eye imaginal disc where it induces cell death. However, if this situation is prevented by genetic engineering, these cells form little wings in the eye region an indication that this gene also controls wing formation.
5 — As demonstrated by Bachiller *et al.* (1994) there is a conservation of a functional hierarchy between mammalian and insect homeobox genes.

Some of the Earliest Fossil Insects had Four Wings and Two Winglets Located on the First Thoracic Segment

The thorax of an insect consists of three segments: pro, meso and metathorax. In most insects the two pairs of wings are located in the mesothorax and the metathorax. The prothorax is wingless.

A geological survey in central France provided the most extensive and best preserved collection of extinct orders of insects which inhabited the forests of the Upper Carboniferous (Kukalova 1970).

In these insects the head was small, the eyes were large and the thoracic segments were nearly equal in size. The prothorax possessed a pair of lobes or winglets which may have acted as immovable "foresails". According to Kukalova the lobes "were undoubtedly homologous with the functional wings" but their basal articulation was unlike that of the large four wings.

The prothoracic winglets were most distinct in the species *Stenodictya lobata* (Fig. 9.1) which was found in these deposits (300 million years old) (Gullan and Cranston 2000). The prothoracic wings had venation resembling that of the flight wings (Grimaldi and Engel 2005).

In most respects the earliest insects have traits which place them close to the mayflies and the dragonflies.

Extra Wing-like Appendages in the First Thoracic Segment of Living Insects Share the Genetic Program of Normal Wings

Treehoppers are insects, noted by the presence of a structure, called the helmet, which resides on the top of the animal's thorax and extends dorsally mimicking thorns and other features.

Significant is that the helmet is an extension of the dorsal portion of the first segment of the three-segmented thorax shared by all insects. Early developmental stages disclose that the helmet forms from paired buds (much like wings) and is attached to the thorax by articulations. Its anatomy provides strong evidence that the helmet is a modified wing on the first thoracic segment (Moczec 2011) (Fig. 9.10).

Prud'Homme *et al.* (2011) could show that the helmet shared a genetic program with the wings. Transcription factors including *nubbin* were found in the developing helmet and its expression parallels that of wings. Two other genes *Distal-less* (*Dll*) and *homothorax* (*hth*) were also expressed in the developing helmet indicating that the treehopper's helmet is a prothorax wing homologue. Additional work revealed that the *Scr* gene prevents wing formation in the prothorax through the repression of *nubbin*.

Hence, insects could have had six wings, as they have six legs, but the formation of wings on the first thoracic segment was partially or totally inhibited in most species during evolution. The necessary genes to produce the wing are present but they are repressed. This is a situation like that in echinoderms which have eye genes, but no eyes.

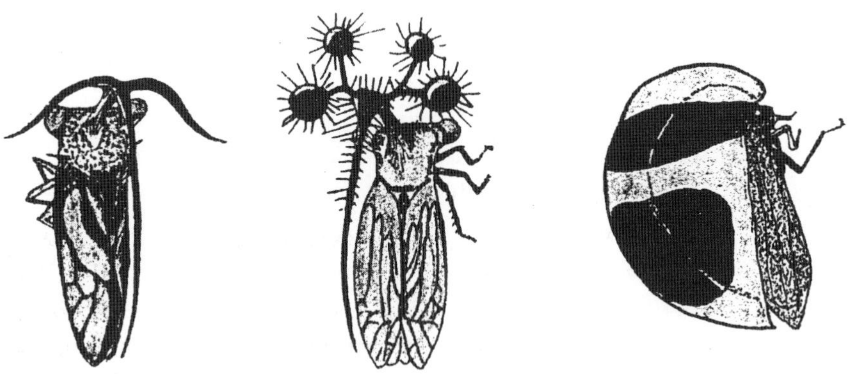

Figure 9.10 Insect helmets considered to be modified wings. (**1**) *Stilocentrus anchora*, (**2**) *Bocidium globulare*, (**3**) *Membracis foliosa*. Insects from Brazil, called treehoppers, produce highly diverse structures named helmets. These have the same genetic program as wings.

There are Flightless Birds But there are no Flightless Bats

The characteristic of flightless birds is either the loss or great reduction of wings, in some cases accompanied by considerable increase in body size. An ostrich may attain 150 kg.

All flightless birds are thought to be descendants of flying ancestors. This is obvious in some species, such as flightless rails and coots which are very closely related to other modern birds that fly. The wing, although sometimes much reduced, is clearly of the form used by flying birds (Fig. 9.11).

Two traits have genetic significance: (1) Reduced wings have preserved the same characteristics of normal ones indicating that they are the product of the same molecular cascade. (2) The inability to fly is not directly correlated with the type of habitat.

Birds found on islands, like the kagu of New Caledonia and the kakapo of New Zealand, are flightless. Living on land, but also flightless are rheas, emus, cassowaries and ostriches. Water-birds, including penguins and the extinct Great auk, also belong to this group.

The presence of a well formed wing does not imply flight capability. Inhabiting the conifer forests of New Zealand the owl parrot (*Strigops habroptilus*) is an example of a bird with large and well formed wings that is unable to fly (del Hoyo *et al.* 1997). Another case of a flightless bird having large wings (measuring 25 cm) is the Flightless Cormorant (*Phalacrocorax harrisi*) of the Galapagos Islands (del Hoyo *et al.* 1992). The kagu which looks like a small heron (*Rhynochetos jubatus*) is a third case of a bird with large wings but unable to fly. It inhabits the forests of New Caledonia and has rounded wings of normal size. It is flightless because of its reduced wing musculature, although it is able to glide when moving downhill (Perrins 2003) (Fig. 9.11).

Kiwis have taken flightlessness to extremes. They evolved on New Zealand around 30 million years ago and the genus *Apteryx* is represented by three living species inhabiting forest and scrub. Their vestigial wings are buried in their feathers, and externally their tails have entirely disappeared. There is no sternum, to which in other birds the flight muscles are attached.

The bats with their 977 species have been on the planet for not less than 40 million years. Their wingspans vary widely ranging from 1.5 meters, in the large flying fox, to as little as 15 cm in the long-nosed bat. Remarkably, all of them have wings of the same type and all fly with great efficiency. No bat has been described which lacks wings or has reduced ones (Burnie 2004).

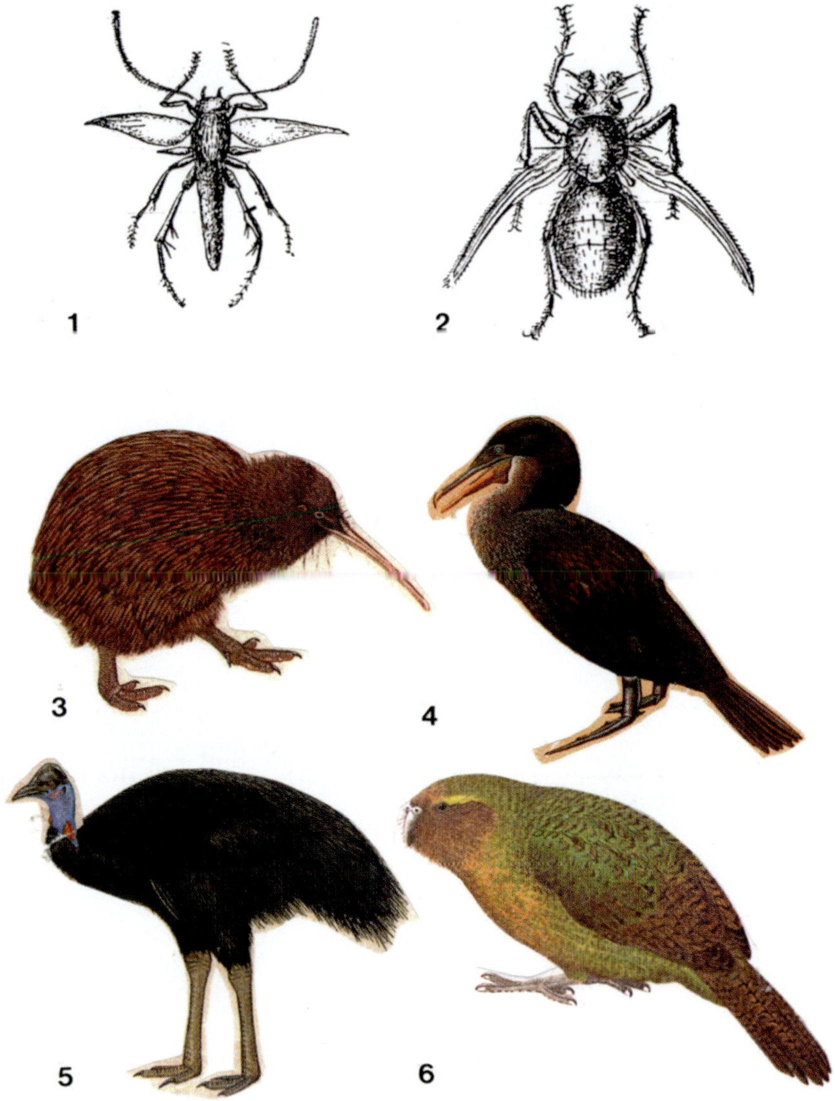

Figure 9.11 Flightless insects and birds. (**1**) *Embryonopsis halticella*. Butterfly with reduced wings unable to fly. (**2**) *Amalopteryx maritima*. Diptera with reduced wings unable to fly. (**3**) *Apteryx australis*, Brown Kiwi. Flightless female with much reduced wings. (**4**) Phalacrocorax harrisi, Flightless Cormorant. The wings measure 25 cm but the bird is flightless. (**5**) *Casuarius bennetti*, Dwarf Cassowary. Bird with rudimentary wings considered to be flightless. (**6**) *Strigops habroptilus*, Kakapo. This parrot has large wings but is unable to fly.

What Characterizes Flight Periodicity is the Re-emergence of Coherent Flight Capacity Following Long Periods of Latency

The periodicity of flight is characterized by the following features: (1) Flight arises without immediate predecessors in five major groups (six if the squids are included). (2) Few intermediate forms seem to have been involved or it appeared suddenly. (3) It tends to occur in a restricted class or order. (4) It consists of a coherent package of structures and functions several of them being obligatory components in all the groups. (5) The evolutionary position of the organism is not directly related to the complexity of the phenomenon. The flight of birds is no more complex than that of insects. (6) The number of wings is not correlated to the evolutionary position of the species. Birds have only two wings, whereas insects and fishes may have four. (7) Female and male insects may have wings or no wings. Well known genes decide their formation or elimination. (8) Flight occurs both in air, water and the interface that separates water from air. Active flying in water is particularly evident in rays that beat their pectoral fins up and down like the wings of birds. (9) There seems to be no obligatory relationship to a specific type of environment, since fishes also fly in a medium (air) adverse to their normal respiration, which takes place in water. Moreover, all mammal orders, except the bats, remained terrestrial and aquatic, and among all the aquatic and terrestrial invertebrates, only the insects conquered the air.

Flight is a latent process that apparently erupts from nowhere. The genetic and molecular evidence supports the conclusion that wing genes, that had remained repressed for a given period of time, were suddenly activated producing wings or that particular DNA sequences recombined leading to the production of ancestral molecular cascades (Figs. 9.4 and 9.7).

Gliding is also a Periodic Event

Surprisingly regular gliding appears also as a latent process which exhibits a periodicity similar to that of flight.

No gene analysis is known from gliding species, but the membranes formed in gliding mammals are much similar in shape, distribution and form to those of bats. The Philippine colugo *Cynocephalus volans* has a very large membrane that

Table 9.1 Periodic Table of Flight and Gliding

Increasing grade of complexity	ABSENCE OF FLIGHT IN AIR	FLIGHT IN AIR OR WATER	GLIDING: REGULAR AND OPTIONAL
1	INVERTEBRATES ALL PHYLA WINGLESS EXCEPT THE ARTHROPODA	INVERTEBRATES INSECTS FLYING Most species in air but some in water SQUIDS water and air	INVERTEBRATES INSECTS Some species Optional
2	INSECTS (ARTHROPODA) Some species wingless Silverfish Fleas	FISH Most species in water Flying rays	FISH Regular *Exocoetus* AMPHIBIANS Regular *Rhacophorus*
3	FISH Most species flightless in air AMPHIBIANS All species flight-less in air	FISH Flying in water and air Bony fish	REPTILES Regular *Draco* BIRDS Some species Optional
4	EXTANT REP-TILES All species flight-less in air	FOSSIL REP-TILES Flying in air Pterosaurs	MARSUPIALS Regular *Petaurus*
5	BIRDS which have wings but are flightless Kiwis Ostriches	BIRDS Most species in air	PLACENTALS DERMOPTERA Regular *Cynocephalus*
6	MAMMALS Many species flightless in air and water	MAMMALS BATS All species in air CETACEANS SEALS Most species in water	PLACENTALS RODENTIA Regular *Petaurista*

Flight in air and flight in water cannot be easily distinguished. Both represent active movements, in a medium of variable density, by the use of additional organs such as wings or fins. The periodicity is revealed by the sudden eruption of flight in the most unrelated groups of animals. Besides, this property was not allowed to develop in all invertebrate phyla except the arthropods and within these only in the insects. The expression or repression of flight capacity is known to be decided by genes that control wing development. Gliding turns out to be also a periodic event reemerging after periods of latency. In the same way as flight became a widespread event in insects, where it emerged in most species, gliding also became of general occurrence among mammals. Yet at the side of gliding species there are extremely similar ones that do not glide (Original).

stretches from the side of the animal's neck to the tips of the fingers and toes and continues to the very tip of the tail. When gliding with its expanded limbs this animal is most similar to a bat since the body membrane also involves the tail (Macdonald 2002) (Fig. 9.6).

The Periodic Table of Flight and Gliding

Three vertical columns are present in Table 9.1.

Column 1 Absence of Flight in Air

This includes the numerous invertebrate groups which are wingless and the many orders of vertebrates that are also not able to fly.

Column 2 Flight in Air or Water

The first group are the flying insects and the last the bats and marine mammals.

Column 3 Gliding: Regular and Optional

From insects to rodents.

Chapter 10

High Mental Ability has Resurged without Previous Announcement

Our Mind is a Layer Cake of Stapled Concepts Going Long Back in Time

We tend to think that our mental attitude and decision making relies mainly or solely on the latest information that we acquired during our upbringing and adult life. But our mind carries the imprints, and the simple concepts, that we as a species acquired throughout time.

Without noticing it, in our present daily behavior we use many of the typical processes and ways of thinking that were part of our daily experience in Pre-History, Antiquity, the Middle Ages, the Voyages of Discovery, the Industrial Revolution and the Space Age, only to mention some main landmarks.

We still eat like humans did in the caves of Lascaux when we gorged on raw meat burned on fire. At present we call this activity preparing grilled ribs. Today our eyes shine with delight and our appetite is satiated in the same way that it was 15,000 years ago. The pleasure of picking fruits and roots are processes equally imprinted in our mind.

Later the emergence of agriculture and animal husbandry led to the creation of hierarchic societies. Oral transmission became substituted by the memorization of symbols. Fixed relations between figures and numbers started to occupy our mind in the form of geometry. For the first time ethical problems also became a preoccupation. All these novel situations became such an integral part of our mind, that we take them for granted.

The Middle Ages left in our brains the mark of obscurantism and of submission, which resurfaces in our present society under several disguised forms.

The finding of new continents and the demonstration that our planet was a sphere rotating free in space, had a strong liberating impact. Suddenly the power of the individual was asserted and our attitudes and traditions were revised.

But as industrialization and colonialism expanded, the need to impose ourselves as the masters of other ethnic groups became the main preoccupation. The 1600s, 1700s and early 1800s saw the generally accepted use of slaves, child labor and women's degradation as sex objects. These ways of thinking also continue to be part of our mind although in more sophisticated forms.

The Space Age has opened the planetary system to travelling, not only by long voyages to distant celestial bodies but soon for week-end trips into space. A new layer of experiences is being added to our mind.

What is of relevance, is that, without noticing it, we use in our daily thinking the ideas that were established in earlier centuries. Some of the most evident are the concepts of: instinct, intelligence, and selection. These are terms that belong mainly to the 1800s when research on animal behavior and evolution was based on an ignorance of genetics, microbiology, biochemistry, biophysics and crystallography. These were scientific disciplines that did not yet exist.

At present these concepts are being superseded. Some like instinct and intelligence, were coupled to the assumption of human biological supremacy. Others still prevail, not because they are necessarily valid, but because they fit the ideology of globalization. This demands a chaotic behavior of economic competition instead of ordered social development. As a consequence randomness and selection are upheld as the mechanisms of biological events.

Imprint after imprint, building layer after layer, our mind uses daily the thoughts, the reflexes and the geometric relationships that we acquired since the dawn of our life as a species.

The Origin and Development of the Brain has been Based on "Severely Biased" Ideas

A clear cut example of what has just been stated is the search for the origin of the brain.

It started with the question of how the human brain differs from that of chimpanzees and gorillas, as well as from other animals. This theme was intensely debated since evolution was established as a biological phenomenon in 1809 by Lamarck. Anatomical differences, as well as size differences, have been the main source of a controversy which raged between 1840 and 1940. Antiquated arguments continue to be used to this day. As Striedter (2004) states, the analysis of the origin of the vertebrate brain remains unresolved because its interpretation is "severely biased".

As an example, he points out the different types of phylogenetic trees that have been conceived throughout time. He asserts that: "The most insidious idea in the study of brain evolution is the very old notion that biological evolution proceeded in a linear and progressive manner, from lower to higher forms of organization and with *Homo sapiens* at the very top of the so called phylogenetic scale". This linear and progressive idea of brain evolution dominated throughout the 19th and most of the twentieth century since medically oriented neuroscientists and anthropologists continued to support it (Campbell and Hodos 1991).

A second type of approach has concretized in the "Traditional Tree" in which humans are at the top and other animal groups are placed as side branches.

A third form depicts vertebrates distributed in a "Bush or Tumbleweed Tree" in which it is stressed that nonhuman lineages do not represent blind alleys but are only the outermost terminal branchlets of this phylogeny.

"Cladograms" are a fourth solution, now generally used by evolutionary biologists, placing both living and extinct species along the top of a diagram. Such cladograms have also been much criticized because they lead to quite different conclusions depending on what kind of molecular information is considered most relevant (Carroll 1988).

"Intelligence Tests" were Made by People with Little Knowledge of Intelligence — A Concept Difficult to Define

If one puts in doubt the difficulty of defining intelligence, a search in Webster's New Twentieth Dictionary (1976), will dispel such an impression. One is confronted with several and quite different definitions. Intelligence is "(a) the ability to learn or understand from experience; ability to acquire and retain knowledge; mental ability; (b) the ability to respond quickly and successfully to a new situation; use of the faculty of reason in solving problems, directing conduct, etc. effectively; (c) in psychology, measured success in using these abilities to perform certain tasks".

Actually "intelligence tests" measure mainly what is covered by the third definition under (c), *i.e.* the ability to use properly a certain amount of knowledge acquired in a selected area based on given social values.

The second definition (b) covers clearly "business success". It is a measure of how fast one prospers within the economic system.

The first definition (a) is closer to the process that occurs in scientific endeavor, as well as in the arts. In them intelligence appears as an ability to bring together, in a coherent way, phenomena that seemed unrelated. It is essentially a unifying ability. It has nothing to do with economic success, encyclopedic knowledge or social status. This is attested by the fact that some scientists, as well as artists, made their great contributions at a tender age, some were quite poor individuals throughout their life, and others had not access to great libraries.

Before the Brain Arrived in Evolution Elementary Memory and Reactivity were Present in Living Organisms

Unicellular organisms, such as protozoans, have no cellular structures comparable to the nervous system found in multicellular animals. Nonetheless they are capable of detecting stimuli and responding adaptively. Some, even exhibit a rudimentary form of memory. Eckert and Randall (1978) ask "How is this possible without a nervous system?"

When a *Paramecium* bumps into an obstacle, it swims backward. If it gets a stimulus from the posterior end it swims more rapidly. Experiments using intracellular microelectrodes revealed that electrical charges and calcium concentration are involved in these responses.

Another protozoa, *Stentor*, when mechanically disturbed, contracts its body with the aid of an internal system of microfilaments. It actually remembers this disturbance because the response decreases exponentially with the number of trials. Microelectrodes were also used in this experiment.

Eckert and Randall (1978) conclude: "It is evident from these two examples that the surface membrane of a unicellular organism can perform the functions of sensory reception, simple integration of receptor signals, and control of the motor response and that a rather simple modification of membrane function can give rise to an elementary form of memory. Thus the rudiments of the sensory-neuromotor systems of metazoans are present already in the electrophysiological properties of the Protozoa. The similarity between the electrical properties of nerve cells and protozoa suggests that these properties are very general and very ancient. This lends support to the view that the evolution of nervous systems has taken place, primarily, by the development of increasingly sophisticated neural circuits rather than through changes in fundamental properties of nerve cells" (Fig. 10.1).

We are so attached to the idea that our brain is such an exceptional organ that we tend to consider it as coming from nowhere or having an evolutionary origin that puts it apart. As Withers (1992) points out, the brain has had an evolution, like any other organ.

The origin of the brain is also much more modest than we could imagine. The organization of sensory receptors, neurons and effector cells — which build the nervous system — started in the skin. These cells were originally located in the epidermis of animals such as the coelenterates. Later they aggregated and migrated building the clusters of neurons that formed ganglia and subsequently small brains. In higher organisms the central nervous system developed into an anterior concentration of neural tissue (the large brain) that became connected with the longitudinal nerve cords. This differentiation already occurred in spiders.

Periodicity at the Mental Level Seems Unexpected but Mental Processes are Equally Directed by the Evolution of DNA Sequences

Llinás (2002) states that "the scientific basis for mindness requires a rigorous evolutionary perspective" and this is corroborated by the emerging evidence revealing that mental ability follows periodicity decided by the evolution of DNA.

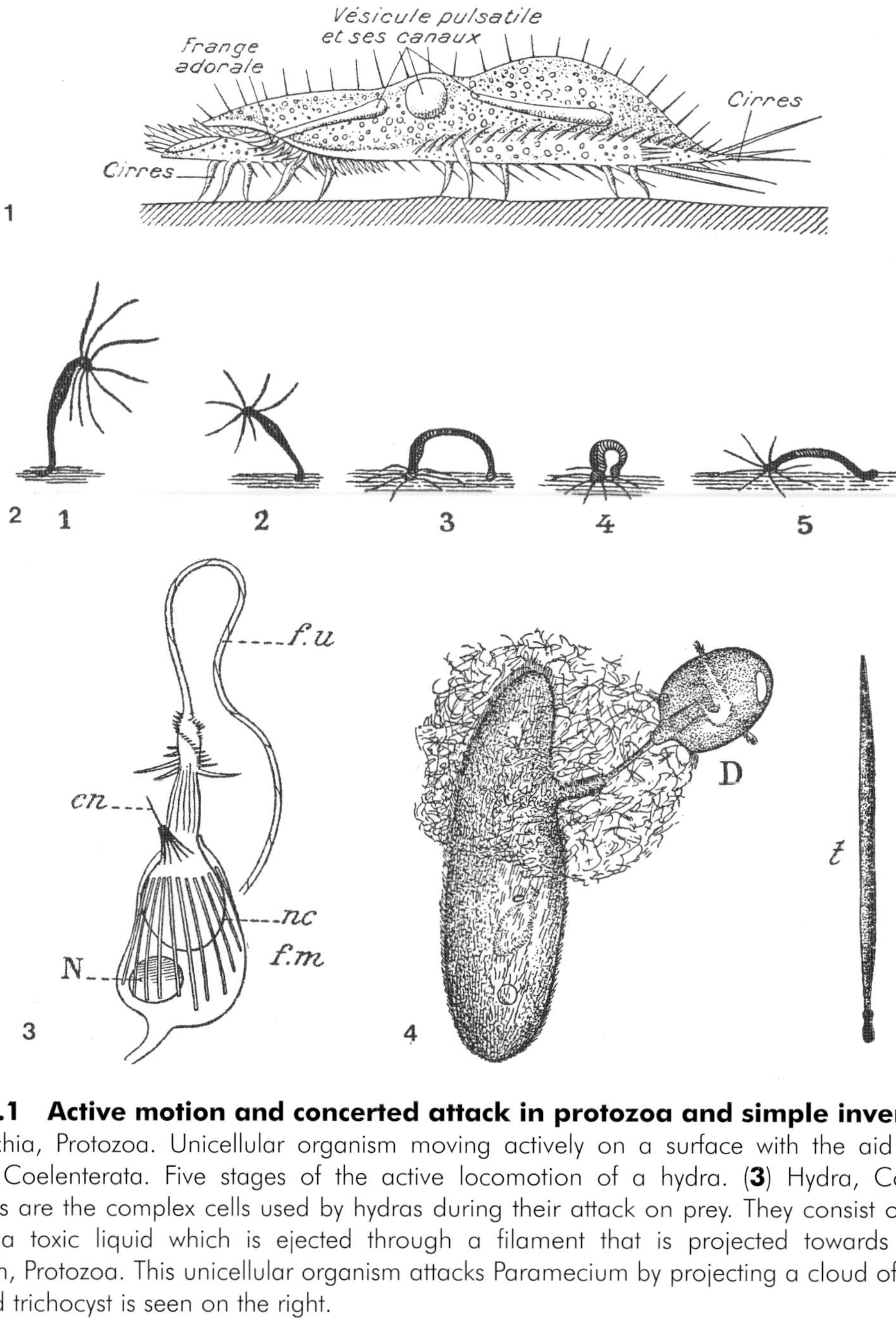

Figure 10.1 Active motion and concerted attack in protozoa and simple invertebrates. (**1**) Stylonychia, Protozoa. Unicellular organism moving actively on a surface with the aid of its cirri. (**2**) Hydra, Coelenterata. Five stages of the active locomotion of a hydra. (**3**) Hydra, Coelenterata. Nematocysts are the complex cells used by hydras during their attack on prey. They consist of a capsule containing a toxic liquid which is ejected through a filament that is projected towards the victim. (**4**) Didinium, Protozoa. This unicellular organism attacks Paramecium by projecting a cloud of trichocysts. An enlarged trichocyst is seen on the right.

To start with the brain develops under the control of well defined DNA sequences. *Hox* genes have been found in the last two decades to be the determinants of the body division into segments in flies, humans, worms, plants and other organisms in which they were investigated. These genes are aggregated in a cluster along the chromosome's DNA built by about 180 base pairs. The order present in the embryo and the adult can be traced directly to the order occurring within the large sequence of DNA. The *Hox cluster* of Drosophila becomes, in humans, divided into four separated DNA segments *Hox A, B, C* and *D*, but in all of them the order of the genes within the cluster and their relationship to the body axis is maintained.

Important is that the *Hox* genes also decide the formation of part of the brain.

Like in other vertebrates the A to D *Hox* genes are present in birds. The study of the brain development in the chicken revealed that a combination of *Hox* genes and related genes, are expressed in rhombomeres, a series of seven transient bulges in the developing rhombencephalon corresponding to cranial sensory and motor nerves. They are known from the chick as well as mammals. This is in agreement with mutations of *Hox* genes in mice that alter the position of specific cranial nerves or prevent their formation.

In the zebrafish gene expression patterns clearly define rhombomeres. Again, mutations in these genes disrupt region-specific neural differentiation in the hindbrain (Zhang *et al.* 1993, Lumsden and Keynes 1989, Narita and Rijli 2009, Purves *et al.* 2012).

The Present Molecular Study of the Brain Opens a New Era of Mental Cognition with Unpredictable Consequences

Molecular biology has advanced so rapidly, in the last ten years, that the brain has become its novel area of research, the experimental development being explosive.

Llinás (2002) summarizes in a nutshell the present status of brain research and neuroscience. He starts by stating that "there must be a fundamental reorientation of perspective in order to approach the neurobiological nature of the mind" and emphasizes that "Many types of neurons in the nervous system are endowed with particular types of intrinsic electrical activity that imbue them with particular functional properties". It is this "intrinsic activity" and "electrical glue" that allow the brain to organize itself functionally and architecturally during development and are at the root of cognition.

Neurological research revealed that in humans the general circuitry of the brain, present at birth, is not fundamentally modified during normal maturation but rather is only fine-tuned.

Molecular biology is at present fully concentrated in dissecting the brain at the molecular level. Numerous laboratories participate in this effort (Inacio *et al.* 2015, Kaltschmidt and Kaltschmidt 2015, Ben-Yakov *et al.* 2015). Memory and other vital brain functions are becoming closer to being manipulated and this research will add to a better control of mental diseases, such as Alzheimer's but also will create ethical problems connected with directing human behavior.

The Traditional Classification of Insect Sociality — Hierarchical Ranking Based on Intermediate Stages Between Solitary Behavior and Complex Societies

Costa (1997) exposes the limitations of the traditional division of insects into six main groups based on their degree of sociality:

1 — *Solitary*, or nonsocial, which is characteristic of most insect species (e.g. praying mantis).
2 — *Subsocial*, parental care of immatured individuals for at least part of lifecycle (e.g. dung and carrion beetles, some treehoppers).
3 — *Communal*, members of the same generation live together but there is no brood care (e.g. gregarious caterpillars and sawflies, some bees).
4 — *Quasisocial*, members of the same generation live together and cooperate in brood care (e.g. some bees, some wasps).
5 — *Semisocial*, members of the same generation live together, cooperate in brood care and there is reproductive division of labor (e.g. many wasp species).
6 — *Eusocial*, overlapping generations, cooperate in brood care and there is reproductive division of labor (e.g. all ants, all termites, many bees, many wasps).

According to Costa (1997) this hierarchical classification has little meaning since other unrelated animal groups display similar social traits such as spiders and naked mole rats. He emphasizes that instead of basing the concept of sociality on different types of selection it must be considered in a broader meaning.

However, this initial division is valuable because it shows that although there are intermediate stages of sociality, this phenomenon became explosive when it reached certain orders. It is not surprising that the highly complex societies are called "evolutionary marvels".

It is important to note that bees have species that belong to three different categories: communal, quasisocial and eusocial, showing that, like in other types of periodicity, simpler forms of the same trait, occur in related species.

Also to be noted is that trail-based cooperative foraging, typical of caterpillar societies, is restricted to a handful of species in scattered families. This suggests that this type of social behavior may have evolved independently several times, as is known to be the case in complex societies, which reinforces the evidence on the sudden emergence of this behavior.

Besides, the species of caterpillars that are solitary exhibit the posturing, flicking and defensive regurgitation found in social species, as well as trail marking. Thus, social behavior, or a solitary condition, appeared and disappeared without previous announcement.

Formation and Evolution of Societies in Bees, Ants, Termites and Humans — Our Society is Becoming Closer to that of Bees

Bees, ants, and termites are social insects that have developed: (1) a highly organized society, (2) a well-defined language, (3) possess dialects, (4) decide the fate of their progeny by differential fertilization, (5) have an agriculture consisting of the cultivation of fungi, (6) raise other species such as aphids, and use their sugary products in a manner akin to humans raising cattle and collecting their milk. Yet, they are not directly related to mammals (Fig. 10.2).

The honey-bees are considered to be the only animals other than man that communicate by a symbolic language (Dreller and Kirchnert 1993). Moreover, bees and ants, like birds, are known to use celestial navigation. They follow with different degrees of accuracy the movements of the sun and its relationship to the horizon.

The honeybee (*Apis mellifera*) is the best known social insect. It originated in Africa being a recent invader of temperate regions. Its colonies reach between 20,000 and 80,000 individuals. This figure is of the same magnitude as the population of many human towns. The queen may lay 1,000 eggs per day. The workers nurse the larvae and construct a royal cell for the queen. A large amount of time is spent in determining the colony's needs. The workers are able to change tasks which enables the society to adapt to novel environmental conditions.

Termites live in galleries constructed in wood or soil. They differ from bees in that workers are sterile individuals of both sexes and the reproductive male is a permanent member of the colony. Soldiers have large heads and mandibles participating in the defense of the society.

A termite's nest is a complex edifice. The central chimney is a ventilation shaft that helps to keep the temperature constant. Outside changes, of as much as 13 °C between day and night, do not affect the inside temperature. Other compartments

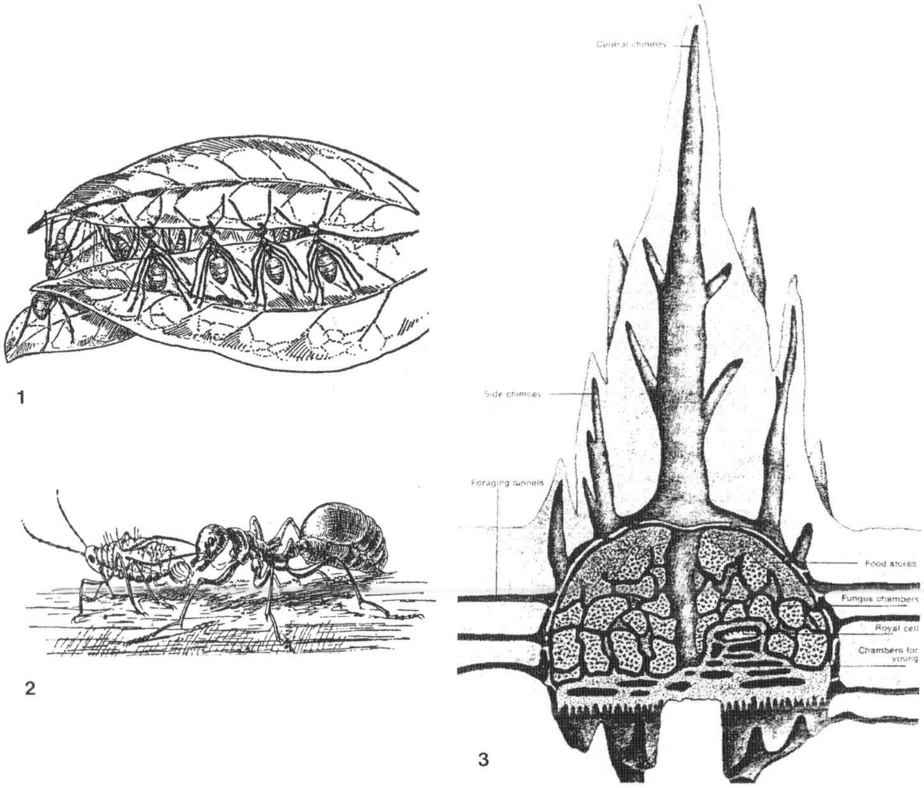

Figure 10.2 Mental ability in insects. (**1**) Oecophylla longinoda, **African weaver ant**. Ant weavers use their feet and mandibles to combine the borders of leaves. Simultaneously, other ants carry larvae, which have in their mouth a liquid, that seals the leaves together. (**2**) **Ant farming**. *Formica sp.* Ants use aphids (insects), which excrete a sugary substance, as a source of nutrition. They even move them to appropriate sites where the aphids can better feed on the sap of plants. (**3**) **Nest architecture in termites**. *Hodotermes meridionalis*. The complexity of the termites' nest is evident from this cross-section. The central chimney is a ventilation shaft that helps to keep the temperature constant. Outside changes of as much as 13°C occurring between day and night do not affect the inside temperature. Other compartments function as: foraging tunnels, food stores, chambers where fungi are cultivated, the royal cell and chambers for the young termites.

function as: foraging tunnels, food stores, chambers where fungi are cultivated, the royal cell and rooms for the young termites (Fig. 10.2).

Ants have a social organization that resembles that of termites. Like bees and wasps their workers are sterile females, and they may form as well a soldier caste.

Human societies exhibit several of the dominant features found in insect societies. The sterilization of members of the community, although it is disgusting, occurs in human societies. The European "castrati" of the 1600s were men that were castrated when young so as to retain their soprano voice for singing in church choirs, this made them sterile for the rest of their life. The same was with the "eunuchs" who were castrated men in charge of an Oriental harem. In Sweden, women and men considered socially undesirable were sterilized (by the end of the 1940s about 2000 people per year) (Tunlid 2004).

The distribution of tasks among the members of the society is, like in insects, a general feature of human social behavior. Castes take many forms in our society. They are still prevalent in India, where the same term is used for humans as is employed for insect diversification. In European societies castes still exist today in a more attenuated form under the designation of "social classes".

Human societies are far from having evolved as far as bee societies. These have had at their disposal much longer time to improve their behavior. Bees have long ago developed a mechanism that enables them to create different types of individuals within their population. This is achieved by fertilizing their eggs, leading to females, or allowing them to develop without fertilization resulting in males. In the first case the individuals have a complete set of chromosomes in the last case half of the chromosomes are enough to build a perfect animal (Gullan and Cranston 2000).

Human society is also not static. Artificial fertilization of human eggs is now a current procedure. Technologies are being devised that allow to manipulate human DNA to redesign the genome of future individuals. The technique, developed in the U.S.A. with the code name *CRISPR/Cas9*, is based on a bacterial protein and a nuclease which enables researchers to generate permanent mutations leading to directed and precise DNA engineering (Sternberg and Doudna 2015).

We are not on our way to build a bee-like human society — we are already there.

Octopuses Display the Highest Mental Ability Among Advanced Invertebrates which is Linked to the Recruitment of Novel Genes

Octopuses have the capacity to learn to distinguish between different geometric figures and can perform tasks in a way comparable to mammals. They can recognize by sight 29 different shapes. Can be trained to distinguish between a cross and a square and 27 intermediate forms. They were given rewards of food when they attacked one of the symbols. The geometry became evident when it turned out that the animals tended to classify shapes along a dimension that seemed to depend upon the relation of outline to area (Wells 1983) (Fig. 10.3).

Octopuses have revealed another unexpected capacity. Like primates they can invent tools. By diving in the Indonesian sea Finn *et al.* (2009) discovered that the octopus *Amphioctopus marginatus* was able to collect empty coconut shells and to assemble their halves as a shelter. It also manipulated and arranged the shells so that the concave surfaces were uppermost.

Figure 10.3 Counting in birds and monkeys. (**1**) *Fulica americana*, American Coot. (**2**) *Columba livia*, Rock dove. (**3**) *Macaca mulatta*, Rhesus monkey. **Problem solving in insects and molluscs**. (**4**) *Apis sp.*, Insects. Different symbols used in experiments on form perception by the bee Apis. The shapes in the top row are not distinguished from each other, but are readily distinguished from those in the second row. (**5**) *Octopus sp.*, Molluscs. Octopuses can distinguish by sight between 29 shapes shown above. They can be trained to distinguish between a cross (above left) and a square (below right) and 27 intermediate forms. The numbers refer to the order in which the octopuses ranked the shapes. This turns out to be directly related to the transition from cross to square.

Molecular evidence has now been gathered on the brains of cephalopod molluscs (squids, cuttlefishes and octopuses), since they are known to possess complex neural and sensory structures. They have the most highly centralized brains of any group of invertebrates characterized by: (1) eyes and visual systems with optic lobes; (2) a specialized center of the brain that controls learning and memory and (3) unique vasculature supporting such complexity of the central nervous system (Yoshida *et al.* 2015). Using various molecular technologies to localize the expression of selected genes they found three types of genomic innovations: (1) Recruitment of novel genes into morphogenetic pathways. (2) Recombination of various coding and regulatory regions of different genes. (3) Duplication and divergence of genes.

Bees have also been found to be able to distinguish between different symbols. *Apis* sp. could not sort out squares from circles but readily realized that they were different from crosses and Y formed figures (Chapman 1969) (Fig. 10.3).

Migration in Butterflies, Birds, Whales and Humans — an Example of Innovation and Improvisation

Insects with their microscopic brains are able to migrate and with high precision. The best studied case is the North American monarch butterfly *Danaus plexippus*. Its autumn migration comprises a journey of several thousand kilometers to the Gulf of Mexico. With the onset of spring the butterflies return north towards the Great Lakes of the U.S.A. They migrate in masses of many thousands. "So predictable is this movement with respect to direction, time of year and the vast numbers involved that many towns along the monarchs' flight path celebrate their arrival with festivals" (Baker 1980).

Spiny lobsters (Crustaceans) migrate between winter and summer residences. Among the fishes, salmons and eels are well known for their long journeys between sea and freshwater and reptiles (e.g. turtles) cross the oceans to return to the same beaches to lay their eggs. Amphibians are the only vertebrates that have a most reduced form of migration extending to less than 1 mile between feeding and breeding sites (Baker 1980).

In birds, migration has been studied extensively. They use as cues the position of the sun, the magnetic field, the landscape, and other sources of information (Boyle and Conway 2007).

Migration tended to be regarded as a natural event. It should be realized that when birds emerged in evolution they did not start necessarily by migrating. This was an elaborate process that depended on the acquisition of: (1) much empirical experience, (2) assembling of learning and (3) combination of dispersed information. No bird in the Cretaceous (130 million years ago), or even several millions of years later, is expected to have started to fly from the North to the South Polar regions, every year, as the Arctic Tern (*Sterna paradisaea*) does during migration, covering up to 15,000 km. Another species, the short-tailed shearwaters (*Puffinus tenuirostris*) makes an annual migration between the North Pacific and its breeding grounds in South Australia which requires a round trip of more than 30,000 km (Pough *et al.* 2005). Such long and hazardous routes are not expected to have been discovered and established overnight.

Research in many areas, which has included the chemistry and physics of the bird's ultramicroscopic anatomy, has disclosed that migration represents one of the best examples of innovation and improvisation. The birds have been obliged to solve problems to follow new routes, as the geography of the earth, its magnetic field, and other parameters changed drastically as the years passed.

Baleen whales, that are among the largest living animals, migrate 20,000 km between polar and tropical waters each year. They are well known for their high mental ability which includes communication by a variety of sounds.

Hence, the size and complexity of the brain do not seem to be decisive factors in determining this form of behavior since it occurs, with high precision, in butterflies, birds and whales.

Humans are also known to have moved to novel geographic areas for thousands of years, populating other continents and continue to do so at present as attested by the number of Middle East refugees that are entering Europe in 2016.

Use of Tools and Independent Learning in Birds is on a Level Similar to Great Apes

Birds are capable of the most exquisite forms of behavior and can surmount difficult situations surpassing mammals in many respects.

The last decades have seen an upsurge in the study of cognition in birds.

Tools have been defined as external objects, used in the attainment of an immediate goal, by extending the function of the hand, mouth, beak or claw (Lawick-Goodall 1970). Examples are: (1) A vulture breaking an egg by hitting it with a stone. (2) Insertion of twigs in crevices by the Woodpecker finch of the Galapagos islands. (3) Leaf manufacture in Caledonian crows. These are true tool cases which are different from borderline situations such as a gull dropping an egg from a rock.

Other well known examples are: (1) Birds-of-paradise use their feet as tools to hold and manipulate food. They easily extract the edible part from capsular fruits harvesting them in a most effective way. (2) The Goldfinch (*Carduelis carduelis*) when presented with food at the end of a string can use its feet and bill to successively lift it close to its mouth, finally eating the prey.

It turned out that the use of tools in birds was much more common than previously thought. Over 120 cases of tool use were reported in 104 species with 39 assigned to the true tool category (Lefebvre *et al*. 2002).

Independent learning is equally impressive. The Hooded crow (*Corvus cornix*) after seeing Eskimos setting lines through holes on the ice for catching fish, rapidly learned the device and alone used the same lines to catch the fish. Another case of independent learning is the way Blue tits started opening milk bottles that had their tops covered with foil. They easily perforated the capsules to get a meal and the novel technique spread rapidly among the other members of the species. Like in mammals, the storing of food, as a reserve for the winter, is a regular habit among the Nutcracker and the Siberian jay (Perrins 1976).

The crows (Corvidae) are so mentally advanced that they have been compared to the great apes. Many species cache food during periods of seasonal abundance. Western Scrub jays are able to remember where they have hidden it but are also aware of its perishable condition and take adequate measures to eat it in time.

Ravens obtain insect larvae from holes in wood. Their hook tools are made by trimming twigs in a complex series of steps. Most corvids form long-term pairs that extend throughout the life of the two sexes and show many other forms of social behavior.

In many ways the cognitive capacity of corvids is on a level similar to that of the great apes including: causal reasoning, imagination, flexibility and prospection. There is an impressive similarity of high mental ability in the two distant groups of animals (Emery and Clayton 2004, Seed *et al.* 2009).

African grey parrots (*Psittacus erithacus*) are known for their sophisticated cognitive skills and the ability to solve reasoning tasks in other domains. An experiment using the shaking of containers revealed that they possess ape-like reasoning skills (Schloegl *et al.* 2012).

Birds that strike a submerged prey must take into account prey position, movement, depth, distance and other factors. All these calculations are made by herons, kingfishers, little egrets and other species coping with refraction over a wide range of prey positions by making corrections to the incident angles (Fig. 10.4) (Katzir 1993, Kral 2003).

In a review comparing the social cognition of primates and birds Marler (1996) states that: "there are more similarities than differences between birds and primates".

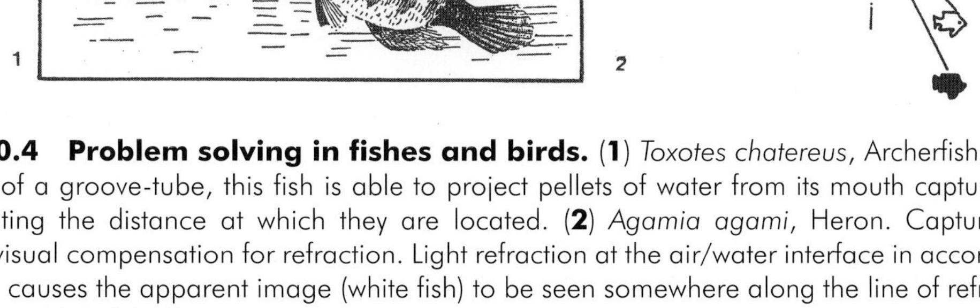

Figure 10.4 Problem solving in fishes and birds. (**1**) *Toxotes chatereus*, Archerfish from Java. By means of a groove-tube, this fish is able to project pellets of water from its mouth capturing insects by calculating the distance at which they are located. (**2**) *Agamia agami*, Heron. Capture in water demands visual compensation for refraction. Light refraction at the air/water interface in accordance with Snell's law causes the apparent image (white fish) to be seen somewhere along the line of refraction, i.e. above the real fish (black).

Independently of the birds, fishes developed the same capacity of problem solving. The archerfish (*Toxotes chatereus*) is famous for this ability. It "shoots" insect prey, off overhanging vegetation, using a precisely aimed jet of water of up to 1.5 m (Burnie 2004) (Fig.10.4).

There is now information available on behavior complexity among amphibians and reptiles. Frogs are known for their vocal communication abilities and salamanders have sophisticated pheromonal communication. Toads learn to anticipate and avoid toxic salt solutions and are also capable of maze learning. Reptiles exhibit parental care (crocodilians) and monitor lizards may have counting-like skills and extract prey hidden in logs by using their forelimbs (*Varanus beccarii*) (Burghardt 2013).

Singing in Birds, Whales and Humans has a Molecular Basis

Singing demands mental innovation based on order, because any succession of sounds that are melodic exclude randomness — they follow a well-defined mathematical sequence known since antiquity. This is why listening to the nightingale or a canary results in a sensation of harmony and mental pleasure.

The molecular basis of singing and musical ability is best known in birds and humans. Like human youngsters, songbirds learn to vocalize by imitating the sounds of their elders. When zebra finches learn to sing, or adult canaries change their song, there is an increase in the protein produced by the *FoxP2* gene in the bird's brain. If the amount of this protein is reduced, by using virus-mediated RNA interference, there is a disruption in bird song which becomes more variable and imprecise than that of controls (Haesler *et al.* 2007).

Other molecules such as hormones participate in the song process (Shen *et al.* 1995). Neural conversion of androgen to estrogen by the enzyme aromatase is an important step in the development and expression of masculine behavior in birds. The distribution of the aromatase messenger RNA has been mapped in the brain of finches and found to be widely expressed. The presence of estrogen receptors in the brain of several avian species has turned out to be unique to songbirds (Gahr *et al.* 1993).

Genetic aberrations of the human gene *FoxP2* impair speech production. The protein encoded by this gene is essential for full articulation of the human language. Mutations of *FoxP2* cause verbal disorders. This is actually the same gene that controls song production in birds and whose variation in activity in a bird's brain leads to song disorder (Haesler *et al.* 2007).

Humans are not the only mammals that sing and that can produce melodies. The humpback whale *Megaptera novaeangliae* has been studied extensively using deep-sea hydrophones combined with satellite tracking. These cetaceans migrate to tropical waters during the winter where they reproduce. Male mating strategy involves singing, as is the case in birds, as well as in humans under certain conditions. Lone whale males repeat long complex songs each lasting for about 10 min in an unbroken series. They can sing incessantly for over 24 h. They follow a hierarchical model in which "each song consists of a series of themes repeated in a specific order, the themes in turn are made up of phrases repeated a variable number of times" (Macdonald 2002). Songs from different years are quite distinct from one another. Besides, the songs of different humpback populations are independent. They can even adopt a new song from a different whale locality. Macdonald (2002) concludes that the humpbacks "learn the detailed acoustic structure of their songs". The parallel with human musical ability is striking.

Spirals Produced by Spiders and by Humans — Drugs Lead in Both Cases to the Drawing of Irregular Geometric Figures

Spirals are curves that can be defined by mathematical equations. This implies that they follow a rigid pattern. They are of several types: Archimedes spiral, logarithmic spiral, Fibonacci spiral and three-dimensional spirals which build helices, such as the well known DNA molecule.

A spiral is a geometric curve, that conveys an inherent order in the arrangement of matter because, before they were formed in living organisms, they occurred in galaxies where their shape was directed by magnetic fields acting on their atomic and molecular construction (Li and Henning 2011). It is thus not surprising that they emerge in the construction of plants and animals since these consist of the same chemical elements that are found in the stars that assemble in galaxies. Besides, spirals are already part of the body of protozoa.

Later in evolution, when the brain arrived, it led to the production in space of spirals. This is specially evident in spiders and humans.

Spiders — with a brain of a size close to that of insects — have a refined geometrical mind. They can draw their geometric figures by actually building them in space. As they produce their webs several types of spirals are constructed with the utmost accuracy. They can also repair their webs re-establishing the original pattern.

A species that produces such a type of web is *Gasteracantha falcicornmis* (Preston-Mafham 1996). Surprisingly the spider starts by building a triangle, not a spiral. Hence, it must have a triangle in its mind before it concretizes it in space. A human cannot draw a triangle on a sheet of paper without previously having it in his (or her) mind.

The web in the form of a spiral is finished by following eight main stages leading to a temporary spiral. This is followed by a second type the sticky spiral (Fig. 10.5).

In other species there is a central hub followed by a strengthening zone and a free zone, but the main area of the web is covered by the sticky spiral. Webs may take many forms depending on whether they are stationary or used as tools to trap insects.

Foelix (1982), who studied spider behavior extensively, reached the conclusion that there has been an evolution in the building of the orb web. Two groups of spiders have formed simple radial webs but created also more complex structures like the spiral form. Some species, such as *Ixeuticus robustus* build irregular traps. Thus spiders use different solutions to obtain food. They build different types of tools that may take the form of spirals but may have a different geometry or no geometry at all (as when jumping). Their minds differ with the species to which they belong. The fact that the pattern of behavior is species-bound indicates that the geometric pattern has a genetic, *i.e.* a molecular basis.

Experiments with drugs, using spiders in captivity, have shown that under their influence they are mentally disturbed, the result being the building of an inaccurate spiral. Age also affects this mental procedure. These results lend support to an explanation in which the spiral is conceived by the spider's mind before it is produced as a web. The situation has its parallel in humans, who under the

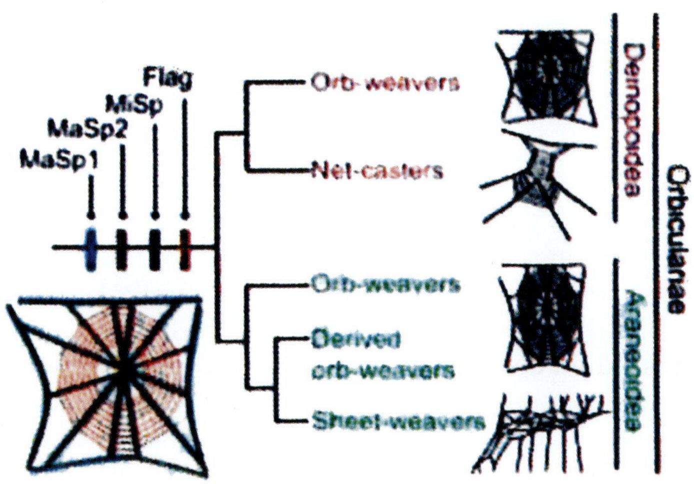

Figure 10.5 Spider architecture. The silks used in spider orb webs are composed of proteins (Flag, Misp and others). The study of the DNAs and silk genes of different groups of spiders support the single origin of orb webs.

effects of drugs or due to old age, also get their minds disrupted producing irregular writing and behaving in unpredictable ways.

The Same Brain Chemicals are Present in Spiders and Humans

The mind of a spider, located in its central nervous system, consists of nerve cells activated by brain chemical signals. Some are the same which occur in our brain. These are: (1) The amino acid *GABA*, which is restricted to the central nervous system and which is an inhibitory neurotransmitter. (2) Serotonin, which is a neurotransmitter is also present in the vertebrate brain. (3) Octopamine, also a neurotransmitter found in insects and vertebrates known for its role in behavior.

As Barth (2002) points out there are even structural similarities between the spider's brain and the human brain. One of them is the distribution of blood vessels which follows an identical pattern in both structures.

Gene Comparison Supports the Single Origin of Orb Webs

It had long been believed that the first spiders (about 220 million years ago) produced only one silk protein but present living spiders produce a number of different silks which they use in specific tasks: one silk cushions their eggs, another protects their egg cases, a third entraps their prey, a fourth builds the frame lines of the web and still another is employed to build the spiral lines (Brunetta and Craig 2010). Garb *et al.* (2006) concluded that "The spider orb web is an impressive example of animal architecture".

There are differences between the adhesive mechanisms of capture spirals spun by Araneoids (aqueous glue) and Deinopoids (dry fibrils). The two types of webs were considered as having occurred separately being an example of what is called convergence. However, the analysis of silk genes has led to a different conclusion. Complementary DNAs from representatives of a pet-casting spider and an orb weaver were characterized. Two proteins *MiSp* and *Flag* are exclusively known from Araneoids. But Araneoids and Deinopoids were found to be united by their shared use of the protein *MaSp2*. The analysis of the silk genes supports the single origin of orb webs (Garb *et al.* 2006) (Fig. 10.5).

Abstract Thinking in Spiders is Accompanied by a Dynamic Mind

Specialists who studied spiders for many years were puzzled by the building of different types of webs. The result was a series of contradictory interpretations of their origin and evolution. Some considered the spiral web as a late arrival resulting from experience accumulated during previous building of simpler webs. Others were inclined to think that the spiral appeared first and was later modified.

What is independent of the various interpretations, is the finding that there has been an enormous variation in web techniques and solutions to catch prey. This is a demonstration that both processes were far from static but have evolved. An example is the radial web which has led to many different solutions. In its simpler form it is merely a retreat with threads radiating from the opening of the web (Cribellate spiders like *Filistata* and Ecribellate such as *Segestria*). An expansion of this type of web is found in the three-dimensional forms (e.g. *Linyphia*). In another solution the web becomes centric and the spider now sits in the hub. Another invention is the spiral web.

The way the glue was used to trap insects has also evolved. In the simple cobwebs glue is only found in droplets, but in the spiral webs a second sticky silk spiral is laid upon. The Australian spider *Dicrostichus* uses a short thread with a large drop of glue on its end. Orb webs without any sticky spiral are typical for the American species (*Wixia*). As Foelix (1996) states: "These examples show that evolution has not stopped at, but has gone beyond the orb web" and he asks a most pertinent question: "A very basic issue not dealt with previously is how spider webs per se could have evolved". Silk threads are considered to have been originally a means to detect and not to trap prey, sticky silk appearing later as an improvement.

The silk threads produced in the abdominal glands consist of alpha-keratin crystals embedded in a rubber-like matrix of amino acid chains which are linked to the keratin structures. These dry threads form the spokes of the web. The sticky spiral consists of wet threads that have glue droplets surrounding glycoprotein doughnuts located at intervals along its length.

We tend to think that the ability to predict adverse situations and to protect one's property are solely human attributes, but the spiders do it equally well. They coat the threads of their webs with fungicides and bactericides which prevent microorganisms from consuming webs. The spider's mind is able to predict and counteract the adversities of life.

Another Form of Abstract Thinking — Bees, Birds and Chimpanzees can Count

The term counting includes "any discriminations based on the numerosity of a set" (Gallistel 1988). Counting is a cognitive ability that, like other forms of behavior that were attributed solely to the human species, turns out to be part of the mental processes in a rainbow of animals extending from bees, to birds and chimpanzees.

Recently it has also been ascertained that bees can count (Dacke and Srinivasan 2008). Honeybees were trained to receive a food reward after they had passed a specific number of landmarks. The distance to the food reward was varied frequently and randomly whilst keeping the number of landmarks constant. This experimental design encouraged them to find a food source by counting landmarks. The result was that bees could count up to four objects.

Pigeons turn out to be on a par with primates in numerical competence (Scarf *et al.* 2011). The birds were trained to order 35 three-item numerical lists. Each list contained stimuli consisting of one, two or three elements, and subjects were trained to respond to them in ascending order. The birds performed above chance and displayed increasing accuracy. The results demonstrated that birds have numerical competence *i.e.* the ability to acquire an abstract ordinal rule. Pigeons can count to several dozens as found by Hauser (2000).

Other birds were also found to have the ability to count. The American coot (an aquatic bird, *Fulica americana*) distinguishes its own eggs from the eggs that other female coots lay in the same nest. They use a variety of tactics to minimize the adverse reproductive effect of the parasitic behavior of other females of the same species. In many cases, hosts may not be able to distinguish their own from alien eggs. In addition coot eggs vary considerably in color. However, Lyon (2003) found that coots can not only distinguish their own eggs from those of a parasite but can also count them. The counting allowed the females to discard the parasitic eggs (Fig. 10.3).

Chimpanzees (*Pan troglodytes*) and rhesus monkeys (*Macaca mulatta*) can also count (Boysen and Berntson 1989, Brannon and Terrace 1998). Two rhesus monkeys were submitted to tests including seven different types of stimulus sets building 36 numerosity pairs generated from the numbers 1 to 9. The animals could spontaneously distinguish between numbers 1 to 9 and in an ascending order. A juvenile female chimpanzee (5 year old) was trained to count food items and objects by using Arabic numbers. She demonstrated the ability to sum arrays of 0 to 4 food pieces. Her counting and number reasoning abilities were comparable to those of preschool children.

It was also found that chimpanzees could be trained to choose different squares and triangles (McFarland 1981). Earlier experiments had also shown that they can put sticks together to reach food more easily.

Nest Building Demands Architectural Skills Involving Dimensions, Proportions, Comfort and Beauty

It makes no difference if one is inspecting the activity of a simple invertebrate or a complex vertebrate — both build equally highly developed nests following complex architectural patterns.

Ants are able of impressive feats. They may be leaf cutters (*Atta* sp.) which carry large leaf pieces all the way to their nests where they become a source of food after careful processing. Others like the African driver ant (*Dorylus nigricans*) build elaborated nests inhabited by millions of individuals.

There are subterranean termites but three-quarters of the world's termite species build nests on trees and over ground. These last ones are vast earthen towers enclosing numerous galleries and chambers used for specific tasks (such as "fungus gardens").

Among fishes, the seahorse (*Hippocampus guttulatus*) is known for the male carrying its young in a pouch, but other species actively care for their eggs by placing them in a nest, chasing away predators and cleaning the eggs to prevent infection.

The building of nests occurs also in amphibians but is a rare event. It has been reported for the African *Afrixalus*, American *Phyllomedusa*, and Indian *Rhacophorus*. In the three cases the nest is built by leaf folding around the eggs (Biju 2009).

Reptiles, like the Australian python (*Antaresia childreni*) lays a clutch of 8–10 eggs in a hollow tree or an underground chamber. It coils itself protectively round the eggs to regulate their temperature until they hatch.

When it comes to birds one is confronted with spectacular architecture. Some species lay their eggs directly on the ground over pebbles (penguins) others build simple nests composed of a few twigs (doves), but a third group uses all its ingenuity to build impressive structures. The African golden weaver (*Ploceus subaureus*) starts its nest like a spider by first building a cross-bridge between two reeds as a support. Grass stalks are then used to make a ring which will become the circular entrance of the nest. After much work this becomes a perfect sphere consisting of entangled twigs disposed following a well established plan.

Not only geometry but beauty is also an obligatory component. Male bower-birds (Family Ptilorhynchidae) build a special structure, the bower, during their courtship display and decorate it with a wide variety of items such as seeds, grass and moss, or even human artefacts. But they do not stop here to attract females. Brightly colored objects are also added including feathers, flowers and colored bits of plastic and paper (Burnie 2004).

The hazel dormouse (*Muscardinus avellanarius*) hibernates deeply in winter. It rests on a nest for about 7 months about 12 cm in diameter, larger than its summer quarters. This nest is in a burrow or under moss or leaves and also functions as a food storage chamber. Brown rat (*Rattus norvegicus*) females give birth to young in a nest of grass, leaves, paper, rags, or almost any other material, which implies improvisation.

Stone Tool Making Among Monkeys and Apes

The order Primates comprises: prosimians (*e.g.* lemurs), monkeys (242 species) and apes (e.g. gibbons, chimpanzees, gorillas and orang-utans). All three form highly complex social units. Tool making has been extensively studied among monkeys and apes.

Japanese macaques (*Macaca fuscata*) are known to pound stones directly against each other. But the wild bearded capuchin monkeys (*Sapajus libidinosus*) in Brazil, deliberately break stones producing sharp-edged flakes and cores that have the characteristics and morphology of hominid tools. They select quartzite cobbles that they hammer into stones of a mean weight of 600 g with sharp edges. This is "an example of intentional stone breakage by a non-human primate" that produces stone accumulations. As many as 111 capuchin-modified stone artifacts were collected by Proffitt *et al.* (2016).

Unexpectedly the apes perform less well than the monkeys in this ability. Stone flacking during chimpanzee tool use is accidental. It is considered a result of indirect force application during activities such as nut-cracking as is the case with the West African chimpanzee (*Pan troglodytes verus*). Also, when the manufacture of sharp edges was taught to captive bonobos (*Pan paniscus*) the resulting flake assemblage did not replicate the early hominid archaeological record.

Behavior in Plants Starts to Resemble Animal Behavior as they are Investigated at the Level of MicroRNAs — Chemical Communication already Occurs in Bacteria

Using as the title of his work "The Green Plant as an Intelligent Organism" Anthony Trewavas (2006) summarized the data on networks in plants that control information flow in single living cells and in whole plants. Higher plants are known to forage and to acquire resources and to respond actively to environmental changes by decisions that best secure an optimal use of resources. Plants have evolved an integrated complex of hormonal systems that allows them to react to water, light, gravity and other agents in a way positive to their development. Their internal communication includes: proteins, nucleic acids, electrical signals and turgor conditions.

A similar approach was taken by Baluska *et al.* (2006) who entitle their work "Neurobiological View of Plants and Their Body Plan". They emphasized that all metabolic biochemical pathways are conserved in animal and plant cells. Besides, plants develop immunity in ways similar to animals. Higher plants perform complex information processing and use action potentials accompanied by synaptic modes of cell-cell communication. The hormone auxin is transported by synaptic processes behaving like a plant-specific neurotransmitter. The evidence indicates that plants have a complex apparatus which allows storage and processing of information.

Before, it was difficult to accept these comparisons with animal behavior, but the analysis at the RNA level has shown that both messenger RNAs and microRNAs are among the molecules that give plants these behavioral properties.

It has been found that RNA molecules actually circulate throughout a plant, via the phloem, in the pumpkin (*Cucurbita maxima*). In situ hybridization confirmed that the messenger RNA *CmNACP*, which is a member of the *NAC* domain gene family, is present in leaf, stem and root phloem, indicating that this RNA is transported over long distances within the plant. This transport was proven by grafts between pumpkin and cucumber plants. These studies established the existence of a system for the delivery of specific messenger RNA from the plant body to the shoot apex (Ruiz-Medrano *et al.* 1999) (Fig. 10.6).

Small silencing RNAs in plants have also been shown to be mobile and to direct gene modifications in recipient cells. They were identified in *Arabidopsis thaliana* using mutants that block small RNAs biogenesis in this plant. The RNA silencing (carried out by RNAs with only 24 nucleotides) spreads both to neighboring cells and systematically over long distances leading to responses to external stimuli (Molnar *et al.* 2010).

Figure 10.6 Communication in plants. (**1**) *Cucurbita maxima*, Pumpkin. Flowering plant with fruit. (**2**) *Cucumis sativus*, Cucumber. Flowering plant with fruit. Grafts between pumpkin and cucumber established the delivery of specific messenger RNA from the plant body to the shoot apex. Small RNAs were also found to be mobile.

Communication also occurs in bacteria and between bacteria and plants. Pea plants have nodules that contain *Rhizobium* bacteria which fix nitrogen, converting the gas into a form that plants can use. In exchange the bacteria gain a nourishing place to live. This symbiosis is the result of a chemical conversation between the roots and the bacteria.

Another soil dwelling bacterium (*Myxococcus xanthus*) is known to live in small groups. When cells begin to starve they signal their plight by secreting a chemical identified as factor *A* (Fig. 10.7) (Losick and Kaiser 1997).

The Periodicity of Mental Ability

Insects are distributed into 29 orders. In only two of them: order Hymenoptera (ants, bees and wasps) and order Isoptera (termites) did social behavior lead to the formation of highly complex and dynamic societies. Hence, 27 orders were not able to acquire this extreme behavior although they had millions of years of evolution at their disposition like those two orders.

Moreover, biologists were led to the conclusion that in insects "social organization has evolved independently eleven times in the Hymenoptera, but only once in the Isoptera" (McFarland 1981).

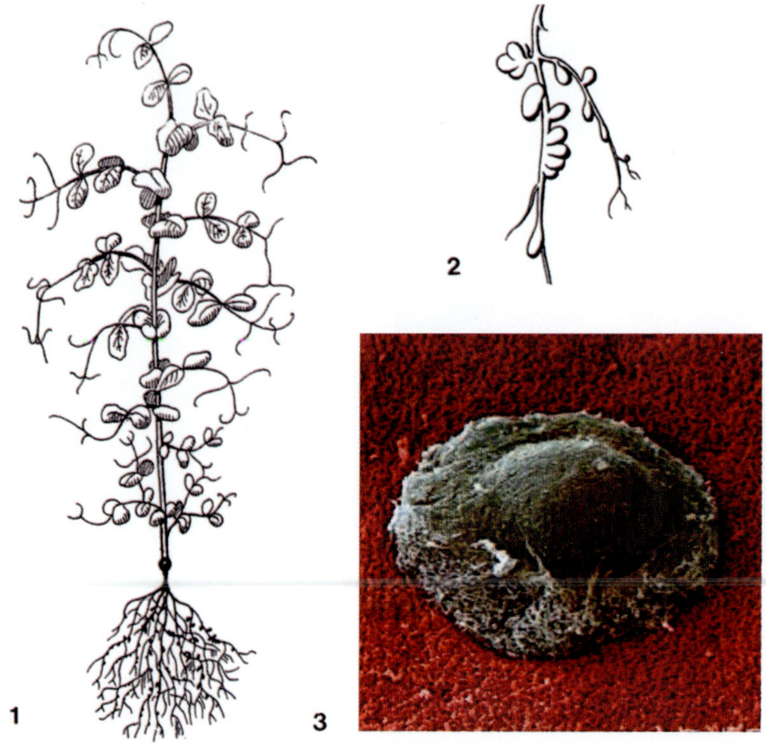

Figure 10.7 Communication in bacteria. (**1**) *Pisum sativum*. Pea plant with nodules in its roots. (**2**) Amplified root of pea plant showing nodules. The nodules contain Rhizobium bacteria, which fix nitrogen, converting the gas into a form plants can use. In exchange, the bacteria gain a nourishing place to live. The symbiosis is the result of a chemical conversation between the roots and the micrcobes. (**3**) *Myxococcus xanthus*. Soil dwelling bacterium. They live in small groups. When cells begin to starve they signal their plight by secreting a chemical identified as factor A.

Although there are intermediate stages between solitary and eusocial behavior the emergence of complex societies in insects erupted in a most unexpected way and not in a form that can be related to the animals' complexity. The orders comprising the Coleoptera (beetles), Diptera (flies) and Lepidoptera (butterflies) are considered by entomologists to be more advanced or equally advanced as those that formed societies.

A feature of importance for the understanding of periodicity is that closely related species do not exhibit the same specific trait. Among the insects species of wasps are: (1) solitary, (2) subsocial (all individuals of the colony are morphologically similar) and (3) eusocial (which have a division of labor in their colonies) (Gullan and Cranston 2000).

Among birds there are finches that are tool-using such as the woodpecker finch (*Cactospiza pallida*) but its close relative the small tree finch (*Camarhynchus parvulus*) does not use tools (Teschke 2011).

Primates come to different solutions depending on the species. Humans are obsessed by mathematics and geometry, whereas chimpanzees specialize in making more than 30 distinct calls including the panthoot which consists of shrieks and roars that can be heard 2 km away. Besides, each adult chimpanzee builds a new, individual tree nest each night for sleeping. Gorillas, instead, have a typical nervous behavior. They bark to deter attacks and walk quietly in single file into thick forest to avoid danger (Burnie 2004).

This diverse mental ability is dependent on genes. The gene *FoxP2* is responsible for singing in both birds and humans and the silk of spirals built by spiders is the product of genes that have been isolated.

High mental ability re-emerges suddenly in octopuses in which it is found to be linked to the recruitment of novel DNA sequences.

The evidence assembled from invertebrates and vertebrates points to the periodicity of mental ability to be the product of DNA evolution.

The Periodic Table of Mental Ability

Table 10.1 consists of 4 vertical columns:

Column 1 Problem Solving

Insects, Arachnids, Molluscs, as well as fish, amphibians, reptiles, birds and mammals show this ability, that is evident in particular species or species groups.

Column 2 Nest Building

This ability is most common in birds but it is rare in certain groups such as fishes. Yet it emerges suddenly in certain species.

Table 10.1 Periodic Table of Mental Ability

Increasing grade of complexity	**PROBLEM SOLVING**	**NEST BUILDING**	**MIGRATION**	**TOOL MAKING**
1	INSECTS ANTS Farming ARACHNIDS SPIDERS Web-repair MOLLUSCS OCTOPUSES 29 shapes	INSECTS AFRICAN ANT *Dorylus* INSECTS TERMITES *Macrotermes*	INSECTS MONARCH BUTTERFLIES LOCUSTS North-South CRUSTACEANS SPINY LOBSTERS Winter and summer residences	INSECTS WEAVER ANTS Leaves sealed ARACHNIDS SPIDERS Webs as traps MOLLUSCS OCTOPUSES Shelters
2	FISH ARCHERFISH Captures insects by projecting water	FISH SEA HORSES Nest building by male	FISH SALMONS EELS Migration from sea to fresh water	FISH No detailed information
3	AMPHIBIANS Maze learning	AMPHIBIANS Nest building by leaf folding	AMPHIBIANS FROGS TOADS Short distance (1 mile) between feeding and breeding sites	AMPHIBIANS No detailed information
4	REPTILES Parental care Counting-like skills	REPTILES PYTHON *Antaresia* Nest building	REPTILES MARINE TURTLES Long distance migration	REPTILES MONITOR LIZARDS Use forelimbs to extract prey
5	BIRDS HERONS Compensation for refraction	BIRDS MOST SPECIES WEAVERS BOWER BIRDS Complex nests with decorations	BIRDS MOST SPECIES From North to South and back	BIRDS RAVENS Insertion of twigs in crevasses
6	MAMMALS CHIMPANZEES Choose different squares	MAMMALS DORMICE CHIMPANZEES Build nests to overnight	MAMMALS ZEBRAS From forests to plains WHALES Between tropical and polar waters	MAMMALS CHIMPANZEES Putting sticks together

Mental ability reemerges in a most unexpected way throughout invertebrates and vertebrates as body complexity increases along the phylogenetic development. Like with vision, where complex organs appeared in simple invertebrates, in mental ability: problem solving, nest building, migration and tool making, attain remarkable development in simple invertebrates as well as complex mammals. The genes involved in some of these mental processes are known: singing in birds and humans (*FoxP2* gene) and web building of spiders (proteins *MiSp*, *Flag* and *MaSp2*) (Original).

Column 3 Migration

Migration is now well established in many animals, be they simple invertebrates or complex mammals. It is also a punctuated event and it may be much more common than it is known today.

Column 4 Tool Making

Studied in detail only recently, this ability becomes more and more evident in animals where it was least expected.

Chapter 11

The Structural and Functional Similarity between Marsupials and Placentals has its Roots in the Plant and Mineral Worlds

The Similarities between Marsupials and Placentals are So Extreme that they have been seen as "Carbon Copies" of One Another

During the Cretaceous period (135-65 million years ago) mammals underwent separation into two distinct groups: The marsupials which carry their young in pouches, and the placentals that retain the young inside the mother's body. The first marsupial fossils are 75 million years old. Placentals are considered to have appeared later. Most of the modern orders were established about 50 million years ago (Halstead 1978, Macdonald 1984).

It has been repeatedly pointed out that the two groups are very much alike. However, this finding has not been examined in terms of periodicity. The reason may lie in the fact that the similarities have not been seen as a result of internally coordinated cellular processes. Another difficulty in interpreting this relationship, has been that the marsupials and the placentals, not only evolved separately, but also on different continents and in different climates where most plant species are different, as is the case in Australia and South America. In consequence the likeness could not be easily attributed to environmental similarity.

The "carbon copy" versions that emerged include species, genera, families and orders. Equivalents are found for members of: carnivora, primates, rodentia, dermoptera, insectivora, edentata, perissodactyla, and artiodactyla (Figs. 11.1 and 11.2).

The evidence coming from fossils already reveals this phenomenon. The large sabretooth cats or tigers had very long upper canines that protruded from the sides of their mouths. In these animals the lower jaw had to open an angle of 90° before it could effectively stab the prey. They emerged in the placentals not only once but several times from independent stocks. Well known genera are *Smilodon* and *Eusmilus*. There also existed a fossil marsupial sabretooth *Thylacosmylus* which was practically identical to the placental species; its upper canines had the same form and shape, it lived at the same time and had the same carnivore habit.

The lions are similarly represented by an equivalent species: the living lion *Panthera leo* (placental) and *Thylacoleo carnifex* (marsupial fossil). The marsupial lion had strong stabbing teeth, and a body and head resembling that of a lion. It was well built for hunting and had carnivore nutrition. Other carnivore equivalents include the hyena, the wolf and the wild cat. Among the most interesting ones is the otter. The European otter (placental) and the water opossum (marsupial) have the same aquatic habit: they swim and catch food in similar ways and are able to close certain organs while under water.

Figure 11.1 Comparison between representative species of placentals with marsupials. (**1**) *Talpa europaea*, European mole (placental). (**2**) *Notoryctes caurinus*, North-western marsupial mole. (**3**) *Muscardinus avellanarius*, dormouse (placental). (**4**) *Dasycercus cristicauda*, crest-tailed mulgara (marsupial). (**5**) *Glaucomys volans*, eastern flying squirrel (placental). (**6**) The marsupial *Petaurus breviceps* has a gliding membrane that is as large as that of the eastern flying squirrel but in this figure the animal is at rest and the membrane is only seen as folds along the lower part of the body.

Among the primates one finds monkeys that resemble several opossum species. And in the rodents several species of mice (placentals) are difficult to distinguish on the basis of shape and habits from the marsupial mice.

The common mole and the pouched mole serve as classical examples. The subterranean way of life, insect feeding, cylindrical body, and reduced eyes and ears make these two species a paradigm of the parallel occurrence of the same structural and functional solutions (Fig 11.1).

The shrew, rhinoceros, and wild boar add to this collection of animals whose structural and functional features are "carbon copies". But, the most remarkable examples are found among the anteaters and the gliding species.

Figure 11.2 Comparison between representative species of placentals with marsupials (*Continued*). (**1**) *Myrmecophaga tridactyla*, giant anteater (placental). (**2**) *Myrmecobius fasciatus*, numbat (marsupial). (**3**) *Marmota flaviventris*, yellow-bellied marmot (placental). (**4**) *Vombatus ursinus*, common wombat (marsupial). (**5**) *Canis lupus*, gray wolf (placental). (**6**) *Thylacinus cynocephalus*, thylacine (marsupial). (**7**) *Leopardus pardalis*, ocelot (placental). (**8**) *Dasyurus hallucatus*, northern quoll (marsupial).

When the placentals are compared with the marsupials nearly all body structures and functions are found to have participated in the establishment of their equivalence. At least the following structures are involved: (1) body configuration, (2) body size, (3) skeletal traits, (4) place of muscle attachment, (5) shape of cranium, (6) shape of mouth, (7) type of tongue, and (8) type of teeth. And at least the following functions are involved: (1) position of the eyes, (2) reduction in eye size, (3) protrusion of the eyes, (4) feeding habits, (5) type of diet, (6) digestive functions, (7) strength of muscles, (8) running capacity, and

(9) gliding capacity. Obviously other structures and functions, though not as prominent, also participated.

Anteaters were Copied in: Monotremes, Marsupials and Placentals

The separation between the monotremes (egg laying mammals) and the viviparous marsupials occurred about 175 million years ago.

The family Tachyglossidae (monotremes) comprises four species which occupy Australia, New Guinea and other large islands (Australasian region). All are recognized by a long narrow snout, barbless spines and sturdy, dome-shaped body. The neck and the tail are hardly seen.

The three species of the long-beaked Echidnas are distinguished from one another by the number of claws and their eating habits (mainly earthworms and crickets).

The fourth species is the short-beaked Echidnas (*Tachyglossus aculeatus*) which is an anteater showing a preference for social insects. With its powerful legs it tears apart underground nests. Once insects are exposed, the sticky tongue collects ants or termites, this is why it is called a myrmecophage (Wilson and Mittermeier 2015).

The same functional solution resurfaces later in the viviparous marsupials. The family Myrmecobiidae has only one species the numbat *Myrmecobius fasciatus*. The fossil record does not shed any light on its emergence as a species which may have occurred in the very late Pleistocene.

The numbat has a sharp snout with a long vermiform tongue that extends in a downward curve. It inhabits the eucalypt forests of Australasia and is the only marsupial that specializes in eating termites and ants. It extracts these insects by pushing its tongue rapidly and repeatedly into the gallery which it digs with its sharp-clawed forefeet. This animal is entirely dependent on termites, requiring between 15,000 to 20,000 each day.

The Edentates diversified 60 million years ago (the beginning of the "Age of Mammals") into several placental orders: (1) anteaters, (2) sloths, (3) armadillos comprising the order Xewarthra and (4) pangolins (order Pholidota). Fossils are well preserved from 30 million years ago when these placental groups expanded. There are four species of Anteaters: Giant anteater (*Myrmecophaga tridactyla*), Northern tamandua (*Tamandua mexicana*), Southern tamandua (*Tamandua tetradactyla*), and Silky anteater (*Cyclopes didactylus*) mainly from Central and South America (Fig. 11.2).

The anteaters feed exclusively on social insects (primarily ants and termites). Giant anteaters are terrestrial but other species are mostly arboreal. All anteaters can dig with claws (the giant anteaters eating the largest ants and the silky anteaters the smallest ones). Their snouts are disproportionately long and their tongues are even longer than their heads. In all anteaters, the tongues are covered by sticky saliva. Their stomachs do not secrete hydrochloric acid but depend instead for the digestion on the formic acid content of the ants they consume.

All species avoid large-jawed ants and termite soldiers that bite their tongue, yet they get sufficient nutrition by ingesting as much as 35,000 ants a day.

The Pangolins belong to another placental order but although they live in Africa and Asia they specialize in eating ants and termites like South American anteaters. They also probe the nests of their prey with long narrow tongues. The family covers 7 species of the genus *Manis* from the Giant pangolin (*M. gigantea*) to the Malayan pangolin (*M. javanica*).

The tongue of the Giant pangolin can be extended for 40 cm and viscous saliva is secreted onto the tongue by an enormous salivary gland which sits in the chest. The captured ants are ground up in the specialized, horny stomach. The pangolins use their powerful claws to demolish the nests of ground termites and ants and the giant pangolin may take 200,000 ants a night.

The similarities of the structures, functions and behavior are impressive, recurring from the monotremes to the placentals.

Gliding was Produced Again and Again in the Marsupials and Placentals

These species were briefly mentioned on Chapter 9, in connection with flight, but are now described in detail to allow an exact comparison.

Among the Marsupials, the Greater Gliders belong to the genus *Petauroides* (family Pseudocheiridae). There are three species: Central Greater Glider (*P. armillatus*), Northern Greater Glider (*P. minor*) and Southern Greater Glider (*P. volans*). All species have well developed claws that allow them to climb up trees. The Greater Gliders possess long and fluffy fur over the body and tail and have a distinctive gliding membrane that allows them to glide between trees. This membrane occurs on each side of the body and consists of two layers of skin. It extends from the elbow of the forearm to the foot.

Another marsupial family is the Petauridae which comprises the Striped Possums, Leadbeater's Possum and Lesser Gliders. All species of the family are arboreal. The claws on the front feet are well developed and sharp which allows them to climb up trees. The fur is very soft.

One species of possum (Leadbeater's) looks superficially like the lesser gliders of the genus *Petaurus*, but it lacks a gliding membrane. Although this species does not glide, it is able to make long jumps, of a meter or more, between tree branches.

The lesser gliders of the genus *Petaurus* comprise 6 species: Northern Glider (*P. abidi*), Yellow-bellied Glider (*P. australis*), Biak Glider (*P. biacensis*), Sugar Glider (*P. breviceps*), Mahogany Glider (*P. gracilis*) and Squirrel Glider (*P. norfolcensis*).

All *Petaurus* species possess a distinctive gliding membrane that allows these animals to glide from tree to tree, it extends from the front paw to the foot. When resting this membrane contracts and is partially hidden. This membrane is built by two layers of skin. The six species differ mainly in body size and color, their diet consists mainly of plant exudates (gum and sap) and nectar and pollen from flowering trees (Fig. 11.1).

A third family of Marsupials is the Acrobatidae which include the Feather-Tailed Gliders and Feather-Tailed Possum also from the Australasian region.

There are two species of gliders: Narrow-toed Feather-tailed Glider (*Acrobates pygmaeus*) and Broad-toed Feather-tailed Glider (*Acrobates frontalis*).

Unlike other species of gliding marsupials in the genera *Petaurus* and *Petauroides*, feather-tailed gliders show no elongation of the major limb bones. Their pelage is soft and dense ventrally and dorsally. Well developed claws are present in the hand which is used in a forceps-like grasp. The gliding membrane extends from the elbow to the knee, and due to the lack of elongation of the limbs is relatively small in area, but *A. frontalis* glides to the impressive distance of 28 meters due to a fringe of longer hairs which increases the surface area of the membrane. These tiny gliders are capable of precise steering descending in a slow spiral around a tree.

Following the Marsupial expansion, again and suddenly, the placentals produced gliding species in two separate groups (Dermoptera and Rodentia) (Wilson *et al*. 2016).

The "Flying Lemur", is not a lemur and does not fly. It only glides and it belongs to a particular order of placental mammals called the Dermoptera which is placed close to the shrews (Macrocelidea) but whose taxonomic location continues to be difficult to define. Colugos as they are also called are supposed to have arisen 60 to 70 million years ago.

There are two species: Malayan colugo (*Cynocephalus variegatus*) and Philippine colugo (*C. volans*). Their gliding membrane stretches from the side of the neck to the tips of the fingers and toes and continues to the very tip of the tail building a large hexagonal figure. This is the result of colugos having limbs of equal length. They also have strong sharp claws which allow them to ascend tree trunks. No other gliding mammal has such an extensive gliding membrane. When this is outspread the colugo assumes the shape of a kite and can execute controlled glides of 136 meters.

Colugos' diet consists mainly of leaves, flowers and sap from tree trunks which they lick with their strong tongue.

Independently of the colugos several large groups of gliding species emerged, but this time, among the Rodents (Rodentia, also placental mammals). These were the squirrels which diversified into different body forms: from ground dwelling marmots to "flying squirrels", which are good gliders.

Several families and many genera have this gliding ability. They are found nearly world wide.

The best known species is the Giant flying squirrel (*Petaurista elegans*). When gliding it stretches out its limbs to extend large patches of furred skin between front and rear legs in a parachute like shape. The large gliding membrane does not involve its long tail. It can glide more than 400 meters and eats mainly flowers, fruits and nuts.

There are no less than 13 other genera of "flying squirrels" belonging to the same family (Sciuridae). These are: *Aeretes, Aeromys, Belomys, Biswamoyopterus, Eupetaurus, Glaucomys, Hylopetes, Iomys, Petaurillus, Petinomys, Pteromys, Pteromyscus* and *Trogopterus*. Within the family Sciuridae there are no less than 273 species in 50 genera but in only 14 genera has the gliding capacity emerged (Macdonald 2002) (Fig. 11.1).

Another family (Anomaluridae) includes the Scaly-Tailed Squirrels of tropical Africa with 3 genera: *Anomalurus, Idiurus* and *Zenkerella*. Their most distinctive feature is a capelike membrane stretched between their four limbs that does not include the tail. The membrane is supported at the front by a rod of cartilage extending from the elbow to the wrist. These squirrels can soar over 100 meters.

Significant is that the family Anomaluridae is not closely related to the other flying squirrels although the membrane is quite similar in shape and extension to that of those species (family Sciuridae) (Wilson *et al.* 2016).

Like the non-gliding possum species which are part of the family Petauridae (marsupials) there is an exception to gliding within the family Anomaluridae (placentals). The Cameroon scaly-tail does not glide (Macdonald 2002). One or several traits are missing that inhibit this process, in agreement with the finding that gliding is the product of the emergence of an array of structures and functions that combine harmoniously.

The DNA of Monotremes has been Sequenced and Compared to that of Marsupials and Placentals — Extreme Gene Permanence

The parallelism of form and function found in monotremes, marsupials and placentals was considered an evolutionary curiosity, because they could hardly be considered to have any close genetic relationship since they diverged millions of years ago. Again, molecular evidence revealed another scenario.

The platypus genome (*Ornithorhynchus anatinus*, Monotremes) was compared to the genomes of other mammals (Warren *et al.* 2008). Conservation patterns of microRNAs in platypus were determined and found to share at least 16 nucleotides with the microRNAs of mice and humans. The protein-coding genes of platypus were also compared with those of marsupials and placentals. The number of these genes in the monotreme is 18,527 which is similar to that of humans 18,600 (placentals) and 20,800 for opossum (marsupials). 82% of the monotreme genes have homologues in these other species.

This extreme gene permanence is reinforced by the work of Putnam *et al.* (2008). The structure and gene content of the DNA of the Florida lancelet (*Branchiostoma floridae*) was analysed by comparing it with humans. The lancelets, or amphioxus, are modern survivors of ancient chordates which had a common ancestor 550 million years ago. The comparative analysis revealed that 17 ancestral gene groups had their location on the same chromosome and were conserved in modern amphioxus and vertebrate genomes. Moreover, 90% of the human genome is encompassed within these gene groups.

Deakin *et al.* (2013) used the increasing number of assembled mammalian genomes to compare their organization across mammalian lineages and to reconstruct the ancestral marsupial chromosome set. They identified large blocks of genes conserved between human and opossum (marsupial) by using fluorescence in situ hybridization.

Hence, the parallelism found between monotremes, marsupials and placentals appears now as a natural result of the similarities of their DNA contents, which suddenly express groups of genes that build identical patterns.

The Marsupials were not Allowed to Fly or to become Marine

Equally important for the understanding of biological periodicity is the repression of genes leading to the elimination of corresponding basic forms and functions.

Most authors have focused on the similarities between the marsupial and the placental species but have bypassed the fact that certain patterns were never allowed to be formed.

The result of this inhibition was the absence of flying marsupials as well as of marine marsupials. On a genetic basis there should have been no hindrance for their formation, since the genes have been conserved from the earliest mammals to humans. Wilson and Mittermeier (2015), in their extensive treatment of all species of marsupials, record that all are terrestrial.

Yet, they could have had another habitat. The platypus (monotremes) has its burrows on land but dives regularly during night into shallow waters to collect food. It moves under water for 40 to 60 seconds to depths of 5 meters. It has even a feature identical to that of placentals. Seals close their ear passages and nostrils when diving and the platypus closes its ears and eyes when underwater. Yet neither the monotremes nor the marsupials became marine.

Coherence Exhibited by Anteaters and Gliding Species

Anteaters possess structures and functions that are involved in the collection, ingestion and digestion of insects. Characteristic of the species analyzed are: (1) long snouts, (2) powerful legs, (3) large claws, (4) a diet consisting mainly of termites and ants, and (5) stomach adaptations to insect digestion. These different traits combine in a coherent activity.

Gliding is also the product of a series of structures and functions that result in harmonious behavior. An impressive number of species show the following characteristics: (1) well developed claws that allow climbing up trees, (2) large gliding membranes, (3) a skin which is usually furred, (4) a diet consisting mainly of flowers and tree exudates and (5) gliding that is directed in a specific way.

Mammal Mimicry and Insect Mimicry are Biologically Identical

Mimicry has been defined as: "Close resemblance of one organism to another or to some object in its environment, as of some insects to the leaves or twigs of plants" (Webster 1976).

The species of marsupials and mammals are as similar to one another as insects from different orders that turn out to be virtually identical. In essence, one is witnessing mammal mimicry. Zoologists have refrained from using the term mimicry for mammals because they live mainly on different continents and do not usually compete for food or environmental niches. The marsupials and the placentals seldom inhabit overlapping areas, whereas the insects that resemble each other usually coexist in the same territory. However, the biological solution has been the same: mammals mimic one another, just as insects do.

Molecular Mimicry is the Result of Key Atoms

At the cell level, proteins with different amino acid sequences, *i.e.* consisting of different atoms, can produce the same pattern and even the same function (Murzin 1993, Schwabe and Travers 1993, Branden and Tooze 1991). One example of the same structure, are 15 proteins with completely different amino acid sequences. All form the eight-stranded α-β-barrel structure, a most regular protein feature. An example of the same function is furnished by the proteins of the staphylococcal nuclease group that were found to have different amino acid sequences, yet perform the same activity.

The molecular mimicry turns out to be the result of key atoms or of specific groups of atoms, both in the case of small and large molecules, that compel the same structure or function to re-emerge. This contributes to explain why organisms with similar genomes but different genetic constitutions can suddenly display the same body solutions, and that very different molecules such as carbonates, cellulose, chitin and proteins, can result in the emergence of similar patterns.

Mineral Pattern — Its Atomic Order may not be Perfect "Yet is not Random". Crystals do not Lie

Mineral symmetry takes various forms: it is internal and external. In both cases it is the sole product of atomic order as demonstrated by X-ray analysis (Wenk and Bulakh 2004).

The internal symmetry is revealed by the existence of symmetry elements: rotation axes, rotoinversion axes, center of symmetry and mirror planes. The presence of these elements is detected, in a well formed crystal by the regular arrangement of the bounding faces.

The number of possible symmetry combinations is not unlimited it results in only 32 crystal classes. These in turn have been grouped into 6 crystal systems: cubic, hexagonal, tetragonal, orthorhombic, monoclinic and triclinic. What characterizes the different systems is the number, and inclination of the axes, some building horizontal, perpendicular or oblique angles. The axes and centers of symmetry are an expression of the atomic order revealed by the X-ray studies.

The external symmetry deals with the form of the three-dimensional crystal. At this level of organization rotation and translation are also evident, but one is dealing with the external morphology of single crystals or of the assembly of crystals into regular units. As Klein and Hurlbut (1985) point out the atomic distribution at times may "not be perfectly ordered and yet also not random". The single or multiple crystal may have recognizable symmetries that can be: twofold (bilateral), threefold, fourfold, fivefold and sixfold.

Crystals exhibit an order that is so patent that it cannot be denied. Crystals do not lie (Fig. 11.3).

The Minerals Exhibit the Mimicry Found in Insects and Mammals without having Genes

Molecular mimicry is equally present in minerals leading to the formation of similar geometric solutions.

The minerals chalcanthite (CuSO45H2O), albite (NaAlSi3O8) and anorthite (CaAl2Si2O8) all form triclinic crystals. The same is the case in zincite (ZnO), bromellite (BeO) and greenockite (CdS), though having different chemical compositions they mimic each other by crystallizing in the hexagonal system (Bloss 1971).

Other minerals that crystallize in the other crystal systems: cubic, tetragonal, orthorhombic and monoclinic also display mimicry.

Copper, platinum and diamond are minerals belonging to the cubic system. They result from the assembly of a single atom (Cu, Pt and C respectively). Other more complex minerals: fluorite (CaF2) and Galena (PbS) also belong to the same system. All these atoms could hardly be more different: carbon (C), platinum (Pt), copper (Cu), fluorine (F), calcium (Ca), lead (Pb) and sulfur (S). Yet their interaction results in the same final pattern.

The tetragonal system is represented by minerals with equally diverse chemical compositions: chalcopyrite, pyrolusite, rutile, scheelite and tin.

Figure 11.3 Minerals constituted by different atomic constellations build similar crystal forms. (**1**) Bilateral symmetry. **Calcite** CaCO3. Twin crystal with rhombohedral form. (**2**) Threefold symmetry. **Tetrahedrite** Cu12Sb4S13. Single crystal. (**3**) Fourfold symmetry. **Andalusite** Al2SiO5. Single crystal. (**4**) Fivefold symmetry. **Quasicrystal** Ho-Mg-Zn. Dodecahedral quasicrystal grown by using excess Mg. Note the clearly defined pentagonal facets. (**5**) Sixfold symmetry. **Emerald** Be3Al2(Si6O18). Green crystal form of beryl. (**6**) Bilateral symmetry. **Argirodite** GeS6Ag8. Single crystal. (**7**) Threefold symmetry. **Boracite** Mg3ClB7O13. Single crystal. (**8**) Fourfold symmetry. **Staurolite** Fe2Al9O6(SiO4)4(O,OH)2. Twin crystal. (**9**) Fivefold symmetry. **Marcasite** FeS2. Association of 5 crystals. (**10**) Sixfold symmetry. **Arsenopyrite** FeAsS. Six crystals formed around a center.

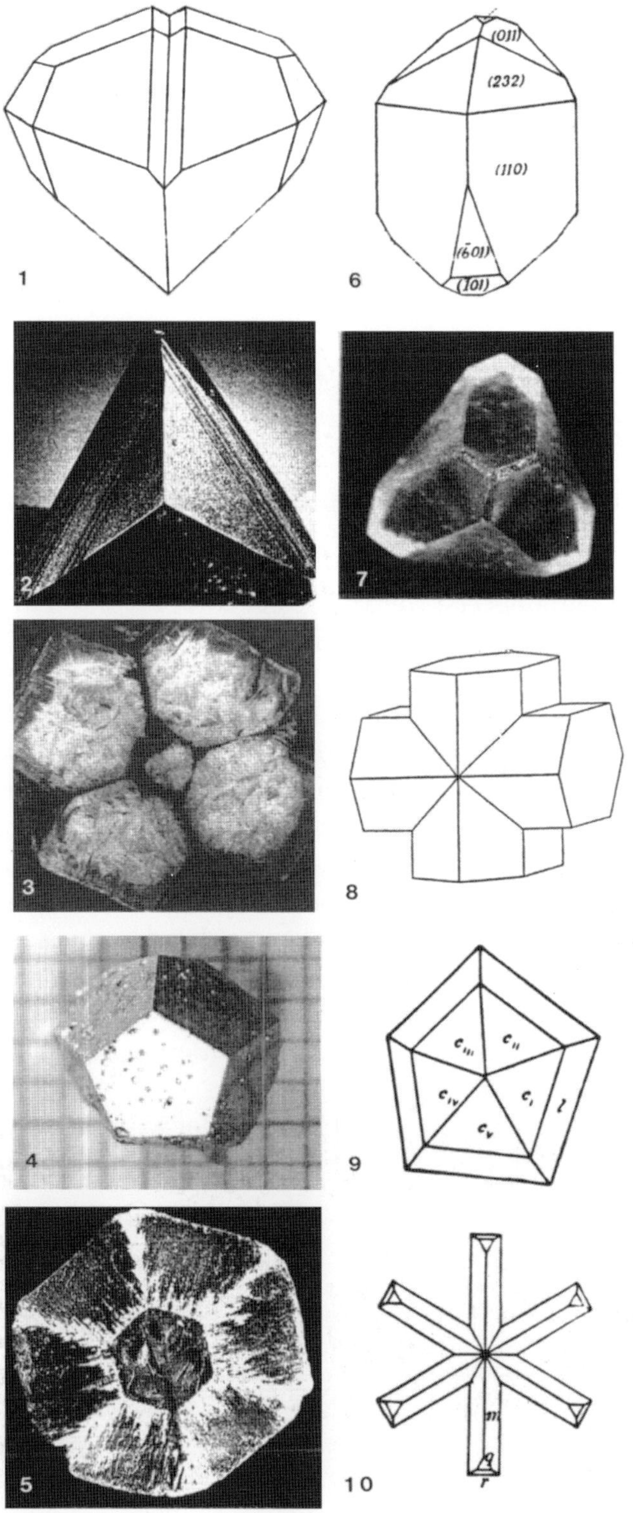

Figure 11.3

The same is true of the orthorhombic system: aragonite, boracite, chrysoberyl, natrolite and sulfur, all have quite different chemical compositions.

The monoclinic system contains as well mimicry: arsenopyrite, gypsum, manganite, muscovite and staurolite exhibit the same symmetry pattern.

Many other minerals could be added to these series, as is the case with the mammals, in which the series is not solely limited to the animals mentioned previously.

The Frustration Experienced by Botanists in Unraveling Plant Evolution has not been Dispelled by the Molecular Analysis

Botanists continue to this day to have great difficulties in establishing the evolutionary relationships between different plant families of flowering plants. Most agree that the Magnoliaceae were early in evolution and that the Liliaceae and the Orchidaceae were late families, but in between confusion reigns (Heywood 1978, Li 1997).

The reason is evident. The differences between plant families are most patent in their flowers, but the main trait of these flowers is their symmetry. What continues to bewilder botanists is that the same type of geometry has arisen, time and again, in quite unrelated families.

Molecular biologists have also clearly expressed their difficulties. Some of the earlier molecular studies, on ribosomal RNA, led to conflicting results (Doyle *et al.* 1994). An extensive phylogenetic analysis of DNA sequences of five genes (from all three plant genomes: mitochondrial, plastid and nuclear) covering 105 species of flowering plants, led also to conflicting results (Qiu *et al.* 1999). Even the more recent analysis of the rates of molecular evolution in flowering plants did not help (Smith and Donoghue 2008).

The reason for this difficulty seems to be that we are still at the molecular level. It is only when the atomic structure will be reached that plant evolution may be elucidated.

The Transfer of Mineral Symmetries to Plants. They Copied the Pattern Displayed by Minerals

The symmetry found in flowers is elucidated by the occurrence among single crystals, and crystal associations, of twofold, threefold, fourfold, fivefold and sixfold symmetries.

The twofold (or bilateral) symmetry of the flower of a snapdragon (*Antirrhinum*) or of an orchid (*Paphiopedilum*), may be compared with a compound crystal of calcite with its bilateral symmetry. The three petals of *Alisma* irradiate from a central region building a threefold symmetry and the same distribution of parts is present in a crystal of boracite. The symmetry of four is displayed by the flowers of *Papaver* as well as by the crystals of andalusite.

An argument used before against such a comparison was that the symmetry of five, so common in plants (e.g. *Rosa*), did not occur in mineral systems. Already it was known that crystals were able to build associations of five components as in marcasite (FeS2). This difficulty was finally removed when quasicrystals were discovered in 1984 (Fig. 11.3).

Metallic glasses were found to embody a new kind of atomic order considered previously impossible. Since then quasicrystals have been produced from other atomic associations consisting of: Ho = Holmium, a rare earth with atomic number 67, Mg = Magnesium and Zn = Zinc (Fig. 11.3). Their fivefold symmetry disconcerted mineralogists.

The sixfold symmetry is displayed by many flowers (e.g. *Narcissus*) and is equally evident in an emerald crystal.

In all cases intermediate solutions between the different symmetries seem to have been excluded. Moreover, within each structure, be it mineral or plant, the parts have the same shape, length and average distance from one another (Fig. 11.4).

The Periodicity of Plant Patterns is Decided by well known DNA Sequences

Plant symmetries are not confined to a single species but recur regularly in unrelated plant families. The bilateral symmetry occurs in the Labiatae, Scrophulariaceae, Orchidaceae, Leguminosae and other families. The threefold symmetry appears mainly in the Annonaceae, Cistaceae, Elatinaceae, Alismataceae, and Commelinaceae; the fourfold symmetry is present in flowers of the Vitaceae, Aquifoliaceae, Cornaceae, Cruciferae, Thymelaeaceae and others. Flowers with fivefold symmetry are most common in the family Rosaceae, but are also present in the Oxalydaceae, Rutaceae, Valerianaceae, Gentianaceae, Gesneriaceae, Solanaceae and other families. The sixfold symmetry emerges in the Liliaceae, Amaryllidaceae, Dioscoreaceae, Iridaceae, Berberidaceae and Ranunculaceae as well as other families.

These symmetries are not only confined to the flowers but may also dominate in the fruits, roots or other plant organs.

Figure 11.4

Like the "carbon copies" described in mammals, the same flower pattern tends to occur in: (1) unrelated families, (2) to predominate in a certain family and (3) within a family there are species with different flower configurations (e.g. Crassulaceae), like the presence or absence of a certain trait in an animal group (Blamey and Grey-Wilson 1989).

The genes responsible for the emergence of these symmetries have been known for several decades (Stubbe 1938).

In the plant *Linaria vulgaris* (common toadflax) the bilateral symmetry of its flowers is transformed into a fivefold symmetry when one of its DNA sequences is modified by methylation of its bases. The same phenomenon occurs in the snapdragon (*Antirrhinum majus*) (Cubas *et al.* 1999) resulting in the same symmetry transformation.

Hence, a simple atomic process (methylation) is occurring in the gene which changes the ensuing molecular cascade in such a drastic way that the leap is from the flower pattern characteristic of one family to that of another completely unrelated one.

The Periodic Table of Equivalence between Marsupials and Placentals

No vertical columns, but parallel rows, are used in this connection. The reason is that the marsupial species can be distributed along a series of patterns that happen to have their counterparts among the placentals (Table 11.1).

The Periodic Table of the Elements is dominated by the formation of vertical columns, but it also includes two parallel rows. These were later created to accommodate the elements within the lanthanides (Nrs. 57 to 69) and the elements within the actinides (Nrs. 89 to 102). Further it was recognized that there was a resemblance between the properties of the two series which led to place them side by side becoming collectively known as the *Rare Earths*. The lanthanides are especially notorious for being chemically similar to each other. Also, the actinides are similar to each other by being all radioactive. Besides, some of the properties found in the elements of lanthanides parallel those of the actinides (Gray 2009).

Figure 11.4 The transfer of mineral symmetries to plants — the same symmetry arising in unrelated families. (**1**) Bilateral symmetry. *Antirrhinum majus*, Scrophulariaceae. (**2**) Threefold symmetry. *Alisma plantago-aquatica*, Alismataceae. (**3**) Fourfold symmetry. *Papaver hybridum*, Papaveraceae. (**4**) Fivefold symmetry. *Borago officinalis*, Boraginaceae. (**5**) Sixfold symmetry. *Narcissus incomparabilis*, Amaryllidaceae. (**6**) Bilateral symmetry. *Cypripedium calceolus*, Orchidaceae. (**7**) Threefold symmetry. *Commelina erecta*, Commelinaceae. (**8**) Fourfold symmetry. *Cornus kousa*, Cornaceae. (**9**) Fivefold symmetry. *Solanum tuberosum*, Solanaceae. (**10**) Sixfold symmetry. *Anemone pulsatilla*, Ranunculaceae.

Table 11.1 Periodic Table of Equivalence between Marsupials and Placentals

Increasing grade of complexity →

MARSUPIAL SERIES

POUCHED MOLE	MULGARA MOUSE	PHASCOG-ALE	OPOSSUM	ANTEATER NUMBAT	GREATER GLIDER	SUGAR GLIDER	WATER OPOSSUM	POUCHED HYENA	TASMANIAN WOLF	MARSUPIAL LION	SABRE-TOOTH TIGER
Notoryctes	*Darcycercus*	*Phascogale*	*Caluromysiops*	*Myrmecobius*	*Petauroides*	*Petaurus*	*Chironectes*	*Borhyaena* Fossil	*Thylacinus*	*Thylacoleo* Fossil	*Thylacosmylus* Fossil

PLACENTAL SERIES

MOLE	MOUSE	SHREW	TARSIER	ANTEATER	SQUIRREL	COLUGO	OTTER	SPOTTED HYENA	WOLF	LION	SABRE-TOOTH TIGER
Talpa	*Mus*	*Sorex*	*Tarsius*	*Tamandua*	*Anomalurus*	*Cynocephalus*	*Lutra*	*Crocuta*	*Canis*	*Panthera*	*Smilodon* Fossil

The similarity between these animal species has been repeatedly described as an intriguing case of "convergence". For years no genetic studies have been available and as a consequence their similarities were considered the product of random events. Such animals, arising on distant continents separated by millions of years, could hardly have the same genes.

The molecular evidence assembled at present discloses a different picture in which the similarities between marsupials and the placentals are found to be an impressive case of periodicity. The similarities of structures, functions and behavior, recur from the monotremes to the placentals making them pure "carbon copies" (species by species). Recently DNA has been sequenced in monotremes, marsupials and placentals indicating that the similarity between these animal species is the natural result of DNA evolution in which groups of genes are suddenly expressed again building identical patterns.

These results find further support in another property that shows periodicity of gene function. Carnivory is displayed mostly by European, Asiatic and North American plant species, but there is also a species that suddenly appeared in Australia which exhibits the same structures and functions that are part of plant carnivory. Besides, this species, Cephalotus, also belongs to a quite different plant family Cephalotaceae. Now it is known that a battery of genes decides these carnivorous traits and that their expression or repression occurs even within the same individual plant. Plants and animals are also now known to have similar genes.

The comparison of the marsupial series and the placental series with that of the lanthanide series and the actinide series of the chemical elements (Table 2.2) may seem farfetched but this is because it is still difficult to think of DNA in pure atomic and electronic terms. The similarities are far from being fortuitous. (1) The chemical elements of both series have common properties that led them to be called rare earths. The marsupials and placentals have the common property that classifies them as mammals. (2) Each series can be compared species by species in animals and in atoms atom by atom. (3) This similarity is not disturbed by the fact that in atoms and mammals there is an increase in complexity from simpler structures to complex ones (Original).

The periodicity evident in the mammals also creates two parallel rows which accommodate the marsupial series and the placental series. Each series has a main feature in common (pouch and placenta respectively) and besides there are impressive similarities, species by species, as the patterns of the animals in the upper and lower rows are compared. They start with the mole and finish with the sabretooth.

The species included are the most representative. Because evolution within the mammalian species is incomplete, the ordering of the various species of mammals, according to increased complexity, is only approximate. Following a most extensive treatment of all mammalian species, in several volumes, Wilson and Mittermeier (2009) were obliged to conclude that: "Unfortunately there is no sure way to determine evolutionary relationships".

Chapter 12

The Periodic Tables Lead to a Law of Biological Periodicity Which has Predictive Power

The "Cambrian Explosion" is an Expression Which Denotes the Lack of Knowledge of the Mechanism Directing Evolution

Geologists and biologists had known for a long time that the simplest invertebrates and algae arose about 1,000 million years ago. What became evident in the last decades is that, as the invertebrates started to diversify, they did not show a slow and progressive process but instead suddenly diversified into the many phyla of invertebrates which are found at present.

This most unexpected result was called the "Cambrian Explosion" and bewildered evolutionary biologists. The event occurred between 541 and 490 million years ago (Fig. 12.1).

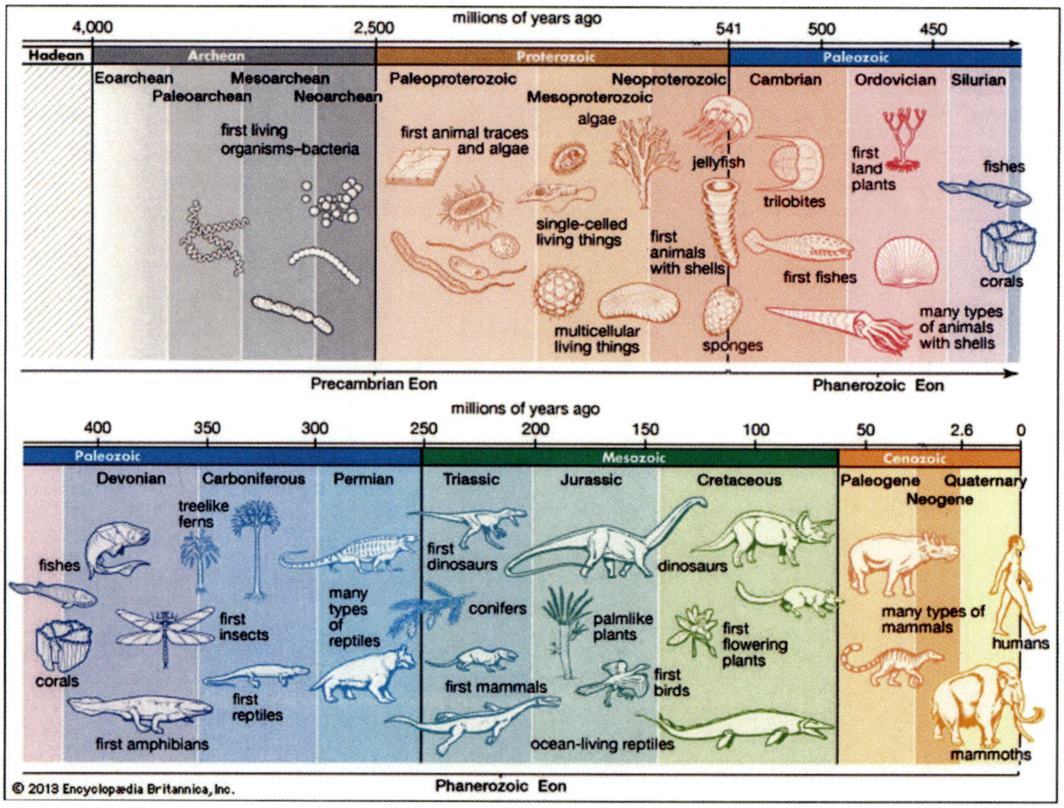

Figure 12.1 The geologic time scale from 4,000 million years ago to the present, showing major evolutionary events at the biological level. The Cambrian period took place between 541 and 490 million years ago.

Within 70 to 80 million years the rate of diversification accelerated enormously and the diversity of life began to resemble that of today. The rapid Cambrian radiation "was noted as early as the 1840s and in 1859 Charles Darwin discussed it as one of the main objections that could be made against the theory of evolution by natural selection" which demanded slow modifications. For a long time this event was ignored to avoid criticizing neo-Darwinism. But in the last decades, as more fossil data supported it, it was revived, "generating extensive scientific debate" (Encyclopedia Britanica).

If one dares to abandon the current interpretation of biological evolution and sees instead evolution as the product of internal atomic DNA transformation, the "Cambrian Explosion" becomes intelligible as a typical example of periodicity in which a given property re-emerges time and again whether the organism likes it or not.

Shapiro (2010) when describing the evolution of mobile DNA stated that "Molecular genetic discoveries, plus a consideration of how mobile DNA rearrangements increase the efficiency of generating functional genomic novelties, make it possible to formulate a 21st century view of interactive evolutionary processes". And he added: "These rapid multi-character changes are fundamentally different from the slowly accumulating small random variations postulated in Darwinian and neo-Darwinian theory".

The Origin of the Eye is not Easily Explainable by Selection. Some Eye Lenses Consist of Pure Minerals

Most conventional biologists who have carried out research on the eye have attributed its evolution to natural selection (Land 2012). For them the eye had to have a biological origin.

However, Gehring (2014) a molecular biologist who produced fruit flies with eyes at the most unexpected body locations by using genetic engineering, has expressed a different opinion.

He argues that it is most difficult to explain the sudden origin of the eye by selection, since at the very beginning selection cannot improve on something that has only an elementary structure. And he states that: "Selection can only act on organs having at least a minimal function".

The evidence available reveals that the eye has not a biological origin but a mineral one.

Crystals, such as calcite ($CaCO_3$), already produce secondary rays oriented in a specific direction. Other transparent minerals, like tourmaline, which are highly complex assemblies of many different atoms, display the same optical property (called birefringence).

Calcite itself is a regular component of the body of living animals and in the living brittle stars the eye lens is made of calcite. Besides, it can be transformed into aragonite. In the marine mollusc chiton the lens of the eye is made of the mineral aragonite. This has the same chemical composition as calcite ($CaCO_3$) but it differs by its high specific gravity and its cleavage (Cronin *et al.* 2014, Klein and Hurlbut 1985, Wenk and Bulakh 2004).

The eye lens of vertebrates consists of packed proteins. These are not assembled at random. They are called crystallins because they actually build a large crystal resulting from their ordered assembly and transparency.

Selection has nothing to do with the assembly of atoms into minerals or the assembly of proteins into crystal structures, these are guided by pure electronic energy states.

DNA and RNA are the Prisoners of 3 Previous Evolutions — their Atomic Internal Construction Made them the Prime Agents of Evolution

First, biological evolution is a terminal and secondary process, not the primordial one (Lima-de-Faria 1988). Matter and energy went through three separate and autonomous evolutions before the biological level was reached. These were: (1) The evolution of the elementary particles (Fig. 2.1). (2) The evolution of the chemical elements (Fig. 2.2). (3) The evolution of the minerals (Chapter 2).

Second, DNA turns out, not to be the product of a disordered assembly of the 118 elements of the Periodic Table, but has been built by putting together a few elements. Only 5 (out of the 118 elements) were employed: hydrogen, oxygen, carbon, nitrogen and phosphorous (Lima-de-Faria 2001, 2003). This holds true also for RNA.

Third, these elements have a unique position in the Periodic Table. Their atomic numbers are: 1 (hydrogen), 6 (carbon), 7 (nitrogen), 8 (oxygen), 15 (phosphorous). They are part of a *particular triangle* near the extreme right end of the Table (close to the halogens and noble gases). With the exception of hydrogen (which is at the origin of the other elements) they are all classified as *nonmetals* (Gray 2009).

These three conditions have canalized in an extreme form all future biological evolution.

Order prevailed at every level of combination, during the formation of simple as well as complex structures, because it had to obey the electronic properties of these atoms and no other ones.

The result was that the electronic evolution of DNA (as well as RNA) became the prime agent and canalizer of biological evolution.

DNA knows Nothing about the Organism that it Produces

DNA not only follows the energetic states of its own elementary particles, but knows nothing about the organism that it creates.

DNA neither "sees" the result of its genetic message (messenger RNA) nor the resulting proteins, as well as other subsequently formed minor molecules that shape the cell's and the organism structures and functions. DNA has evolved and continues to do so, creating its own patterns of base sequences which emerge independently of what effect they may have on the development of the organism.

However, at the same time that it follows the laws of physics and chemistry it also cares for the correct survival of its final product — the living organism — avoiding defaults.

DNA by using repair mechanisms at the DNA, RNA and protein levels ensures that its message is correctly transmitted to the progeny that it never "sees". Repair mechanisms at these three levels are now well established (Cullen 2003, Cakmak 2009).

Among living organisms two examples reveal a similar situation at a higher level of cell organization. The sand wasp (*Ammophila sabulosa*) hunts caterpillars, paralysing them with a sting, and burying them in a burrow depositing on their top a single egg. In this way the wasp ensures the survival and development of its progeny that it never sees.

The same is the case among birds. The Australian brush-turkey (*Alectura lathami*) builds a nest that is a mound consisting of leaves and sand (reaching 5 meters in diameter). The female lays eggs in chambers excavated by the male and the rotting vegetation provides the heat needed for incubation of the eggs which are ignored. The chickens never see their parents (Perrins 2003).

Transposable Elements Generate DNA Novelties by Shaping Genes

Piskurek and Jackson (2012) entitled their article "Transposable Elements: From DNA Parasites to Architects of Metazoan Evolution".

The completion of the human genome had revealed that more than half of our DNA is derived from transposable elements (or transposons). It became also evident that transposons comprise the largest class within most metazoan genomes.

As these authors remind us, the transposons were once categorised as "junk DNA" but are now known to influence genome structure and function by increasing the coding and non-coding genetic repertoire of the host. By stimulating the evolution of genomes transposons generate morphological and functional novelties.

The importance of DNA transposons in the evolution of multicellular organisms was also treated at length by Feschotte and Pritham (2007). They found that these mobile genetic units exhibit a broad diversity in their structure and transposition mechanisms. In addition their movements and accumulation within the genomes "represent a major force shaping the genes of almost all organisms".

RNA Performs Multiple Editing Mechanisms that Change the Information Received from DNA

In his work "The theory of RNA-mediated gene evolution", Morris (2015) describes the observations and experiments that in the last decade revealed that non-coding RNAs function in regulating the transcriptional ability of DNA and gene expression. This novel function of small RNAs is achieved mainly by methylation of DNA. One event is the resulting transformation of DNA bases, such as cytosines to thymines, which ultimately influence the function of the cell.

One of the surprises from the sequencing of the human genome was that roughly 80% is not involved in generating proteins but rather in producing non-coding transcripts.

Many non-coding RNAs are functional in the regulation of gene transcription. Studies in human cells have disclosed that these RNA transcripts interact with chromosome proteins involved in epigenetic cellular processes of regulating chromatin states and gene accessibility.

Changes of the chromatin architecture along the chromosomes can drastically influence gene expression. This modification, as well as the above mentioned DNA methylation, can result in the silencing of a particular gene or gene complex. This problem is now being extensively studied both in plants and animals (Riddihough 2016, Garg and Sharp 2016, Rinn and Guttman 2014, Matzke *et al*. 2015).

Not only DNA but RNA appear now as the directors of evolution.

Biological Periodicity is Characterized by the Following Features

1 Biological periodicity is an event anchored in the mineral world where only simple atomic interactions decide the properties. Among the properties studied, some do not start at the biological level, as previously assumed, but existed before DNA and the cell arrived in evolution. Regeneration is evident in crystals of ammonium oleate (Fig. 7.1), luminescence is pronounced in the

minerals fluorite and scheelite as well as in 19 others (Table 4.1 and Fig. 4.1). Also in vision the ability to deviate light rays (present in lens proteins) was already exhibited by calcite crystals with their double refraction (Fig. 8.1). Only much later, when biological evolution emerged, did these properties appear in bacteria and multicellular organisms as a result of similar atomic processes.

2 The main feature of biological periodicity is its punctuated emergence. It seems to arise from nowhere coming and going without previous announcement. The organisms that exhibit flight, vision, plant carnivory, or any other of the described properties, do not show a relationship to their immediate predecessors. The pterosaurs were not directly related to the flying insects that preceded them, and carnivorous plants appear in plant families that are not closely related.
3 Periodicity emerges with few intermediate forms or abruptly in most cases. High mental ability is present in the building of elaborated societies by ants whereas many other insect species are solitary.
4 It is independent of organism complexity. Vision is present in bacteria in a way that it already resembles that of humans. The placenta emerges independently in plants, invertebrates and mammals.
5 It appears in a well defined organ or on a restricted species group. Plant carnivory is confined only to leaves and only to flowering plants. Luminescence became localized to specific organs and could not extend beyond the fishes.
6 The types of structures and functions that are an obligatory component of a given property (e.g. flight, mental ability, placenta formation) are essentially the same in all the different taxonomic groups that exhibit it. Flight in bats, birds and insects uses a wing whereas a quite different structural solution could have been created (such as jet propulsion in squids).
7 Periodicity occurs at regular intervals but of variable length. This is to be expected since we are dealing with living organisms which are more complex than simple atoms. High mental ability has re-emerged after millions of years but in plant carnivory the interval may have extended to only thousands of years.
8 It is independent of the organism's taxonomic location. The penis is as developed in invertebrates as it is in higher mammals.
9 It is also independent of the environment. Gliding has arisen in animal species living in the tropics as well as in Nordic climates. The marsupials that evolved in Australia (in a different environment and having a different plant diet) became carbon copies of their European placentals. Flight occurs in air but also in water and in the water-air interface.
10 The degree of complexity of a given structure or function does not necessarily follow the general increase in complexity taking place during evolution. Flight

in birds is not more complex than that of flies. Four wings occur in fishes and butterflies but birds have only two. The eye of a fish is not more advanced than that of a cephalopod.

11 Repression, of the genes for a given property, is considered responsible in: plant carnivory (where it appears only in some families of flowering plants), regeneration (where it diminishes sharply in vertebrates) and luminescence (where it was obliterated in the vertebrates following the fishes) (Table 12.1).

Table 12.1 (see facing page)

1. The properties studied show an increase in complexity starting with simple forms and terminating in highly complex ones resulting from combination with additional components. All the chemical elements are derived from the simplest one (hydrogen, atomic Nr. 1) and by combination result in an increasing complexity leading, at present, to Ununoctium (atomic Nr. 118) (Gray 2009). In earlier forms of the Table the most complex element is Nr. 103 (Fig. 2.2).
2. There is a reemergence of the same phenomenon irrespective of the taxonomic location of the organism, being it a simple invertebrate or a complex mammal. Among the elements the same phenomenon is evidenced also irrespective of complexity: Helium (atomic Nr. 2) and Radon (atomic Nr. 86) are both noble gases (Table 2.1).
3. The same property resurfaces following a given interval. In the elements this interval tends to be separated by 8 atoms as is the case for the first two periods of the Table (Alkali Metals and Alkaline Earths). However, in the Transition Metals the interval is much longer than 8 atoms (Scerri 2007). It is expected that at the biological level the interval will be more variable since one is dealing with a more complex level of organization.
4. Atoms having similar properties are usually located along vertical columns, but in addition the different columns are arranged in natural groups due to the similarities between the columns (e.g. Alkali metals and Alkaline Earths). This happens to be the case also at the level of living organisms. In this table luminescence, vision and regeneration, build the first group (the 3 columns start at the mineral or crystal level). A second natural group is formed by particular organs: placenta, penis and mainly wing (flight in air). The third group comprises plant carnivory and mental ability. The fourth is built by the marsupial and placental series.
5. The comparison of plant carnivory with mental ability may seem outrageous. But the detailed analysis of both properties reveals that they actually represent a similar physiological solution in plants and animals. Actually both improve the functions of the organism by sophisticated ways of catching prey which benefits food supply.
6. The most unexpected, yet may be the most significant, is the equivalence found within the Mammals which is found as well within the Rare Earths (Table 2.2).
7. The predictive capacity of the Table of Biological Periodicity will be tested by future research. If it will be modified, due to the acquisition of novel data, that will be a confirmation of its intrinsic value (Original). See Appendix pages 307–311.

Table 12.1 General Table of Biological Periodicity.

Increasing grade of complexity ↓	LUMINESCENCE	VISION	REGENERATION	PLACENTA	PENIS (SINGLE)	FLIGHT (IN AIR)	PLANT CARNIVORY (LEAVES)	MENTAL ABILITY
1	MINERALS EUBACTERIA	MINERAL REFRACTION CYANOBACTERIA	CRYSTALS PROTOZOA	ONYCHOPHORANS BRYOZOANS	FLUKES	INSECTS	DROSOPHYLLACEAE *Drosophyllum*	ANTS TERMITES BEES
2	DINOFLAGELLATES	PLANT CELLS CNIDARIANS	ALGAE SPONGES	SPONGES WORMS FLIES	GNATHOSTOMULIDS GASTROTRICHA	SQUIDS	SARRACENIACEAE *Sarracenia*	SPIDERS OCTOPUSES
3	FUNGI	CRUSTACEANS MOLLUSCS INSECTS	CNIDARIANS	ASCIDIANS FERNS	MOLLUSCS INSECTS	FISH	NEPENTHACEAE *Nepenthes*	ARCHERFISH
4	STARFISH JELLYFISH	FISHES AMPHIBIANS	FLATWORMS	SHARKS TOADS LIZARDS	CAECILIANS LIZARDS	PTEROSAURS	DROSERACEAE *Dionaea*	MONITOR LIZARDS
5	MOLLUSCS INSECTS	REPTILES BIRDS	ECHINODERMS	FLOWERING PLANTS BANDICOOTS	OSTRICHES DUCKS	BIRDS	LENTIBULARIACEAE *Genlisia*	WEAVERBIRDS
6	SEASHORE AND ABYSSAL FISH	MARSUPIALS PLACENTALS	FLOWERING PLANTS	EARLY PLACENTALS LATE PLACENTALS	MARSUPIALS PLACENTALS	BATS	LENTIBULARIACEAE *Utricularia*	CHIMPANZEES

Increasing grade of complexity →

MARSUPIAL SERIES

POUCHED MOLE	MULGARA MOUSE	PHASCOGALE	OPOSSUM	ANT-EATER NUMBAT	GREATER GLIDER	SUGAR GLIDER	WATER OPOSSUM	POUCHED HYENA FOSSIL	TASMANIAN WOLF	MARSUPIAL LION FOSSIL	SABRE-TOOTH TIGER FOSSIL

PLACENTAL SERIES

MOLE	MOUSE	SHREW	TARSIER	ANT-EATER	SQUIRREL	COLUGO	OTTER	SPOTTED HYENA	WOLF	LION	SABRE-TOOTH TIGER FOSSIL

The Molecular Mechanism Responsible for the Emergence of Biological Periodicity

1 Biological periodicity could not occur if the genes of most living organisms were not essentially similar. The sequenced genomes of algae (*Chlamydomonas*), flowering plants (*Arabidopsis*) and humans were compared revealing that their gene numbers are 15,143, 26,341 and circa 23,000 respectively. Besides many gene families can be easily recognized in these 3 types of organisms (Merchant *et al.* 2007).

2 Before the sudden re-emergence of a given property was considered a "paradoxical event". Biological periodicity could not even be contemplated because the genes directly responsible for a given structure and function were not known. Now a single gene, isolated in the test tube, such as *Pax6*, decides vision in nearly all animals.

3 Plants were considered to be genetically far away from animals and as a consequence could not be included in the same periodicity. The placenta which is a general feature of a whole group of mammals, could not be compared to that of plants, but its emergence in these totally unrelated groups is now known to be decided by common genes.

4 The ordered evolution of an organ, such as the eye, has been guided by self-assembly, a pure atomic event, independent of external intervention.

5 Coherence of structures and functions building a harmonious organ with high efficiency are most evident in plant carnivory, flight, mental ability and vision.

6 The novel organs appear as "ready made". Extra pairs of wings in flies and extra pairs of wings in birds are formed, from the onset, with the correct skeleton, articulations, muscles and blood vessels. The extra legs of insects are also organized in the same way consisting of the identical segments and articulations (Bender *et al.* 1983, Lawrence 1992, Gehring 1998).

7 The sudden absence of a property as well as its equally sudden reappearance is due to the repression and the expression of a gene or gene cluster. Gene repression and expression are decided by changes within the DNA of chromosomes but also by the intervention of small RNAs. This is the case in most properties studied.

8 The same genetic constitution which is present within the same animal or within the same plant, allows or blocks the occurrence of a given property. The same individual plant produces side by side carnivorous and non-carnivorous leaves. Among annelids, brachiopods and bryozoans the adults have no eyes whereas the larvae have eyes.

9 The resurfacing of the same property may occur again and again. Bioluminescence is estimated to have evolved independently at least 40 times and plant carnivory evolved also independently on 10 occasions.

The Periodic Law at the Level of Chemical Elements Leads to the Law of Biological Periodicity

Scerri (2007) who treated extensively the periodicity of the chemical elements in his work "The Periodic Table — Its Story and Its Significance" formulated the Law of Periodicity at the chemical level as follows: "The periodic law states that after certain regular and varying intervals the chemical elements show an approximate repetition in their properties". To be noted is that the intervals are regular but "varying" and that the repetition is "approximate". And he adds "periodicity among the elements is neither constant nor exact". However, he emphasizes that there is a fundamental relationship among the elements.

The evidence gathered at the molecular level that unifies living organisms independently of their complexity or phylogenetic location, exposes a periodicity that can also be concretized in the form of a law.

The law of biological periodicity states that after certain regular and varying intervals the living organisms show an approximate repetition in their properties.

The similarity between the periodicity at the atom level and that of living organisms is so striking that the law of biological periodicity is a paraphrase of the law of the chemical elements. This is astonishing and could hardly be expected since we tend to think about biological phenomena following antiquated ways of reasoning far removed from the laws of physics and chemistry.

Now, when the evidence is so compelling, one thinks that it could not be otherwise.

The Predictive Power of the Law of Biological Periodicity

The main value of the Periodic Table, when it was established by Mendeleev, was that it contained empty spaces where atoms were missing. He understood that these were not faults to be disposed of as accidents or exceptions to be concealed "under the carpet". On the contrary he saw them as a sign that there were chemical elements that had not yet been discovered that would fit into these empty squares.

This was Mendeleev's great contribution, he not only predicted the existence of these atoms but also stated clearly the properties that they ought to have. They were soon found corroborating his prediction on the existence of a common order.

As expected, the situation at the organism level is more complex, making prediction less evident, but the following may be asserted.

1. Since periodicity appears as a sole product of atomic and molecular inherent evolution it is expected to extend to other properties than those treated in this work.
2. It is equally expected that, with time, periodicity will become better documented, and as a consequence better understood bringing closer structures and functions that previously seemed to be unrelated.
3. Contrary to the present dominating interpretation of evolution, novel structures and functions are expected to arise from the start by being built by a coherent combination of parts and organs, since they are the product of molecular cascades originating in DNA sequences.
4. Equally the emergence of novel traits is not considered to be the result of numerous random mutations accompanied by long periods of selection. Instead novel traits arise again and again, irrespective of whether the organism likes them or not, because DNA and RNA follow only the laws of physics and chemistry.
5. The electron is a prisoner of its energy states. Electrons can jump from one orbital to another but they cannot fall in between orbitals as already demonstrated by Bohr in 1913 (David *et al.* 2002). There are alternatives due to the existence of movement (otherwise there would have been no evolution) but the energy alternatives of elementary particles are, from the beginning, not of all kinds. The canalization of evolution is present at the very dawn of matter and life.

Prediction is always a difficult process, it only allows inroads into certain aspects of a given phenomenon. As Scerri emphasized "Almost nobody, including Mendeleev, had predicted or even suspected the existence of an entire family of new elements", the inert or noble gases.

References

Abercrombie, M. *et al.* 1990. The New Penguin Dictionary of Biology. Penguin Books, London, UK.

Affolter, M. *et al.* 1990. Homeodomain proteins and the regulation of gene expression. Curr. Opin. Cell Biol. 2: 485–495.

Airth, R.L. 1961. Characteristics of cell-free fungal bioluminescence. In: Light and Life. Eds. W.D. McElroy and B. Glass. Johns-Hopkins Press: 262, Baltimore, U.S.A.

Alaphilippe, F. *et al.* 1960. Sur un cas de Diphallie symétrique chez *Helix aspersa* Müller. J. de Conchyliologie 100: 163–167.

Albert, V.A. *et al.* 1992. Carnivorous plants: phylogeny and structural evolution. Science 257: 1491–1495.

Alberts, B. *et al.* 1983. Molecular Biology of the Cell. Garland Publishing, New York, U.S.A.

Altshuler, D.L. *et al.* 2015. The biophysics of bird flight: functional relationships integrate aerodynamics, morphology, kinematics, muscles, and sensors. Can. J. Zool. 93: 961–975.

Arnold, E.N. 1986. The hemipenis of lacertid lizards (Reptilia: Lacertidae): structure, variation and systematic implications. J. of Natural History 20: 1221–1257.

Aron, M. and Grassé, P. 1939. Precis de Biologie Animale. Masson and Co., Editeurs, Paris, France.

Bachiller, D. *et al.* 1994. Conservation of a functional hierarchy between mammalian and insect *Hox/Hom* genes. EMBO J. 13: 1930–1941.

Baker, R. Ed. 1980. The Mystery of Migration. Macdonald and Jane's, London, UK.

Balusca, F. *et al.* 2006. Neurobiological view of plants and their body plan. In: Communication in Plants, Neuronal Aspects of Plant Life. Eds. F. Baluska *et al.* Springer: 19–32, Berlin, Germany.

Barnes, R.D. 1980. Invertebrate Zoology. Saunders College, Philadelphia, U.S.A.

Barth, F.G. 2002. A Spider's World: Senses and Behaviour. Springer, Berlin, Germany.

Baulcombe, D.C. and Dean, C. 2014. Epigenetic regulation in plant responses to environment. Cold Spring Harbor Perspectives in Biology 2014; 6: a 019471.

Bayer, R.J. *et al.* 1996. Phylogenetic relationships in Sarraceniaceae based on *rbcL* and *ITS* sequences. Systematic Botany 21: 121–134.

Beazley, M. 1980. The Atlas of World Wildlife. Rand McNally and Company, The Netherlands.

Beltrami, G. *et al.* 2010. A sky polarization compass in lizards: the central role of the parietal eye. J. Exp. Biol. 213: 2048–2054.

Bender, W. 2008. MicroRNAs in the *Drosophila bithorax* complex. Genes Dev. 22: 14–19.

Bender, W. *et al.* 1983. Molecular genetics of the bithorax complex in Drosophila melanogaster. Science 221: 23–29.

Ben-Yakov, A. *et al.* 2015. Memory retrieval in mice and men. Cold Spring Harb. Perspect. Biol. doi: 10.1101/cshperspect.a021790.

Berking, S. 1997. Pattern formation in Hydrozoa. Naturwissenschaften 84: 381–388.

Biju, S.D. 2009. A novel nesting behaviour of a treefrog, *Rhacophorus lateralis* in Western Ghats, India. Current Science 97: 433–437.

Birnbaum, K.D. and Alvarado, A.S. 2008. Slicing across kingdoms: regeneration in plants and animals. Cell 132: 697–710.

Björn, L.O. 2015. The diversity of eye optics. In: Photobiology (3rd Ed.) Ed. L.O. Björn: 189–204.

Blakeley, M.P. *et al.* 2008. Neutron crystallography: opportunities, challenges, and limitations. Current Opinion in Structural Biology 18(5): 593–600.

Blakely, M. *et al.* 2008. Quantum model of catalysis based on a mobile proton revealed by subatomic X-ray and neutron diffraction studies of *h-aldose reductase*. Proc. Nat. Acad. Sci., U.S.A. 105(6): 1844–1848.

Blamey, M. and Grey-Wilson, C. 1989. The Illustrated Flora of Britain and Northern Europe. Hodder and Stoughton, London, UK.

Blaringhem, L. 1923. Pasteur et le Transformisme. Masson et Cie, Editeurs, Libraires de l'Académie de Médecine, Paris, France.

Bloss, F.D. 1971. Crystallography and Crystal Chemistry. An Introduction. Holt, Reinhart and Winston, New York, U.S.A.

Bohn, G. 1934. Les Invertebrés (Coelentérés et Vers). Leçons de Zoologie et Biologie Generale. Hermann et Co., Paris, France.

Boncinelli, E. *et al.* 1988. Organization of human homeobox genes. Hum. Reprod. 3: 880–886.

Bonner, J.T. 1952. Morphogenesis. An Essay on Development. Princeton University Press, Princeton, NJ, U.S.A.

Bork, P. and Copley, R. 2001. Filling in the gaps. Nature 409: 818–820.

Boyland, M. *et al.* 1989. *Lux C, D* and *E* genes of the *Vibrio fischeri* luminescence operon code for the reductase, transferase and synthetase enzymes involved in aldehyde biosynthesis. Photochem. Photobiol. 49: 681–688.

Boyle, W.A. and Conway, C.J. 2007. Why migrate? A test of the evolutionary precursor hypothesis. Am. Nat. 169: 344–359.

Boysen, S.T. and Berntson, G.G. 1989. Numerical competence in a chimpanzee (*Pan troglodytes*). J. Comparative Psychology 103: 23–31.

Braam, J. 2005. In touch: plant responses to mechanical stimuli. New Phytologist 165: 373–389.

Braam, J. and Davis, R.W. 1990. Rain-, wind- and touched-induced expression of calmodulin and calmodulin-related genes in *Arabidopsis*. Cell: 357–364.

Brachet, J. 1974. Introduction to Molecular Embryology. Springer-Verlag, Berlin, Germany.

Bradley, D. *et al.* 1997. Inflorescence commitment and architecture in *Arabidopsis*. Science 275: 80–83.

Branden, C. and Tooze, J. 1991. Introduction to Protein Structure. Garland Publishing, Inc., New York, U.S.A.

Brannon, E.M. and Terrace, H.S. 1998. Ordering of the numerosities 1 to 9 by monkeys. Science 282: 746–749.

Brown, T.A. 1999. Genomes. Bios Scientific Publishers, Oxford, UK.

Brunetta, L. and Craig, C.L. 2010. Spider Silk. Yale University Press, New Haven, CT, U.S.A.

Bullar, J.F. 1876. The generative organs of the parasitic Isopoda. J. Anat. and Phys. XI: 118–123.

Burghardt, G.M. 2013. Environmental enrichment and cognitive complexity in reptiles and amphibians: Concepts, review, and implications for captive populations. Applied Animal Behaviour Science 147: 286–298.

Burke, R.D. *et al.* 2006. A genomic view of the sea urchin nervous system. Dev. Biol. 300: 434–460.

Burnie, D. Ed. 2004. Animal. Dorling Kindersley, London, UK.

Cakmak, Y.O. 2009. A review of the potential effect of electroacupuncture and moxibustion on cell repair and survival: the role of heat shock proteins. Acupunct. Med. 27(4): 183–186.

Callaerts, P. *et al.* 2002. HOX genes in the sepiolid squid *Euprymna scolopes*: implications for the evolution of complex body plans. Proc. Natl. Acad. Sci. U.S.A. 99: 2088–2093.

Campbell, A.K. 1988. Chemiluminescence. Principles and Applications in Biology and Medicine. Ellis Horwood, VCH, U.S.A.

Campbell, C.B.G. and Hodos, W. 1991. The *scala naturae* revisited: evolutionary scales and anagenesis in comparative psychology. J. Comp. Psychol. 105: 211–221.

Carrasco, A.E. *et al.* 1984. Cloning of an *X. laevis* gene expressed during early embryogenesis that codes for a peptide region homologous to *Drosopila* homeotic genes. Cell 37: 409–414.

Carroll, R.L. 1988. Vertebrate Paleontology and Evolution. W.H. Freeman and Company, New York, U.S.A.

Carroll, S.B. 2005. Endless Forms Most Beautiful. W.W. Norton and Co., New York, U.S.A.

Carroll, S.B. *et al.* 1995. Homeotic genes and the regulation and evolution of insect wing number. Nature 375: 58–61.

Carter, A.M. 2001. Evolution of the placenta and fetal membranes seen in the light of molecular phylogenetics. Placenta 22: 800–807.

Carter, A.M. and Mess, A. 2007. Evolution of the placenta in eutherian mammals. Placenta 28: 259–262.

Carter, A.M. and Mess, A. 2008. Evolution of the placenta and associated reproductive characters in bats. J. Experimental Zoology (Mol. Dev. Evol.) 310B: 428–449.

Challis, J.R.G. *et al.* Eds. 1979. Prostaglandins. Reproduction in Mammals, Vol. 7. Mechanisms of Hormone Action. Cambridge University Press: 81–116, Cambridge, UK.

Chan, Y.-H.M. and Marshall, W.F. 2012. How cells know the size of their organelles. Science 337: 1186–1189.

Changeux, J.-P. 1985. Neuronal Man. The Biology of Mind. Oxford University Press, Oxford, UK.

Chapman, R.F. 1969. The Insects. Structure and Function. The English Universities Press Ltd., London, UK.

Charlton-Perkins, M. *et al.* 2011. The lens in focus: a comparison of lens development in *Drosophila* and vertebrates. Mol. Genet. Genomics (2011) 286: 189–213.

Charron, J.-B.F. *et al.* 2002. Molecular and structural analyses of a novel temperature stress-induced lipocalin from wheat and *Arabidopsis*. *FEBS* Letters 517: 129–132.

Chen, J.C-H. *et al.* 2012. Direct observation of hydrogen atom dynamics and interactions by ultrahigh resolution neutron protein crystallography. PNAS September 18, 109(38): 15301–15306.

Chen, J.-J. *et al.* 2014. Tissue regeneration after bark girdling: an ideal research tool to investigate plant vascular development and regeneration. Physiologia Plantarum 151: 147–155.

Chesebro, J. *et al.* 2009. Diverging functions of *Scr* between embryonic and post-embryonic development in a hemimetabolous insect *Oncopeltus fasciatus*. Dev. Biol. 329: 142–151.

Choi, Y.S. *et al.* 2003. Genomic structure of the luciferase gene and phylogenetic analysis in the *Hotaria*-group fireflies. Comparative Biochemistry and Physiology, Part B 134: 199–214.

Coenraads, R.R. and Koivula, J.I. 2007. Geologica. H.F. Ullmann, Millenium House Pty, Australia. Swedish translation.

Cohn, M.J. *et al.* 1997. *Hox9* genes and vertebrate limb specification. Nature 387: 97–101.

Colbert, E.H. 1980. Evolution of the Vertebrates. A History of the Backboned Animals through Time. A Wiley-Interscience Publication, John Wiley and Sons, New York, U.S.A.

Colbourne, J.K. *et al.* 2011. The ecoresponsive genome of *Daphnia pulex*. Science 331: 555–561.

Costa, J.T. 1997. Caterpillars as social insects. American Scientist 85: 150–159.

Costa, J.T. 2006. The Other Insect Societies. The Belkmap Press of Harvard University Press, Cambridge, Ma, U.S.A.

Creuzet, S. *et al.* 2005. Patterning the neural crest derivatives during development of the vertebrate head: insights from avian studies. J. Anat. 207: 447–459.

Cronin, T.W. *et al.* 2014. Visual Ecology. Princeton University Press, Princeton, U.S.A.

Cubas, P. *et al.* 1999. An epigenetic mutation responsible for natural variation in floral symmetry. Nature 401: 157–161.

Cullen, B.R. 2003. Nuclear RNA export. J. Cell Sci. 116: 587–597.

Czarkwiani, A. *et al.* 2013. Expression of skeletogenic genes during arm regeneration in the brittle star *Amphiura filiformis*. Gene Expression Patterns 13: 464–472.

Dacke, M. and Srinivasan, M.V. 2008. Evidence for counting in insects. Anim. Cogn. 11: 683–689.

Das, S. 2015. Morphological, molecular, and hormonal basis of limb regeneration across Pancrustacea. Integrative and Comparative Biology. Advance Access, August 20: 1–9.

Davenport, H.W. 1972. Why the stomach does not digest itself. Scientific American 226(1): 87–93.

Davenport, J. 1994. How and why do flying fish fly? Rev. Fish Biol. Fisher. 4: 184–214.

Davenport, J. 2003. Allometric constraints on stability and maximum size in flying fishes: implications for their evolution. J. Fish Biol. 62: 455–463.

David, I. *et al.* 2002. The Cambridge Dictionary of Scientists. Cambridge University Press, Cambridge, UK.

Day, J.C. *et al.* 2004. Evolution of beetle luminescence: the origin of beetle luciferin. Luminescence 19: 8–20.

Deakin, J.E. *et al.* 2013. Reconstruction of the ancestral marsupial karyotype from comparative gene maps. BMC Evolutionary Biology 13: 258.

de Duve, C. 1984. A Guided Tour of the Living Cell, Vol. II. Scientific American Library, Scientific American Books, Inc., New York, U.S.A.

del Hoyo, J. *et al.* Eds. 1992. Handbook of the Birds of the World, Vol. 1. Lynx Edicions, Barcelona, Spain.

del Hoyo, J. *et al.* Eds. 1997. Handbook of the Birds of the World, Vol. 4. Lynx Edicions, Barcelona, Spain.

Denffer, D. von *et al*. 1971. Strasburger's Textbook of Botany. Longman, London, UK.
De Robertis, E. *et al*. 1990. Homeobox genes and the vertebrate body plan. Sci. Am. 1990: 46–52.
Desjardin, D.E. *et al*. 2008. Fungi bioluminescence revisited. Photochem. Photobiol. Sci. 7: 170–182.
de Wet, J.R. *et al*. 1987. Firefly luciferase gene structure and expression in mammalian cells. Molecular and Cellular Biology, Feb. 1987: 725–737.
Dickerson, R.E. and Geis, I. 1979. Chemistry, Matter, and the Universe. An Integrated Approach to General Chemistry. Benjamin/Cummings, Menlo Park, CA, U.S.A.
Dong, P.-X. *et al*. 2008. Cloning, expression and sequence analysis of a luciferase gene from the Chinese firefly *Pyrocoelia pygidialis*. Zoological Research 29: 477–484.
Dorrington, J. 1979. Pituitary and placental hormones. In: Mechanisms of Hormone Action, Reproduction in Mammals, Book 7. Eds. C.R. Austin and R.V. Short. Cambridge University Press: 53–80, Cambridge, UK.
Downing, A.C. 1972. Optical scanning in the lateral eyes of the copepod *Copilia*. Perception 1: 193–207.
Doyle, J.A. *et al*. 1994. Integration of morphological and ribosomal RNA data on the origin of angiosperms. Ann. Missouri Bot. Gard. 81: 419–450.
Dreller, C. and Kirchnert, W.H. 1993. How honeybees perceive the information of the dance language. Naturwissenschaften 80: 319–321.
Duncan, I. 1987. The bithorax complex. Annu. Rev. Genet. 21: 285–319.
Dunlap, P.V. 2009. Bioluminescence microbial. In: Encyclopedia of Microbiology, 3rd Ed. Ed. M. Schoechter. Elsevier: 45–61, Oxford, UK.
Dunlap, P.V. *et al*. 2007. Phylogenetic analysis of host-symbiont specificity and codivergence in bioluminescent symbioses. Cladistics 23: 507–532.
Eckert, R. and Randall, D. 1978. Animal Physiology. W.H. Freeman and Company, San Francisco, CA, U.S.A.
Eilenberg, H. *et al*. 2006. Isolation and characterization of chitinase genes from pitchers of the carnivorous plant *Nepenthes khasiana*. J. Exp. Botany 57: 2775–2784.
Eiraku, M. *et al*. 2011. Self-organizing optic-cup morphogenesis in three-dimensional culture. Nature 472: 51–56.
Ellis, A.G. and Midgley, J.J. 1996. A new plant-animal mutualism involving a plant with sticky leaves and a resident hemipteran insect. Oecologia 106: 478–481.
Emery, N.J. and Clayton, N.S. 2004. The mentality of crows: convergent evolution of intelligence in Corvids and apes. Science 306: 1903–1907.
Eriksson, R. 2015. What do neutrons tell us? /ESS. https://european_spallation_source.se/what-do-neutrons-tell-us2015-08-27.
Esposito, R. *et al*. 2014, 2015. The ascidian pigmented sensory organs: structures and developmental programs. Genesis 53: 15–33 (2014) (2015).
Farley, R.D. 1998. Matrotrophic adaptations and early stages of embryogenesis in the
Desert Scorpion *Paruroctonus mesaensis* (Vaejovidae). J. Morphology 237: 187–211.
Fasken, M.B. and Corbett, A.H. 2005. Process or perish: quality control in mRNA biogenesis. Nat. Struct. Mol. Biol. 12: 482–488.
Feduccia, A. 1999. The Origin and Evolution of Birds. Yale University Press, New Haven, Conn., U.S.A.

Felsenfeld, G. 2014. A brief history of epigenetics. Cold Spring Harb. Perspect. Biol. 2014; 6: a018200.

Ferguson, B.A. *et al.* 2003. Coarse-scale population structure of pathogenic *Armillaria* species in a mixed-conifer forest in the Blue Mountains of northeast Oregon. Can. J. For. Res. 33.

Feschotte, C. and Pritham, E.J. 2007. DNA transposons and the evolution of eukaryotic genomes. Annual Review of Genetics 41: 331–368.

Finn, J.K. *et al.* 2009. Defensive tool use in a coconut-carrying octopus. Curr. Biol. 19(23): R 1069–R 1070.

Fior, J. 2014. Salamander regeneration as a model for developing novel regenerative and anticancer therapies. Journal of Cancer 5: 715–719.

Flemming, A.F. and Blackburn, D.G. 2003. Evolution of placental specializations in viviparous African and South American lizards. J. Experimental Zoology 299A: 33–47.

Foelix, R.F. 1982. Biology of Spiders. Harvard University Press, Cambridge, MA, U.S.A.

Foelix, R.F. 1996. Biology of Spiders. Oxford University Press, Oxford, UK.

Francé, R.H. 1943. La Maravillosa Vida de los Animales. Editorial Labor, Barcelona, Spain.

Frank, L.G. *et al.* 1995. Masculinization costs in hyaenas. Nature 377: 584–585.

Freyer, C. *et al.* 2003. The marsupial placenta: a phylogenetic analysis. J. Experimental Zool. 299A: 59–77.

Frick, J.E. 1998. Evidence of matrotrophy in the viviparous holoturoid echinoderm *Synaptula hydriformis*. Invertebrate Biology 117(2): 169–179.

Friday, A. and Ingram, D.S. Eds. 1985. The Cambridge Encyclopedia of Life Sciences. Cambridge University Press, Cambridge, UK.

Friml, J. 2003. Auxin transport — shaping the plant. Current Opinions in Plant Biology 6: 7–12.

Frohlich, M.W. and Chase, M.W. 2007. After a dozen years of progress the origin of angiosperms is still a great mystery. Nature 450: 1184–1189.

Frohlich, M.W. and Meyerowitz, E.M. 1997. The search for flower homeotic gene homologs in basal angiosperms and gnetales: a potential new source of data on the evolutionary origin of flowers. Int. J. Plant Sci. 158(6 Suppl.): 131–142.

Gagnon, Y.L. 2010. Multifocal spherical fish lenses. Doctoral Thesis, Faculty of Science, Lund University, Lund, Sweden.

Gahr, M. *et al.* 1993. Estrogen receptors in the avian brain: survey reveals general distribution and forebrain areas unique to songbirds. J. Comp. Neurol. 327: 112–122.

Gallie, D.R. and Chang, S.C. 1997. Signal transduction in the carnivorous plant *Sarracenia purpurea*. Plant Physiology 115: 1461–1471.

Gallistel, C.R. 1988. Counting versus subitizing versus the sense of number. Behav. Brain Sci. 11: 585–586.

Gama-Carvalho, M. *et al.* 2014. Regulation of cardiac cell fate by microRNAs: implications for heart regeneration. MPDI. Cells 3(4): 996–1026.

Garb, J.E. *et al.* 2006. Silk genes support the single origin of orb webs. Science 312: 1762.

Garcia-Fernandez, J. and Holland, P.W. 1994. Archetypal organization of the amphioxus *Hox* gene cluster. Nature 370: 563–566.

Garg, S. and Sharp, P.A. 2016. Single-cell variability guided by microRNAs. Science 352: 1390–1391.

Gehring, W.J. 1998. Master Control Genes in Development and Evolution: the Homeobox Story. Yale University Press, New Haven, U.S.A.

Gehring, W.J. 2014. The evolution of vision. *WIREs* Dev. Biol. 2014, 3: 1–40. doi: 10.1002/wdev. 96.

Gehring, W.J. and Ikeo, K. 1999. *Pax6* mastering eye morphogenesis and eye evolution. Trends in Genetics 15: 371–377.

Gehring, W.J. *et al.* 2009. Evolution of the *Hox* gene complex from an evolutionary ground state. Curr. Top. Dev. Biol. 88: 35–61.

Gianini, N.P. 2012. Toward an integrative theory on the origin of bat flight. Evolutionary History of Bats: Fossils, Molecules and Morphology. Eds. G.F. Gunnell and N.B. Simmons. Cambridge Univ. Press, Cambridge, UK.

Gilbert, S.F. 2000. Developmental Biology. Sinauer Associates Publ., Sunderland, MA, U.S.A.

Glover, B.J. 2007. Understanding Flowers and Flowering. Oxford University Press, Oxford, UK.

Gorzelak, P. *et al.* 2011. ^{26}Mg labeling of the sea urchin regenerating spine: Insights into echinoderm biomineralization process. J. Structural Biology 176: 119–126.

Govorin, I.A. 2009. First record of the biphallia phenomenon in male rapa whelk *Rapana venosa* (Mollusca: Gastropoda) in the Black Sea. Mollusca 27: 49–51.

Graf, M.D. and Rochefort, L. 2010. Moss regeneration for fen restoration: field and greenhouse experiments. Restoration Ecology 18: 121–130.

Gray, T. 2009. The Elements. A Visual Exploration of Every Known Atom in the Universe. Black Dog and Leventhal Publishers Inc., New York, U.S.A.

Greenwood, N.N. and Earnshaw, A. 1989. Chemistry of the Elements. Pergamon Press, Oxford, UK.

Greenwood, P.H. and Thomson, K.S. 1960. The pectoral anatomy of *Pantodon buchholzi* Peters (a freshwater flying fish) and the related Osteoglossidae. Proc. Zool. Soc. London 135: 283–301.

Greuet, C. 1965. Structure fine de l'ocelle d'*Erythropsis pavillardi*. Comptes Rendus Acad. Sci. Paris 261: 1904–1907.

Greulach, V.A. 1973. Plant Function and Structure. Macmillan Publishing Co., New York, U.S.A.

Grimaldi, D. and Engel, M.S. 2005. Evolution of the Insects. Cambridge Univ. Press, Cambridge, UK.

Guerrero-Ferreira, R.C. and Nishiguchi, M.K. 2007. Biodiversity among luminescent symbionts from squid of the genera *Uroteuthis*, *Loliolus* and *Euprymna* (Mollusca: Cephalopoda). Cladistics 23: 497–506.

Guilliermond, A. and Mangenot, G. 1941. Precis de Biologie Vegetale. Masson and Co. Editeurs, Paris, France.

Gullan, P.J. *et al.* 2000. The Insects. Blackwell Science, London, UK.

Gullan, P.J. and Cranston, P.S. 2000. The Insects. An Outline of Entomology. Blackwell Science, UK.

Gullan, P.J. and Cranston, P.S. 2005. The Insects. An Outline of Entomology. Blackwell Publishing, London, UK.

Haberlandt, G. 1902. Culturversuche mit isolierten Pflanzenzellen. Sitzungsber. Akademie der Wissenschaften Wien. Mathematisch Naturwissenschaftliche Klasse 111: 69–92.

Hackett, J.D. *et al*. 2004a. Dinoflagellates: a remarkable evolutionary experiment. Amer. J. Botany 91: 1523–1534.

Hackett, J.D. *et al*. 2004b. Migration of the plastid genome to the nucleus in a peridinin dinoflagellate. Current Biology 14: 213–218.

Haddock, S.H.D. *et al*. 2010. Annu. Rev. Mar. Sci. 2: 443– .

Haesler, S. *et al*. 2007. Incomplete and inaccurate vocal imitation after knockdown of *FoxP2* in songbird basal ganglia nucleus area X. PLoS Biol. 5: 2885–2897.

Hakman, I. *et al*. 1985. The development of somatic embryos in tissue cultures initiated from immature embryos of *Picea abies* (Norway Spruce). Plant Science 38: 53–59.

Halder, G.P. *et al*. 1995. Induction of ectopic eyes by targeted expression of the *eyeless* gene in *Drosophila*. Science 267: 1788–1792.

Halliday, T. and Adler, K. Eds. 2004. The New Encyclopedia of Reptiles and Amphibians. Oxford University Press, Oxford, UK.

Halme, A. *et al*. 2010. Retinoids regulate a developmental checkpoint for tissue regeneration in *Drosophila*. Current Biology 20: 458–463.

Halstead, L.B. 1978. The Evolution of the Mammals. Peter Lowe, Eurobook Limited, Oxford, UK.

Hämmerling, J. 1931. Entwicklung und Formbildungsvermögen von *Acetabularia mediterranea*. I. Die normale Entwicklung. Biol. Zbl. 51: 633–647.

Hämmerling, J. 1963. Neo-cytoplasmic interactions in *Acetabularia* and other cells. Annu. Rev. Plant Physiol. 14: 65–92.

Hanström, B. and Johnels, A.G. 1962. Benfiskar. In: Djurens Värld, Band 6, Fiskar: 2. Ed. B. Hanström. Förlagshuset Norden AB, Malmö, Sweden.

Harland, D.P. *et al*. 2012. How jumping spiders see the world. In: How Animals See the World. Eds. O.F. Lazareva *et al*. Oxford University Press: 134–151, Oxford, UK.

Hartmeyer, S. 1998. Carnivory in *Byblis* revisited II: The phenomenon of symbiosis in insect trapping plants. ICPS Newsletter. ICPS home page. Science — December 1998. International Carnivorous Plant Society.

Harvey, E.N. 1952. Bioluminescence. Academic Press, New York, U.S.A.

Hastings, J.W. 2011. Progress and perspectives on bioluminescence: from luminous organisms to molecular mechanisms. In: Chemiluminescence and Bioluminescence: Past, Present and Future. Ed. Aldo Roda, *RSC* Publishing.

Hauser, M.D. 2000. Wild Minds: What Animals Really Think. Holt, New York, U.S.A.

Heald, F. de F. 1898. Conditions for the germination of the spores of bryophytes and pteridophytes. Botanical Gazette 26(1): 25–45.

Hedges, S. 1997. Mongoose's Secret is to Copy its Prey. New Scientist, London, UK.

Heers, A.M. and Dial, K.P. 2012. From extant to extinct: locomotor ontogeny and the evolution of avian flight. Trends in Ecology and Evolution 27: 296–305.

Hejtmancik, J.F. and Shiels, A. 2015. Overview of the lens. Progress in Molecular Biology and Translational Science 134: 119–127.

Herald, F. de F. 1898. A study of regeneration as exhibited by mosses. Botanical Gazette 26: 169–210.

Herbert, A. 2004. The four *Rs* of *RNA*-directed evolution. Nature Genetics 36: 19–25.

Herrera, A.M. *et al*. 2013. Developmental basis of phallus reduction during bird evolution. Current Biology 23: 1–10.

Herring, P. 2002. The Biology of the Deep Ocean. Oxford University Press, Oxford, UK.

Herring, P.J. 1978. Bioluminescence in Action. Academic Press, London, UK.

Herring, P.J. 2007. Sex with the lights on? A review of bioluminescent sexual dimorphism in the sea. J. Mar. Biol. Ass., UK 87: 829–842.

Heslop-Harrison, Y. 1975. Enzyme release in carnivorous plants. In: Lysosomes in Biology and Pathology. Eds. J.T. Dingle and R. Dean. 4: 525–576.

Heywood, V.H. 1978. Flowering Plants of the World. Oxford University Press, Oxford, UK.

Hill, R.W. *et al.* 2004. Animal Physiology. Sinauer Associates, Inc., Sunderland, MA., U.S.A.

Hoch, J.M. and Yuen, B. 2009. An individual barnacle *Semibalanus balanoides* with two penises. J. Crustacean Biology 29: 135–136.

Holland, P. 1992. Homeobox genes in vertebrate evolution. Bioessays 14: 267–273.

Holmes, S. 1979. Henderson's Dictionary of Biological Terms. Longman, London, UK.

Hopkins, P.M. 2001. Limb regeneration in the fiddler crab, *Uca pugilator*: hormonal and growth factor control. Amer. Zool. 41: 389–398.

Hörnschemeyer, T. and Willkommen, J. 2007. The contribution of flight system characters to the reconstruction of the phylogeny of the Pterygota. Arthropod Systematics and Phylogeny 65: 15–23.

Horst, D.J. van der 2003. Insect adipokinetik hormones: release and integration of flight energy metabolism. Comparative Biochemistry and Physiology, Part *B* 136: 217–226.

Hou, Z. *et al.* 2009. Adaptive history of single copy genes highly expressed in the term human placenta. Genomics 93: 33–41.

Houg, A. 2010. Pictured: "Ugly" Chick Born with Four Wings. The Telegraph (2010-11-23) (15 May 2010).

Huebner, E. and Lococo, D.J. 1994. Oogenesis in a placental viviparous onychophoran. Tissue and Cell 26(6): 867–889.

Inácio, A.R. *et al.* 2015. Endogenous IFN-β signaling exerts anti-inflammatory actions in experimentally induced focal cerebral ischemia. J. of Neuroinflammation 12: 211.

International Chicken Genome Sequencing Consortium (incl. L. Andersson and H. Ellegren) 2004. Sequence and comparative analysis of the chicken genome provide unique perspectives on vertebrate evolution. Nature 432: 695–715.

Isaac, A. *et al.* 1998. *Tbx* genes and limb identity in chick embryo development. Development 125: 1867–1875.

Isayama, T. *et al.* 2006. Differences in the pharmacological activation of visual opsins. Visual Neuroscience 23(6): 899–908.

Ishikawa, M. *et al.* 2011. *Physcomitrella* cyclin-dependent kinase *A* links cell cycle reactivation to other cellular changes during reprogramming of leaf cells. Plant Cell 23: 2924–2938.

Jaffe, H.W. 1988. Introduction to Crystal Chemistry. Student Edition. Cambridge University Press, Cambridge, UK.

Jex, A.R. *et al.* 2011. *Ascaris suum* draft genome. Nature 479: 529–533.

Johnson, G.P. and Renzaglia, K.S. 2008. Embryology of *Ceratopteris richardii* (Pteridaceae, tribe Ceratopterideae) with emphasis on placental development. J. Plant Research 121: 581–592.

Johnson-Murray, J.L. 1987. The comparative myology of the gliding membranes of *Acrobates*, *Petauroides* and *Petaurus* contrasted with the cutaneous myology of *Hemibelideus* and *Pseudocheirus* (Marsupialia: Phalangeridae) and with selected gliding Rodentia (Sciuridae and Anamoluridae). Australian J. Zoology 35: 101–113.

Joyce, B.L. *et al.* 2014. Morphology and ploidy level determination of *Pteris vittata* callus during induction and regeneration. BMC Biotechnology 14: 96.

Kalia, R.K. *et al.* 2007. Plantlet regeneration from fascicular buds of seedling shoot apices of *Pinus roxburghii.* Sarg. Biologia Plantarum 51: 653–659.

Kaltschmidt, B. and Kaltschmidt, C. 2015. *NF*-KAPPA*B* in long-term memory and structural plasticity in the adult mammalian brain. Fontiers in Molecular Neuroscience 8(article 69): 1–11.

Kamimura, Y. 2006. Right-handed penises of the earwig *Labidura riparia* (Insecta, Dermaptera, Labiduridae): evolutionary relationships between structural and behavioral asymmetries. J. of Morphology 267: 1381–1389.

Kaneko-Ishino, T. and Ishino, F. 2010. Retrotransposon silencing by DNA methylation contributed to the evolution of placentation and genomic imprinting in mammals. Develop. Growth Differ. 52: 533–543.

Kardong, K.V. 1995. Vertebrates. Wm. C. Brown Publishers, Oxford, UK.

Katzir, G. 1993. Visual mechanisms of prey capture in water birds. In: Vision, Brain, and Behavior in Birds. A Bradford Book. Eds. H.P. Zeigler and H.-J. Bischof. The MIT Press: 301–315, Cambridge, Massachusetts, U.S.A.

Kawakami, A. 2010. Stem cell system in tissue regeneration in fish. Development, Growth and Differentiation 52: 77–87.

Khadra, Y.B. *et al.* 2014. Homeobox genes expressed during echinoderm arm regeneration. Biochem. Genet. 52: 166–180.

Khadra, Y.B. *et al.* 2015. Wound repair during arm regeneration in the red starfish *Echinaster sepositus*. Wound Repair and Regeneration 23: 611–622.

King, B.L. *et al.* 2011. A natural deletion of the *HoxC* cluster in elasmobranch fishes. Science 334: 1517.

King, J.E. 1964. Seals of the World. The British Museum (Natural History), London, UK.

Kizil, C. *et al.* 2011. Adult neurogenesis and brain regeneration in zebrafish. Developmental Neurobiology. Published on line 18 May 2011: 429–461.

Klein, C. and Hurlbut, Jr. C.S. 1985. Manual of Mineralogy (After J.D. Dana). John Wiley and Sons, New York, U.S.A.

Knight, M.R. *et al.* 1991. Transgenic plant aequorin reports the effects of touch and cold-shock and elicitors on cytoplasmic calcium. Nature 352: 524–526.

Köhler, H.-R. *et al.* 2007. Sex steroid receptor evolution and signalling in aquatic invertebrates. Ecotoxicology 16: 131–143.

Korniushin, A.V. and Glaubrecht, M. 2003. Novel reproductive modes in freshwater clams: brooding and larval morphology in Southeast Asian taxa of *Corbicula* (Mollusca, Bivalvia, Corbiculidae). Acta Zoologica (Stockholm) 84: 293–315.

Kossiakoff, A.A. 1982. Protein dynamics investigated by the neutron diffraction — hydrogen exchange technique. Nature 296: 713–721.

Koyanagi, M. *et al.* 2008. Jellyfish vision starts with *cAMP* signaling mediated by opsin-*Gs* cascade. *PNAS* 105: 15576–15580.

Kral, K. 2003. Behavioural-analytical studies of the role of head movements in depth perception in insects, birds and mammals. Behav. Processes 64: 1–12.

Kramer, J. 1979. Orchids. Harry N. Abrams, Inc., Publishers, New York, U.S.A.

Kraus, P. and Lufkin, T. 2006. *Dlx* homeobox gene control of mammalian limb and craniofacial development. Am. J. Med. Genet. Part A 140A: 1366–1374.

Kudo, R.R. 1971. Protozoology. Charles C. Thomas, Illinois, U.S.A.

Kukalova, J. 1970. Revisional study of the order Palaeodictyoptera in the Upper Carboniferous shales of Commentry, France. Part III. Psyche 77: 1–44.

Kumar, J.P. 2012. Building an ommatidium one cell at a time. Developmental Dynamics 241: 136–149.

Lackie, J.M. and Dow, J.A.T. Eds. 1995 The Dictionary of Cell Biology. Academic; Harcourt Brace and Company, London, UK.

Land, M.F. 2012. The evolution of lenses. Ophthalmic and Physiological Optics 32(2012): 449–460.

Land, M.F. and Fernald, R.D. 1992. The evolution of eyes. Annu. Rev. Neurosci. 1992, 15: 1–29.

Lawick-Goodall, J. van 1970. Tool-using in primates and other vertebrates. Adv. Study Behav. 3: 195–249.

Lawrence, P.A. 1992. The Making of a Fly. The Genetics of Animal Design. Blackwell Scientific Publications, Oxford, U.K.

Lefebvre, L. *et al.* 2002. Tools and brains in birds. Behaviour 139: 939–973.

Lehninger, A.L. 1975. Biochemistry. Worth Publishers, Inc., New York, U.S.A.

Lewis, E.B. 1978. A gene complex controlling segmentation in *Drosophila*. Nature 276: 565–570.

Lewis, E.B. 1992. Clusters of master control genes regulate the development of higher organisms. The 1991 Albert Lasker Medical Awards. J. Am. Med. Assoc. 267: 1524–1531.

Li, C.-H. *et al.* 2008. Somatic embryogenesis and plant regeneration in elite genotypes of *Picea koraiensis*. Plant Biotechnol. Rep. 2: 259–265.

Li, H.-B. and Henning, T. 2011. The alignment of molecular cloud magnetic fields with the spiral arms in *M33*. Nature 479: 499–501.

Li, Q. *et al.* 2015. Regeneration across metazoan phylogeny: lessons from model organisms. J. of Genetics and Genomics 42: 57–70.

Li, Wen-Hsing 1997. Molecular Evolution. Sinauer Associates, Inc., Publishers, Sunderland, MA, U.S.A.

Ligrone, R. and Duckett, J.G. 2011. Morphology versus molecules in moss phylogeny: new insights (or controversies) from placental and vascular anatomy in *Oedipodium griffithianum*. Plant Systematics and Evolution 296: 275–282.

Lima-de-Faria, A. 1973. Equations defining the position of ribosomal cistrons in the eukaryotic chromosome. Nature New Biology 241: 136–139.

Lima-de-Faria, A. 1988. Evolution without Selection. Form and Function by Autoevolution. Elsevier, Amsterdam, New York, U.S.A.

Lima-de-Faria, A. 1995. Biological Periodicity. Its Molecular Mechanism and Evolutionary Implications. JAI Press Inc., Greenwich, Connecticut, U.S.A. JAI Press is at present part of Elsevier, Amsterdam, The Netherlands.

Lima-de-Faria, A. 1997. The atomic basis of biological symmetry and periodicity. BioSystems 43: 115–135.

Lima-de-Faria, A. 2001. Genetic mechanisms involved in the periodicity of flight. Caryologia 54: 189–208.

Lima-de-Faria, A. 2003. One Hundred Years of Chromosome Research and what Remains to be Learned. Kluwer Academic Publishers, Dordrecht, London, UK.

Lima-de-Faria, A. *et al.* 1977. Fusion of human cells with Haplopappus protoplasts by means of Sendai virus. Hereditas 87: 57–66.

Lima-de-Faria, A. *et al.* 1983. Chromosome divisions and nuclear fusion in cell hybrids *Homo-Haplopappus*. Memorias da Academia das Ciencias de Lisboa 1985, 26: 235–244.

Llinás, R.R. 2002. I of the Vortex. From Neurons to Self. A Bradford Book, The *MIT* Press, Cambridge, MA, U.S.A.

Lodish, H. *et al.* 2000. Molecular Cell Biology. Media Connected, W.H. Freeman and Company, New York, U.S.A.

Logan, M. *et al.* 1998. Differential regulation of T-box and homeobox transcription factors suggests roles in controlling chick limb-type identity. Development 125: 1825–1835.

Losick, R. and Kaiser, D. 1997. Why and how bacteria communicate. Scientific American, February 1997: 52–57.

Louisy, P. 2015. European and Mediterranean Marine Fish. Identification Guide. Les Editions Eugen Ulmer, Paris, France.

Lovern, M.B. *et al.* 2004. Effects of testosterone on the development of neuromuscular systems and their target tissues involved in courtship and copulation in green anoles (*Anolis carolinensis*). Hormomes and Behavior 45: 295–305.

Lu, P. *et al.* 1996. Identification of a meristem L1 layer-specific gene in Arabidopsis that is expressed during embryonic pattern formation and defines a new class of homeobox genes. Plant Cell 8: 2155–2168.

Lumsden, A. and Keynes, R. 1989. Segmental patterns of neuronal development in the chick hindbrain. Nature 337: 424–428.

Lush, M.E. and Piotrowski, T. 2014. Sensory hair cell regeneration in the zebrafish lateral line. Developmental Dynamics 243: 1187–1202.

Lyon, B.E. 2003. Egg recognition and counting reduce costs of avian conspecific brood parasitism. Nature 422: 495–499.

Macdonald, D. Ed. 1984. The Encyclopedia of Mammals: Vols. 1 and 2. George Allen & Unwin, London, UK.

Macdonald, D. Ed. 2002. The New Encyclopedia of Mammals. Oxford University Press, Oxford, UK.

Maginnis, T.L. 2006. The costs of autotomy and regeneration in animals: a review and framework for future research. Behavioral Ecology. Advance Access publication 19 June 2006: 857–872.

Margulis, L. and Schwartz, K. 1982. Five Kingdoms. W.H. Freeman and Co., San Francisco, CA., U.S.A.

Marler, P. 1996. Social cognition. Are primates smarter than birds? In: Current Ornithology, Vol. 13. Eds. V. Jr Nolan and E.D. Ketterson. Plenum Press: 1–32, New York, U.S.A.

Martin, G. *et al.* 1989. Epidermal focussing and the light microenvironment within leaves of *Medicago sativa*. Physiologia Plantarum 76: 485–492.

Matzke, M. and Matzke, A.J.U. 2003. RNAi extends its reach. Science 301: 1060–1061.

Matzke, M.A. *et al.* 2015. RNA-directed DNA methylation: the evolution of a complex epigenetic pathway in flowering plants. Annu. Rev. Plant Biol. 66: 243–267.

Mazurs, E.G. 1974. Graphic Representations of the Periodic System During One Hundred Years. The University of Alabama Press, University, Alabama, U.S.A.

Mccampbell, K.K. and Wingert, R.A. 2014. New tides: using zebrafish to study renal regeneration. Translational Research 163: 109–122.

McCracken, K.G. 2000. The 20-cm spiny penis of the Argentine Lake Duck (*Oxyura vittata*). The AUK 117: 820–825.

McElroy, W.D. 1964. Insect bioluminescence. In: The Physiology of Insecta, Vol 1. Ed. M. Rockstein. Academic Press: 463–508, New York, U.S.A.

McFarland, D. Ed. 1981. The Oxford Companion to Animal Behaviour. Oxford Univ. Press, Oxford, UK.

McGinnis, W. *et al.* 1984a. A conserved DNA sequence in homeotic genes of the *Drosophila* Antennapedia and bithorax complex. Nature 308: 428–433.

McGinnis, W. *et al.* 1984b. Molecular cloning and chromosome mapping of a mouse DNA sequence homologous to homeotic genes of *Drosophila*. Cell 38: 675–680.

McGinnis, W. and Kuziora, M. 1994. The molecular architects of body design. Sci. Am. 270: 36–42.

McPherson, S. 2010. Carnivorous Plants and their Habitats. Redfern Natural History Productions, Dorset, UK.

Mebs, D. 2002. Venomous and Poisonous Animals. *CRC* Press, Boca Raton, U.S.A.

Meier, R. *et al.* 1999. Ovoviviparity and viviparity in the Diptera. Biol. Rev. 74: 199–258.

Merchant, S.S. *et al.* 2007. The *Chlamydomonas* genome reveals the evolution of key animal and plant functions. Science 318: 245–251.

Mess, A. and Carter, A.M. 2007. Evolution of the placenta during the early radiation of placental mammals. Comparative Biochemistry and Physiology. Part A 148: 769–779.

Meyer-Rochow, V.B. 2007. Glowworms: a review of *Arachnocampa* spp. and kin. Luminescence 22: 251–265.

Moczec, A.P. 2011. The origins of novelty. Nature 473: 34–35.

Molnar, A. *et al.* 2010. Small silencing RNAs in plants are mobile and direct epigenetic modification in recipient cells. Science 328: 872–875.

Moosbrugger, M. *et al.* 2012. The placental analogue and the pattern of sexual reproduction in the cheilostome bryozoan *Bicellariella ciliata* (Gymnolaemata). Frontiers in Zoology 9: 29.

Morgan, Th.H. 1934. Embryology and Genetics. Columbia University Press, New York, U.S.A.

Morris, K.V. 2015. The theory of RNA-mediated gene evolution. Epigenetics 10(1): 1–5.

Mossman, H.W. 1937. Comparative morphogenesis of the fetal membranes and accessory uterine structures. Carnegie Inst. Washington. Contr. Embryol. 26: 129–246.

Mulvey, J. 1979. The new frontier of particle physics. Nature 278: 403–409.

Murzin, A.G. 1993. Can homologous proteins evolve different enzymatic activities? Trends in Biochemical Sciences 18(11): 403–405.

Nagl, W. 1962. Über Endopolyploidie, Restitutionskernbildung und Kernstrukturen im Suspensor von Angiospermen und einer Gymnosperme. Österr. Bot. Zeitschrift 109(4/5): 431–494.

Nagl, W. 1973. The angiosperm suspensor and the mammalian trophoblast: organs with similar cell structure and function? Soc. Bot. Fr. Mém. Coll. Morphol. 1973: 289–302.

Nagl, W. 1976. Zellkern und Zellzyklen. Verlag Eugen Ulmer, Stuttgart, Germany.

Napier, J.R. and Napier, P.H. 1985. The Natural History of the Primates. British Museum (Natural History), London, UK.

Narita, Y. and Rijli, F.M. 2009. *Hox* genes in neural patterning and circuit formation in the mouse hindbrain. Curr. Top. Dev. Biol. 88: 139–167.

Nelson, D.R. 1986. Quasicrystals. Sci. Am. 255(2): 33–41.

Neufeld, C.J. and Palmer, A.R. 2008. Precisely proportioned: intertidal barnacles alter penis form to suit coastal wave action. Proc. R. Soc. B 275: 1081–1087.

Neumann, C.J. and Nüsslein-Volhard, C. 2000. Patterning of the zebrafish retina by a wave of sonic hedgehog activity. Science 289: 2137–2139.

Niklitschek, A. 1943. Técnica de la Vida. Iniciación al Estudio de la Biologia. Iberia-Joaquin Gil, Barcelona, Spain.

Nilson, J.H. *et al.* 1991. Different combinations of regulatory elements may explain why placenta-specific expression of the glycoprotein hormone α-subunit gene occurs only in primates and horses. Biology of Reproduction 44: 231–237.

Nilsson, D.-E. 1988. A new type of imaging optics in compound eyes. Nature 332: 76–78.

Norberg, U.M. 1990. Vertebrate Flight. Springer-Verlag, Berlin, Germany.

Nye, H.L.D. *et al.* 2003. Regeneration of the urodele limb: a review. Developmental Dynamics 226: 280–294.

Oba, Y. *et al.* 2010. Identification of a functional luciferase gene in the non-luminous diurnal firefly, *Lucidina biplagiata*. Insect Molecular Biology 19(6): 737–743.

Oba, Y. *et al.* 2011. The terrestrial bioluminescent animals of Japan. Zoological Science 28: 771–789.

O'Dor, R. *et al.* 2012. Squid Rocket Science: How Squid Launch into Air. Deep-Sea Research II (2012), http://dx.doi.org/10.1016/j.dsr2.2012.07.002

Ogé, L. *et al.* 2008. Protein repair L-isoaspartyl methyltransferase1 is involved in both seed longevity and germination vigor in *Arabidopsis*. Plant Cell 20(11): 3022–3037.

Ohno, S. 1972. So much "Junk" DNA in our genome. In: Evolution of Genetic Systems. Ed. H.H. Smith. Gordon and Breach: 366–370, New York, U.S.A.

Ohuchi, H. *et al.* 1998. Correlation of wing-leg identity in ectopic FGF-induced chimeric limbs with the differential expression of chick of *Tbx5* and *Tbx4*. Development 125: 51–60.

Olberg, R.M. 2012. Visual control of prey-capture flight in dragonflies. Current Opinion in Neurobiology 22: 267–271.

Onuma, Y. *et al.* 2002. Conservation of Pax 6 function and upstream activation by Notch signaling in eye development of frogs and flies. Proc. Nat. Acad. Sci. U.S.A. 99(4): 2020–2025.

Orr, R. 1986. Mammals of Britain and Europe. Peerage Books, London, UK.

Ostrom, J.H. 1986. The cursorial origin of avian flight. In: The Origin of Birds and the Evolution of Flight. Ed. K. Padian. Mem. Calif. Acad. Sci. 8:73–81.

Ostrovsky, A.N. 2013. From incipient to substantial: Evolution of placentotrophy in a phylum of aquatic colonial invertebrates. Evolution 67–5: 1368–1382.

Ow, D.W. *et al.* 1986. Transient and stable expression of the firefly luciferase gene in plant cells and transgenic plants. Science 234: 856–859.

Padamsee, H.S. 2003. Unifying the Universe. IoP. Institute of Physics Publishing, Bristol, U.S.A.

Pagels, H.R. 1982. The Cosmic Code. Michael Joseph, London, UK.

Paterson, J.R. *et al.* 2011. Acute vision in the giant Cambrian predator *Anomalocaris* and the origin of compound eyes. Nature 480: 237–240.

Paxton, J.R. and Eschmeyer, W.N. Eds. 1998. Encyclopedia of Fishes. Natural World. Academic, San Diego, CA., U.S.A.

Peifer, M. *et al.* 1987. The bithorax complex: control of segmental identity. Genes Dev. 1: 891–898.

Perrins, C. 1976. Bird Life. Elsevier, Phaidon, Oxford, UK.

Perrins, C. Ed. 2003. The New Encyclopedia of Birds. Oxford University Press, Oxford, UK.

Petie, R. 2012. Visually guided swimming in box jellyfish. Doctoral Thesis in Integrative Zoology. Lund University, Lund, Sweden.

Pires, M.N. *et al.* 2011. Why do placentas evolve? An evaluation of the life-history facilitation hypothesis in the fish genus *Poeciliopsis*.

Piskurek, O. and Jackson, D.J. 2012. Transposable elements: from DNA parasites to architects of metazoan evolution. Genes 3: 409–422.

Plachno, B.J. *et al.* 2006. Fluorescence labelling of phosphatase activity in digestive glands of carnivorous plants. Plant Biol. 8: 813–820.

Poran, N.S. and Coss, R.G. 1990. Development of antisnake defenses in Californian ground squirrels, I. Behaviour 112: 222–245.

Pough, F.H. *et al.* 2005. Vertebrate Life. Pearson Prentice Hall, Pearson Education International, Upper Saddle River, N.J., U.S.A.

Preston-Mafham, R. 1996. Den Stora Boken om Spindlar. Quarto Publishing, Streiffert Förlag AB, Stockholm, Sweden.

Price, E.R. 2010. Dietary lipid composition and avian migratory flight performance: development of a theoretical framework for avian fat storage. Comparative Biochemistry and Physiology, Part *A* 157: 297–309.

Proffitt, T. *et al.* 2016. Wild monkeys flake stone tools. Nature 539: 85–88.

Provini, P. 2013. Study on the pelvic system of birds and on the origin of flight. Revue de Paléobiologie, Genève 32: 405–411.

Prud'Homme, B. *et al.* 2011. Body plan innovation in treehoppers through the evolution of an extra wing-like appendage. Nature 473: 83–86.

Pulianmackal, A.J. *et al.* 2014. Competence and regulatory interactions during regeneration in plants. Frontiers in Plant Science 5, article 142: 1–16.

Purushothaman, S. *et al.* 2015. Transcriptomic and proteomic analyses of *Amphiura filiformis* arm tissue-undergoing regeneration. J. Proteomics 112: 113–124.

Purves, D. *et al.* Eds. 2012. Neuroscience. Sinauer Associates, Sunderland, MA, U.S.A.

Putnam, N.H. *et al.* 2008. The amphioxus genome and the evolution of the chordate karyotype. Nature 453: 1064–1072.

Qiu, Yin-Long *et al.* 1999. The earliest angiosperms: evidence from mitochondrial, plastid and nuclear genomes. Nature 402(6760): 404–407.

Quiring, R. *et al.* 1994. Homology of the *eyeless* gene of *Drosophila* to the *Small eye* gene in mice and *Aniridia* in humans. Science 265: 785–789.

Rabaud, E. 1932. Zoologie Biologique. Gauthier-Villars Editeurs, Paris, France.

Rabaud, E. 1934. Zoologie Biologique (Fascicule 3). Gauthier-Villars, Paris, France.

Raible, F. *et al*. 2006. Opsins and clusters of sensory G-protein-coupled receptors in the sea urchin genome. Dev. Biol. 300: 461–475.

Randel, N. and Jekely, G. 2016. Phototaxis and the origin of visual eyes. Philosophical Transactions *B*, The Royal Society Publishing 2016: 1–11.

Raven, P.H. *et al*. 1999. Biology of Plants. 6th Edition. W.H. Freeman, U.S.A.

Rayner, J.M.V. 1986. Pleuston: animals which move in water and air. Endeavour, New Series 10(2): 58–64.

Reinardy, H.C. *et al*. 2015. Tissue regeneration and biomineralization in sea urchins: role of *Notch* signaling and presence of stem cell markers. *PLOS ONE* 10(8): e0133860. doi:10.1371/journal.pone.0133860.

Riddihough, G. 2016. Signals in RNA. Science 352: 1407–1416.

Rinn, J. and Guttman, M. 2014. RNA and dynamic nuclear organization. Science 345: 1240.

Rintelen, T. von and Glaubrecht, M. 2005. Anatomy of an adaptive radiation: a unique reproductive strategy in the endemic freshwater gastropod *Tylomelania* (Cerithioidea: Pachychilidae) on Sulawesi, Indonesia and its biogeographical implications. Biol. J. of the Linnean Soc. 85: 513–542.

Robertson, A.B. *et al*. 2009. Base excision repair: the long and short of it. Cell Mol. Life Sci. 66: 981–993.

Romer, A.S. and Parsons, T.S. 1978. The Vertebrate Body. W.B. Saunders Co., Philadelphia, U.S.A.

Romoser, W.S. 1973. The Science of Entomology. Macmillan Publ. Co., New York, U.S.A.

Ruiz-Medrano, R. *et al*. 1999. Phloem long-distance transport of *CmNACP* mRNA: implications for supracellular regulation in plants. Development 126: 4405–4419.

Russel-Hunter, W.D. 1979. A Life of Invertebrates. Macmillan Publishing Co., New York, U.S.A.

Ryan, C.A. 1973. Proteolytic enzymes and their inhibitors in plants. Ann. Rev. Plant Physiol. 24: 173–196.

Rydén, L.G. and Hunt, L.T. 1993. Evolution of protein complexity: the blue coppercontaining oxidases and related proteins. J. Mol. Evol. 36: 41–66.

Sabri, A. *et al*. 2011. Placenta-like structure of the aphid endoparasitic wasp Aphidius ervi: A strategy of optimal resources acquisition. PLoS One 6(4): e18847.

Salvini-Plawen, L.V. and Mayr, E. 1977. On the evolution of photoreceptors and eyes. Evol. Biology 10: 207–263.

Sanderson, R.T. 1967. Inorganic Chemistry. Reinhold Publishing Co., New York, U.S.A.

Sawaya, R.J. and Sazima, I. 2003. A new species of *Tantilla* (Serpentes; Colubridae) from southeastern Brazil. Herpetologica 59: 119–126.

Scarf, D. *et al*. 2011. Pigeons on par with primates in numerical competence. Science 334: 1664.

Scerri, E.R. 2007. The Periodic Table. Oxford University Press, New York, U.S.A.

Schloegl, C. *et al*. 2012. Grey parrots use inferential reasoning based on acoustic cues alone. Proc. R. Soc. *B* 279: 4135–4142.

Schnitzler, C.E. *et al*. 2012. Genomic organization, evolution, and expression of photoprotein and opsin genes in *Mnemiopsis leidyi*: a new view of ctenophore photocytes. *BMC* Biology 10: 107.

Schuerges, N. *et al.* 2016. Cyanobacteria use micro-optics to sense light direction. e-Life 2016; 5:e 12620. *DOI*: 10.7554/eLife.12620.

Schwab, I.R. and O'Connor, G.R. 2005. The lonely eye. Br. J. Ophtalmol. 89: 256.

Schwabe, J.W.R. and Travers, A.A. 1993. DNA-binding proteins: What is evolution playing at? Current Biology 3(9): 628–630.

Seed, A. *et al.* 2009. Intelligence in Corvids and apes: a case of convergent evolution? Ethology 115: 401–419.

Seifert, A.W. *et al.* 2012. The influence of fundamental traits on mechanisms controlling appendage regeneration. Biol. Rev. 87: 330–345.

Semyonov, D.V. and Dunayev, Y.A. 1989. Publication's name in Russian letters. Vol. LXVIII(10): 56–63.

Sena, G. and Birnbaum, K.D. 2010. Built to rebuild: in search of organizing principles in plant regeneration. Current Opinion in Genetics and Development 20: 460–465.

Sereno, P.C. and Chenggang, R. 1992. Early evolution of avian flight and perching: new evidence from the lower Cretaceous of China. Science 255: 845–848.

Serov, O. and Serova, I. 2004. Genes and chromosomes: control of development. Annals of the Brazilian Academy of Sciences 76: 529–540.

Sewell, M.A. *et al.* 2006. Evidence for matrotrophy in the viviparous sea cucumber *Leptosynapta clarki*: A role for the genital haemal sinus. Invertebrate Reproduction and Development 49(4): 225–236.

Shapiro, J.A. 2010. Mobile DNA and evolution in the 21st century. Shapiro *Mobile DNA* 2010, 1: 4: 1–14.

Shen, P. *et al.* 1995. An atlas of aromatase mRNA expression in the zebra finch brain. J. Comp. Neurol. 360: 172–184.

Shepherd, W. 1939. Science Marches on. Harrap, London, UK.

Shimonura, O. 2006. Bioluminescence. Chemical Principles and Methods. World Scientific, New Jersey, U.S.A.

Simpson, G.G. and Beck, W.S. 1965. Life. An Introduction to Biology. Harcourt, Brace and World Inc., New York, U.S.A.

Slack, A. 1979. Carnivorous Plants. Ebury Press, London, UK.

Smith, S.A. and Donoghue, M.J. 2008. Rates of molecular evolution are linked to life history in flowering plants. Science 322: 86- ?.

Spector, D.L. 1984. Dinoflagellate nuclei. In: Dinoflagellates. Ed. D.L. Spector. Academic Press: 107–147, Orlando, CA., U.S.A.

Spielman, M. *et al.* 2001. The epigenetic basis of gender in flowering plants and mammals. Trends in Genetics 17(12): 705–711.

Spudich, J.L. 2006. The multitalented microbial sensory rhodopsins. Trends Microbiology 2006, 14: 480–487.

Stavenga, D.G. 1989. Pigments in compound eyes. In: Facets of Vision. Eds. D.G. Stavenga and R.C. Hardie. Springer-Verlag: 152–172, Berlin, Germany.

Stavenga, D.G. and Kuiper, J.W. 1977. Insect pupil mechanisms. I. On the pigment migration in the retinula cells of Hymenoptera (suborder Apocrita). J. Comp. Physiol. 113: 55–72.

Stenkamp, D.L. 2015. Development of the vertebrate eye and retina. Progress in Molecular Biology and Translational Science 134: 397–414.

Sternberg, S.H. and Doudna, J.A. 2015. Expanding the biologist's toolkit with *CRISPR-Cas9*. Molecular Cell 58: 568–574.

Stewart, C.N. Jr. 2006. Go with the glow: fluorescent proteins to light transgenic organisms. Trends in Biotechnology 24(4): 155–162.

Striedter, G.F. 2004. Brain evolution. In: The Human Nervous System. Eds. G. Paxinos and J.K. Mai. Elsevier: 3–21, New York, U.S.A.

Stubbe, H. 1938. Genmutation. Handbuch der Vererbungswissenschaft. ZF, Berlin, Germany.

Stubbe, H. 1966. Genetik und Zytologie von *Antirrhinum L.* Fischer, Jena, Germany.

Su, Y.H. and Zhang, X.S. 2014. The hormonal control of regeneration in plants. Current Topics in Developmental Biology 108: 35–69.

Suarez-Castillo, E.C. *et al.* 2004. Ependymin, a gene involved in regeneration and neuroplasticity in vertebrates, is overexpressed during regeneration in the echinoderm *Holothuria glaberrima*. Gene 334: 133–143.

Sugimoto, K. *et al.* 2011. Regeneration in plants and animals: dedifferentiation, transdifferentiation or just differentiation? Trends in Cell Biology 21: 212–218.

Sun, L.N. *et al.* 2013. Cloning and expression analysis of *Wnt6* and *Hox6* during intestinal regeneration in the sea cucumber *Apostichopus japonicus*. Genetics and Molecular Research 12: 5321–5334.

Tachibana, M. *et al.* 2013. Human embryonic stem cells derived by somatic cell nuclear transfer. Cell 153: 1–11.

Taiz, L. and Zeiger, E. 1998. Plant Physiology. Sauer Associates Inc., Publishers, Sunderland, MA, U.S.A.

Takahashi, K. *et al.* 1974. Specific inhibition of acid proteases from brain, kidney, skeletal muscle, and insectivorous plants by diazoacetyl-DL-norleucine methyl ester and by pepstatin. J. Biochem. 76: 897–899.

Terry, L.J. *et al.* 2007. Crossing the nuclear envelope: hierarchical regulation of nucleocytoplasmic transport. Science 318: 1412–1416.

Teschke, I. 2011. Sometimes tool use is not the key: no evidence for cognitive adaptive specializations in tool-using woodpecker finches. Animal Behaviour 82: 945–956.

Thompson, M.B. and Speake, B.K. 2006. A review of the evolution of viviparity in lizards: structure, function and physiology of the placenta. J. Comp. Physiol. *B* 176: 179–189.

Titushin, M.S. *et al.* 2011. Protein-protein complexation in bioluminescence. Protein Cell 2(12): 957–972.

Tokarz, R.R. 1988. Copulatory behaviour of the lizard *Anolis sagrei*: alternation of hemipenis use. Anim. Behav. 36: 1518–1524.

Tokarz, R.R. 1989. Pattern of hemipenis use in the male lizard *Anolis sagrei* after unilateral castration. The J. Experimental Zoology 250: 93–99.

Tökés, Z.A. *et al.* 1974. Digestive enzymes secreted by the carnivorous plant *Nepenthes* macferlanei L. Planta 119: 39–46.

Tornini, V.A. and Poss, K.D. 2014. Keeping at arm's length during regeneration. Developmental Cell 29: 139–143.

Trewanas, A. 2006. The green plant as an intelligent organism. In: Communication in Plants, Neuronal Aspects of Plant Life. Eds. F. Baluska *et al.* Springer: 1–18, Berlin, Germany.

Tunlid, A. 2004. Ärftlighetsforsknings gränser. Lunds Universitet, Ugglan, Minervaserien 11.

Uddin, M. *et al.* 2008. Distinct genomic structures of adaptation in pre- and postnatal environments during human evolution. Proc. Natl. Acad. Sci. U.S.A. 105: 3215–3220.

Ullrich-Lüter, E.M. *et al.* 2011. Unique system of photoreceptors in sea urchin tube feet. *PNAS* 108: 8367–8372.

Vallès, Y. and Gosliner, T.M. 2006. Shedding light onto the genera (Mollusca, Nudibranchia) *Kaloplocamus* and *Plocamopherus* with description of new species belonging to these unique bioluminescent Dorids. The Veliger 48(3): 178–205.

van Wyk, B.-E. and Wink, M. 2004. Medicinal Plants of the World. Briza Publications, Pretoria, South Africa.

Venkitaraman, A.R. 2005. Aborting the birth of cancer. Nature 434: 829–830.

Vogel, P. 2005. The current molecular phylogeny of eutherian mammals challenges previous interpretations of placental evolution. Placenta 26: 591–596.

Vogelmann, T.C. *et al.* 1989. Photosynthetic light gradients and spectral régime within leaves of *Medicago sativa*. Phil. Trans. R. Soc. Lond. B 323: 411–421.

von Bayer, H.C. 1992. Taming the Atom. Random House, New York, U.S.A.

Wada, S. *et al.* 2012. Expression of UV-sensitive parapinopsin in the iguana parietal eyes and its implication in UV-sensitivity in vertebrate pineal-related organs. PLoS ONE 7(6): e39003: 7.

Walker, C. 1974. Introduction: the world of birds. In: The World Atlas of Birds. Mitchell Beazley Publishers: 10–33, London, UK.

Wampler, J.E. 1978. Measurements and physical characteristics of luminescence. In: Bioluminescence in Action. Ed. P.J. Herring. Academic Press: 1–48, New York, U.S.A.

Warren, W.C. *et al.* 2008. Genome analysis of the platypus reveals unique signatures of evolution. Nature 453: 175–183.

Webster, N. 1976. Webster's New Twentieth Century Dictionary of the English Language Unabridged. Collins World, U.S.A.

Weinberg, S. 2015. To Explain the World. Penguin Random House, UK.

Weitz, H.J. 2004. Naturally bioluminescent fungi. Mycologist 18: 4.

Wellik, D.M. 2009. *Hox* genes and vertebrate axial pattern. Curr. Top. Dev. Biol. 88: 257–277.

Wells, M. 1983. Cephalopods do it differently. New Scientist 100: 332–337.

Wenk, H.-R. and Bulakh, A. 2004. Minerals. Their Constitution and Origin. Cambridge University Press, Cambridge, UK.

Widder, E.A. 2010. Bioluminescence in the ocean: origins of biological, chemical, and ecological diversity. Science 328: 704–708.

Wildman, D.E. 2011. Review: Toward an integrated evolutionary understanding of the mammalian placenta. Placenta 32. Supplement B, Trophoblast Research 25: 5142–5145.

Wildman, D.E. *et al.* 2006. Evolution of the mammalian placenta revealed by phylogenetic analysis. P.N.A.S., U.S.A., 103: 3203–3208.

Williams, S.E. *et al.* 1994. Relationships of Droseraceae: A cladistic analysis of rbcL sequence and morphological data. American Journal of Botany 81: 1027–1037.

Williford, A. *et al.* 2004. Evolution of a novel function: nutritive milk in the viviparous cockroach, *Diploptera* punctata. Evolution and Development 6–2: 67–77.

Wilmut, I. *et al.* 1997. Viable offspring derived from fetal and adult mammalian cells. Nature 385: 810–813.

Wilson, D.E. and Mittermeier, R.A. Eds. 2009. Handbook of the Mammals of the World, Vol. 1. Carnivores. Lynx Edicions, Barcelona, Spain.

Wilson, D.E. and Mittermeier, R.A. Eds. 2014. Handbook of the Mammals of the World, Vol. 4. Sea Mammals. Lynx Edicions, Barcelona, Spain.

Wilson, D.E. and Mittermeier, R.A. Eds. 2015. Handbook of the Mammals of the World, Vol. 5. Monotremes and Marsupials. Lynx Edicions, Barcelona, Spain.

Wilson, D.E. *et al.* Eds. 2016. Handbook of the Mammals of the World, Vol. 6. Lagomorphs and Rodents I. Lynx Edicions, Barcelona, Spain.

Withers, P.C. 1992. Comparative Animal Physiology. Saunders College, New York, U.S.A.

Wooding, P. and Burton, G. 2008. Comparative Placentation. Springer, Berlin, Germany.

Wourms, J.P. 1993. Maximization of evolutionary trends for placental viviparity in the spadenose shark *Scoliodon laticaudus*. Environmental Biology of Fishes 38: 269–294.

Wourms, J.P. *et al.* 1988. The maternal-embryonic relationship in viviparous fishes. Fish Physiology XIB: 1–134.

Xiao, L. *et al.* 2012. Transcriptome of protoplasts reprogrammed into stem cells in *Physcomitrella patens*. *PLoS One* 7, e 35961.

Xu, L. and Huang, H. 2014. Genetic and epigenetic controls of plant regeneration. Current Topics in Dev. Biol. 108: 1–33.

Xu, X.M. and Møller, S.G. 2011. The value of *Arabidopsis* research in understanding human disease states. Current Opinion in Biotechnology 22: 300–307.

Yang, G. *et al.* 2010. Molecular characterization of two novel milk proteins in the tsetse fly (Glossina morsitans morsitans). Insect Molecular Biology 19(2): 253–262.

Yashina, S. *et al.* 2012. Regeneration of whole fertile plants from 30,000-y-old fruit tissue buried in Siberian permafrost. PNAS 109: 4008–4013.

Yoshida, M.A. *et al.* 2015. Molecular evidence for convergence and parallelism in evolution of complex brains of cephalopod molluscs: insights from visual systems. Integrative and Comparative Biology 55: 1070–1083.

Young, N.D. *et al.* 2011. The *Medicago* genome provides insight into the evolution of rhizobial symbioses. Nature 480: 520–523.

Zaniolo, G. *et al.* 1998. Brood pouch differentiation in *Botrylloides violaceus* a viviparous ascidian (Tunicata). Invertebrate Reproduction and Development 33(1): 11–23.

Zenkteler, M. and Relska-Roszak, D. 2003. Bidirectional pollination of angiosperm and gymnosperm ovules. Acta Biologica Cracoviensia, Series Botanica 45: 77–81.

Zhang, U. *et al.* 1993. Ectopic *HoxA-1* induces rhombomere transformation in mouse hindbrain. Development 120: 2431–2442.

Zhou, Z. 2004. The origin and early evolution of birds: discoveries, diputes, and perspectives from fossil evidence. Naturwissenschaften 91: 455–471.

Zoncu, R. *et al.* 2011. *mTORC1* senses lysosomal amino acids through an inside-out mechanism that requires the vascular H^+-ATPase. Science 334: 678–683.

Zybina, T.G. and Zybina, E.V. 2005. Cell reproduction and genome multiplication in the proliferative and invasive trophoblast cell populations of mammalian placenta. Cell Biology International 29: 1071–1083.

Sources of Figures

Fig. 1.1 From "*LUM*" Lunds Universitets Meddelande 2016. (Nr. 2) (Page 6).

Fig. 1.2 Eriksson, R. 2015. What do Neutrons Tell Us?/ESS. https://europeanspallationsource.se/what-do-neutrons-tell-us_2015-08-27.

Fig. 1.3 von Baeyer, H.C. 1992. Taming the Atom. Random House, New York, U.S.A. (Plate following page 102).

Fig. 2.1 Padamsee, H.S. 2003. Unifying the Universe. Institute of Physics Publishing, Bristol and Philadelphia, U.S.A. (Fig. 2.17, page 134).

Fig. 2.2 (**1**) Sanderson, R.T. 1967. Inorganic Chemistry. Reinhold Publ. Co., New York, U.S.A. (Page 14). (**2**) After Benfy, T. 1964. Spiral periodic chart. In: Chemistry. Benfy, T. (Editor) page 14, Amer. Chemical Soc. Washington, D.C., 37(6): 14. From Jaffe, H.W. 1988. Introduction to Crystal Chemistry. Cambridge Univ. Press, Cambridge, UK (Fig. page 11).

Fig. 3.1 (**1**) Wettstein, R. 1944. Tratado de Botanica Sistematica. Labor, Barcelona, Spain (Fig. 604, page 843). (**2**) Wettstein, R. 1944. Tratado de Botanica Sistematica. Labor, Barcelona, Spain (Fig. 470, page 666). (**3**) Wettstein, R. 1944. Tratado de Botanica Sistematica. Labor, Barcelona, Spain (Fig. 470, page 666). (**4**) Gola, G. *et al.* 1943. Tratado de Botanica. Labor, Barcelona, Spain (Fig. 84, page 95). (**5**) Guilliermond, A. and Mangenot, G. 1941. Precis de Biologie Vegetale. Masson Ed., Paris, France (Fig. 184, page 424).

Fig. 3.2 (**1**) http://www.best-carnivorous-plants.com/CP-Photos/Byblis-gigantea-Copyright-Rosti. (**2**) https://upload.wikimedia.org/wikipedia/commons/thumb/9/9a/Roridula-dentata-01.jp. (**3**) Guilliermond, A. and Mangenot, G. 1941. Precis de Biologie Vegetale. Masson Ed., Paris, France (Fig. 185, page 425). (**4**) and (**5**) Gola, G. *et al.* 1943. Tratado de Botanica. Labor, Barcelona, Spain (Fig. 338, page 409). (**6**) Guilliermond, A. and Mangenot, G. 1941. Precis de Biologie Vegetale. Masson Ed., Paris, France (Fig. 186, page 426). (**7**) Gola, G. *et al.* 1943. Tratado de Botanica. Labor, Barcelona, Spain (Fig. 229, page 220).

Fig. 3.3 (**1**) Guilliermond, A. and Mangenot, G. 1941. Precis de Biologie Vegetale. Masson Ed., Paris, France (Fig. 183, page 423). (**2**) Frisch, K.V. 1938. Noi e la Vita. U. Hoepli, Milano, Italy (Fig. 41, page 73). (**3**) Gola, G. *et al.* 1943. Tratado de Botanica. Labor, Barcelona, Spain (Fig. 340, page 412). (**4**) Guilliermond, A. and Mangenot, G. 1941. Precis de Biologie Vegetale. Masson, Ed., Paris, France (Fig. 188, page 428). (**5**) Wettstein, R. 1944. Tratado de Botanica Sistematica. Labor, Barcelona, Spain (Fig. 470, page 666).

Fig. 3.4 McPherson, S. 2010. Carnivorous Plants and their Habitats, Vol. 1. Redfern Natural History Productions, London, UK (Fig. 39, page 75).

Fig. 3.5 (**1**) and (**2**) and (**3**) Wettstein, R. 1944. Tratado de Botanica Sistematica. Labor, Barcelona, Spain (Fig. 604, page 843). (**4**) Francé, R.H. 1943. La Maravillosa Vida de las Plantas. Labor, Barcelona, Spain (Fig. 55, page 302). (**5**) and (**6**) Wettstein, R. 1944. Tratado de Botanica Sistematica. Labor, Barcelona, Spain (Fig. 602, page 841).

Fig. 3.6 (**1**) https://upload.wikimedia.org/wikipedia/commons/thumb/7/7e/Carica-papaya-001.JP (**2**) https://upload.wikimedia.org/wikipedia/commons/d/da/Ghana-pineapple-field.jpg

Fig. 3.7 (**1**) Strasburger, E. 1943. Tratado de Botanica. Manuel Marin, Barcelona, Spain (Fig. 210, page 164). (**2**) Palhinha, R.T. and Cunha, A.G. 1939. Curso de Botanica. J. Rodrigues Editors, Lisboa, Portugal (Fig. 218, page 585). (**3**) http://luirig.altervista.org/cpm/albums/bot-units63.aristolochia-pallida21621.jpg (**4**) https://upload.wikimedia.org/wikipedia/commons/c/cf/Aristolochia-rotunda.jpg.

Fig. 3.8 (**1**) Guilliermond, A. and Mangenot, G. 1941. Precis de Biologie Vegetale. Masson Ed., Paris, France (Fig. 285, page 593). (**2**) Bohn, G. 1935. Les Invertébrés. Leçons de Zoologie et Biologie Generale, IV. Hermann Editeurs, Paris, France (Fig. 68, page 106). (**3**) Schäffer, C. 1942. Curso y Practicas de Biologia. Labor, Barcelona, Spain (Fig. 384, page 313). (**4**) Pierantoni, U. 1944. Tratado de Zoologia. Editorial Labor, Barcelona, Spain (Fig. 945, page 866). (**5**) Aron, M. and Grassé, P. 1939. Precis de Biologie Animale. Masson Ed., Paris, France (Fig. 663, page 1039). (**6**) Francé, R.H. 1943. La Maravillosa Vida de los Animales. Labor, Barcelona, Spain (Fig. 59, page 368).

Fig. 3.9 (**1**) Francé, R.H. 1943. La Maravillosa Vida de las Plantas. Editorial Labor, Barcelona, Spain (Fig. 2, page 6). (**2**) Guilliermond, A. and Mangenot, G. 1941. Precis de Biologie Vegetale. Masson Ed., Paris, France (Fig. 290, page 601). (**3**) Guilliermond, A. and Mangenot, G. 1941. Precis de Biologie Vegetale. Masson Ed., Paris, France (Fig. 291, page 602).

Fig. 3.10 (**1**) After A. Levan. From Müntzing, A. 1961. Genetic Research. LT. Förlag, Stockholm, Sweden (Fig. 162, page 230). (**2**) https://upload.wikimedia.org/wikipedia/commons/0/04/Herbstzeitlose.jpg (**3**) https://upload.wikimedia.org/wikipedia/commons/thumb/a/ad/Dendrobates-auratus (**4**) https://upload.wikimedia.org/wikipedia/commons/thumb/1/1f/L14cobra.jpg/1024px. (**5**) Gola, G. *et al.* 1943. Tratado de Botanica. Editorial Labor, Barcelona, Spain (Fig. 227, page 218). (**6**) http://www.iaeeperth2012.org/wp-content/uploads2015/01/Real-Human-Stomach.jpg.

Fig. 3.11 (**1**) Wilson, D.E. and Mittermeier, R.A. (Editors) 2009. Handbook of the Mammals of the World, Vol. 1. Lynx Edicions, Barcelona, Spain (Plate 17, Nr. 5, page 309). (**2**) Sandars, E. 1937. A Beast Book for the Pocket. Oxford University Press, London, UK (Fig. page 98). (**3**) From *Flickr*, Photographer Calle Lidström. (**4**) From *Flickr*, Photographer Philip Hay. (**5**) From *Flickr*, Photographer Franco Folini.

Fig. 4.1 (**1**) and (**2**) Author Parent Géry, Scheelite sous UVC (Chine) 4.jpg. https:commons.wikimedia.org/wiki/File:Scheelite-sous-UVC(Chine)4.jpg (**3**) and (**4**) George Elling Collection, England. Rob Lavinsky. https://commons.wikimedia.org/wiki/File:Fluorite,calcite,scheelite,quartz-sousU.

Fig. 4.2 Wampler, J.E. 1978. Measurements and physical characteristics of luminescence. In: Bioluminescence in Action. Herring, P.J. (Editor). Academic Press: 1-48, New York, U.S.A.

Fig. 4.3 (**1**) Combes, R. 1938. La biologie vegetale. In: La Biologie. (Palais de la Decouverte), Masson Ed., Paris, France (Fig. page 21). (**2**) Pierantoni, U. 1944. Tratado de Zoologia. Editorial Labor, Barcelona, Spain (Fig. 78, page 107). (**3**) Pierantoni, U. 1944. Tratado de Zoologia. Editorial Labor, Barcelona, Spain (Fig. 204, page 322). (**4**) Francé, R.H. 1943. La Maravillosa Vida de los Animales. Editorial Labor, Barcelona, Spain (Plate XVI, page 193). (**5**) Pierantoni, U. 1944. Tratado de Zoologia. Editorial Labor, Barcelona, Spain (Fig. 78, page 107).

Fig. 4.4 (**1**) Aron, M. and Grassé, P. 1939. Precis de Biologie Animale. Masson Ed., Paris, France (Fig. 557, page 903). (**2**) Frisch, K.V. 1938. Noi e la Vita. U. Hoepli, Milano, Italy (Plate III). (**3**) Frisch, K.V. 1938. Noi e la Vita. U. Hoepli, Milano, Italy (Plate III). (**4**) After Pierantoni. From Aron, M. and Grassé, P. 1939. Precis de Biologie Animale. Masson Ed., Paris, France (Fig. 21, page 530). (**5**) Aron, M. and Grassé, P. 1939. Precis de Biologie Animale. Masson Ed., Paris, France (Fig. 559, page 904).

Fig. 4.5 Ow, D.W. *et al.* 1986. Transient and stable expression of the firefly luciferase gene in plant cells and transgenic plants. Science, New Series 234(4778): 856-859 (Fig. 5, page 858).

Fig. 5.1 (**1**) Perrier, R. 1936. Cours Elementaire de Zoologie. Masson Ed., Paris, France (Fig. 650, page 687). (**2**) Rabaud, E. 1934. Zoologie Biologique. Fascicule III. Gauthier-Villars, Paris, France (Fig. page 660). (**3**) Frisch, K.V. 1938. Noi e la Vita. Ed. U. Hoepli, Milano, Italy (Fig. 182, page 340). (**4**) Frisch, K.V. 1938. Noi e la Vita. Ed. U. Hoepli, Milano, Italy (Fig. 184, page 344).

Fig. 5.2 Wooding, P. and Burton, G. 2008. Comparative Placentation. Structures, Functions and Evolution. Springer, Berlin, Germany (Fig. 9.1, page 241).

Fig. 5.3 (**1**) Nagl, W. 1962. Über Endopolyploidie, Restitutionskernbildung und Kernstrukturen im Suspensor von Angiospermen und einer Gymnosperme. Österr. Bot. Zeitschrift 109: 431-493 (Fig. 9, page 447). (**2**) Pierantoni, U. 1944. Tratado de Zoologia. Editorial Labor, Barcelona, Spain (Fig. 452, page 530). (**3**) https://upload.wikimedia.org/wikipedia/commons/thumb/c/c9/Abedus-herberti.jpg/48 (**4**) https:upload.wikimedia.org/wikipedia/commons/1/12/Lethocerus-americanus23.jpg

Fig. 5.4 Spielman, M. *et al.* 2001. The epigenetic basis of gender in flowering plants and mammals. Trends in Genetics 17(12): 705-711 (Fig. page 706).

Fig. 6.1 (**1**) After Turner and Bagnara 1976. From Romer, A.S. and Parsons, T.S. 1978. The Vertebrate Body. Saunders Company, Philadelphia, U.S.A. (Fig. page 300). (**2**) Romer, A.S. and Parsons, T.S. 1978. The Vertebrate Body. Saunders Company, Philadelphia, U.S.A. (Fig. 136, page 154). (**3**) Frank, L.G. *et al.* 1995. Masculinization costs in hyaenas. Nature 377: 584-585 (Fig. page 585).

Fig. 6.2 (**1**) After Noble and Noble 1976. From Barnes, R.D. 1980. Invertebrate Zoology. Saunders College, Philadelphia, U.S.A. (Fig. page 233). (**2**) Bullar, J.F. 1876. The generative organs of the parasitic isopoda. J. Anat. Phys. XI: 118-123 (Plate IV, page 122). (**3**) Perrier, R. 1936. Cours Elementaire de Zoologie. Masson Ed., Paris, France (Fig. 294, page 392). (**4**) Barnes, R.D. 1980. Invertebrate Zoology. Saunders College, Philadelphia, U.S.A. (Fig. 14.32, page 703).

Fig. 6.3 (**1**) Govorin, I.A. 2009. First record of the biphallia phenomenon in male rapa whelk *Rapana venosa* (Mollusca, Gastropoda) in Black Sea. Mollusca 27(1): 49-51 (Fig. 1, page 50). (**2**) Govorin, I.A. 2009. First record of the biphallia phenomenon in male rapa whelk *Rapana venosa* (Mollusca, Gastropoda) in Black Sea. Mollusca 27(1): 49-51 (Fig. 2, page 50). (**3**) Frisch, K.V. 1938. Noi e la Vita. Ulrico Hoepli Editor, Milano, Italy (Fig. 152, page 291).

Fig. 6.4 (**1**) Sawaya, R.J. and Sazima, I. 2003. A new species of Tantilla (Serpentes: Colubridae) from southeastern Brazil. Herpetologica 59(1): 119-126 (Fig.3, page 123). (**2**) Semyonov, D.V. and Dunayev, Y.A. 1989. Morphology of the hemipenis and classification of lizards of the genus *Phrynocephalus* (Reptilia: Agamidae). Title of journal in

Russian. Volume 68(10): 56-64 (Fig. 2, page 59). (**3**) Tokarz, R.R. 1988. Copulatory behaviour of the lizard *Anolis sagrei*: alternation of hemipenis use. Anim. Behav. 36: 1518-1524 (Fig. 1, page 1519).

Fig. 6.5 Herrera, A.M. *et al.* 2013. Developmental basis of phallus reduction during bird evolution. Current Biology 23: 1-10 (Fig. 4F, page 6).

Fig. 7.1 (**1**) After Pasteur 1857. From Blaringhem, L. 1923. Pasteur et le Transformisme. Masson Ed., Paris, France (Fig. page 53). (**2**) After Lehman. From Bonner, J.T. 1952. Morphogenesis. Princeton Univ. Press, Princeton, NJ, U.S.A. (**3**) After Gruber. From Aron, M. and Grassé, P. 1939. Precis de Biologie Animale. Masson Ed., Paris, France (Fig. 114, page 230). (**4**) After Morgan, from Korschelt. From Aron, M. and Grassé, P. 1939. Precis de Biologie Animale. Masson Ed., Paris, France (Fig. 116, page 231). (**5**) After Lütken. From Bohn, G. 1935. Les Invertébrés IV. Actualités Scient. et Industrielles No. 242. Hermann Ed., Paris, France (Fig. 70, page IV-110). (**6**) After Richters. From Aron, M. and Grassé, P. 1939. Precis de Biologie Animale. Masson Ed., Paris, France (Fig. 118, page 233).

Fig. 7.2 From Sena, G. and Birnbaum, K.D. 2010. Built to rebuild: in search of organizing principles in plant regeneration. Current Opinion in Genetics and Development 20: 460-465 (Fig. 1, page 461). Copyright Elsevier.

Fig. 7.3 From Li, Q. *et al.* 2015. Regeneration across metazoan phylogeny: lessons from model organisms. J. Genetics and Genomics 42: 57-70 (Fig. 3, page 61). Copyright Elsevier.

Fig. 7.4 (**1**) After Brecher. From Bohn, G. 1934. Leçons de Zoologie Generale V (155). Hermann Ed., Paris, France (Fig. 5, page V-17). (**2**) After Bordage. From Bohn, G. 1934. Leçons de Zoologie Generale V (155). Hermann Ed., Paris, France (Fig. 4, page V-16). (**3**) After Torniep and Korschelt. From Aron, M. and Grassé, P. 1939. Precis de Biologie Animale. Masson Ed., Paris, France (Fig. 123, page 242). (**4**) After Korschelt. From Aron, M. and Grassé, P. 1939. Precis de Biologie Animale. Masson Ed., Paris, France (Fig. 119, page 234).

Fig. 7.5 From Li, Q. *et al.* 2015. Regeneration across metazoan phylogeny: lessons from model organisms. J. Genetics and Genomics 42: 57-70 (Fig. 5, page 65). Copyright Elsevier.

Fig. 7.6 (**1**) After Child. From Buchsbaum, R. 1951. Animals without Backbones. Penguin Books, UK (Fig. page 146). (**2**) After Stoppel. From Palhinha, R.T. and da Cunha, G.A. 1939. Curso de Botanica. Rodrigues Ed., Lisboa, Portugal (Fig. 97, page 355). (**3**) After Loeb, from Morgan 1901. From Bonner, J.T. 1952. Morphogenesis. Princeton Univ. Press, Princeton, NJ, U.S.A.

Fig. 7.7 From Xu, L. and Huang, H. 2014. Genetic and epigenetic controls of plant regeneration. Current Topics in Developmental Biology 108: 1-33 (Fig. 1.2, page 6). Copyright Elsevier.

Fig. 8.1 (**1**) https://www.thorlabs.de/Images/GuideImages/116_GlanThompsonPolar_4.jpg (**2**) Cabrera, A. 1937. Historia Natural. Geologia. Instituto Gallach, Barcelona, Spain (Fig. page 151). (**3**) Klein, C. and Hurlbut, Jr. C.S. 1985. Manual of Mineralogy (After James D. Dana). First Editions 1857-1900. John Wiley and Sons, New York, U.S.A. (Fig. 7.6, page 237).

Fig. 8.2 Schuerges, N. *et al.* 2016. Cyanobacteria use micro-optics to sense light direction. e-Life 2016; 5:e 12620. *DOI*: 10.7554/eLife.12620.

Fig. 8.3 (**1**) Niklitschek, A. 1943. Tecnica de la Vida. Iberia-Joaquin Gil, Barcelona, Spain (Fig. page 320). (**2**) Guilliermond, A. and Mangenot, G. 1941. Precis de Biologie Vegetale. Masson Ed., Paris, France (Fig. 294, page 609). (**3**) Vogelmann, T.C. *et al.* 1989. Photosynthetic light gradients and spectral régime within leaves of *Medicago sativa*. Phil. Trans. R. Soc. London *B* 323: 411-421 (Fig. page 419). (**4**) Rabaud, E. 1932. Zoologie Biologique. Gauthier-Villars Ed., Paris, France (Fig. 116, page 154).

Fig. 8.4 (**1**) Frisch, K.V. 1938. Noi e la Vita. U. Hoepli, Milano, Italy (Fig. 90, page 166). (**2**) Rabaud, E. 1932. Zoologie Biologique. Gauthier-Villars Ed., Paris, France (Fig. 107, page 146). (**3**) Rabaud, E. 1932. Zoologie Biologique. Gauthier-Villars Ed., Paris, France (Fig. 108, page 147). (**4**) Rabaud, E. 1932. Zoologie Biologique. Gauthier-Villars Ed., Paris, France (Fig. 100, page 139).

Fig. 8.5 (**1**) Bohn, G. 1934. Vertebres Inférieures, VI. Hermann and Co. Ed., Paris, France (Fig. 34, page VI-47). (**2**) Rabaud, E. 1932. Zoologie Biologique. Gauthier-Villars Ed., Paris, France (Fig. 115, page 153). (**3**) Rabaud, E. 1932. Zoologie Biologique. Gauthier-Villars Ed., Paris, France (Fig. 113, page 151). (**4**) https://upload.wikimedia.org/wikipedia/commons/thumb/4/4e/GrandCanyon_Nat.Park._Desert_Spiny_Lizard (**5**) https://upload.wikimedia.org/wikipedia/commons/thumb/c/cO/Podarcis_sicula_rb.jpg/

Fig. 8.6 Gehring, W.J. and Ikeo, K. 1999. Pax6 mastering eye morphogenesis and eye evolution. Trends in Genetics 15(9): 371-377 (Fig. 1, page 372).

Fig. 8.7 (**1**) Rioja, E. 1929. Los Animales Marinos. Editorial Labor, Barcelona, Spain (Fig. 44, page 106). (**2**) Pierantoni, U. 1944. Tratado de Zoologia. Editorial Labor, Barcelona, Spain (Fig. 731, page 700). (**3**) Gehring, W.J. 1998. Master Control Genes in Development and Evolution. Yale University Press, London, UK (The three figures from Plate 7).

Fig. 9.1 (**1**) Kukalova, J. 1970. Revisional study of the order Palaeodictyoptera in the Upper Carboniferous shales of Commentry, France. Part III. Psyche 77: 1–44 (Fig. 50, facing page 1). (**2**) Robert, P.A. 1936. Les Insectes, I. Delachaux et Nestle, Neuchatel, Switzerland (Plate 26). (**3**) Eliasson, C.U. *et al.* 2005. Fjärilar: Dagfjärilar. Nationalnyckeln till Sveriges Flora och Fauna. Artdatabanken, Sveriges Lantbruksuniversitet, Sweden (Fig. page 289). (**4**) Douwes, P. *et al.* 2012. Steklar: Myror-Getingar. Nationalnyckeln till Sveriges Flora och Fauna. Artdatabanken, Sveriges Lantbruksuniversitet, Sweden (Fig. page 47).

Fig. 9.2 (**1**) Matthew, P. Martyniuk — Own Work. Created: 26 October 2014. https://commons.wikimedia.org/wiki/File:Pterodactylus_holotype_fly_mmartyniuk.png (**2**) Smith, K. 2012. The ground breaker. Nature 489: 22–25.

Fig. 9.3 (**1**) del Hoyo, J. *et al.* (Editors) 1992. Handbook of the Birds of the World, Vol. 1. Lynx Edicions, Barcelona, Spain (Photo page 355). (**2**) del Hoyo, J. *et al.* (Editors) 1996. Handbook of the Birds of the World, Vol. 3. Lynx Edicions, Barcelona, Spain (Photo page 575). (**3**) File: Piki Wiki Israel 11327 Wildlife and Plants of Israel. JPG. Uploaded by MathKnight-at-Tau Created 19 August 2013. https://commons.wikimedia.org/wiki/Bat

Fig. 9.4 Swartz, S.M. and Konow, N. 2015. Advances in the study of bat flight: the wing and the wind. Can. J. Zool. 93: 977–990 (Fig. 1, page 978).

Fig. 9.5 (**1**) Nature 2012. Vol. 487: 409. Deep-Sea Research. (Fig. page 409). Pt// http://dx.doi.org/10.1016/j.dsr2.2012.07.002(2012). (**2**) https://commons.wikimedia.org/wiki/Category:Cheilopogon_melanurus. (**3**) File: Tsukushi TU. jpg Uploaded by

Izuzuki Created 30 July 2013. Izuzuki — http://www.izuzuki.com/ (**4**) https://upload.wikimedia.org/wikipedia/commons/5/58/Eagle_ray_jb.jpg

Fig. 9.6 (**1**) Bramwell, M. (Editor) 1980. The Atlas of World Wildlife. Mitchell Beazley, London, UK (Fig. page 116). (**2**) File: *Petaurista elegans*. jpg — Wikimedia Commons https://commons.wikimedia.org/wiki/File:Petaurista_elegans.jpg (**3**) https://commons.wikimedia.org/wiki/File:Petaurus_australis.jpg (**4**) https://commons.wikimedia.org/wiki/File:Dermopt%C3%A8re3.jpg?uselang=sv

Fig. 9.7 Ross, H.H. 1962. How to Collect and Preserve Insects. Natural History Survey Division, Harlow B. Mills, Chief. State of Illinois, Department of Registration and Education (Fig. 18, page 30). http://www.biodiversitylibrary.org/item/105840

Fig. 9.8 (**1**) After Lutz. From Barnes, R.D. 1980. Invertebrate Zoology. Holt-Saunders Int. Editions, Philadelphia, U.S.A. (Fig. 16–12, page 849). (**2**) Aron, M. and Grassé, P. 1939. Precis de Biologie Animale. Masson Ed., Paris, France (Fig. 482, page 824). (**3**) Aron, M. and Grassé, P. 1939. Precis de Biologie Animale. Masson Ed., Paris, France (Fig. 498, page 836). (**4**) Pierantoni, U. 1944. Tratado de Zoologia. Editorial Labor, Barcelona, Spain (Fig. 567, page 610).

Fig. 9.9 (**1**) Lawrence, P.A. 1992. The Making of a Fly. Oxford. Blackwell Scientific Publications, London, UK (Fig. 5.1 Plate). (**2**) After Rodriguez-Esteban, C. *et al.* 1999. T-box genes Thx4 and Thx5 regulate limb outgrowth and identity. Nature 398: 814–818 (Fig. 3, page 816). Photographs courtesy of J.C. Izpisua-Belmonte. From Gilbert, S.F. 2000. Developmental Biology. Sinauer Ass., U.S.A. (Fig. 16.6, page 508).

Fig. 9.10 Cuénot, L. 1932. La Genese des Especes Animales. Felix Alcan, Paris, France (Fig. 97, page 440).

Fig. 9.11 (**1**) Cuénot, L. 1932. La Genese des Especes Animales. Felix Alcan, Paris, France (Fig. 101(A), page 476). (**2**) Cuénot, L. 1932. La Genese des Especes Animales. Felix Alcan, Paris, France (Fig. 101(B), page 476). (**3**) del Hoyo, J. *et al.* (Editors) 1992. Handbook of the Birds of the World, Vol. 1. Lynx Edicions, Barcelona, Spain (Plate 5, Nr. 1, page 109). (**4**) del Hoyo, J. *et al.* (Editors) 1992. Handbook of the Birds of the World, Vol. 1. Lynx Edicions, Barcelona, Spain (Plate 23, Nr. 39, page 353). (**5**) del Hoyo, J. *et al.* (Editors) 1992. Handbook of the Birds of the World, Vol. 1. Lynx Edicions, Barcelona, Spain (Plate 3, Nr. 2, page 97). (**6**) del Hoyo, J. *et al.* (Editors) 1997. Handbook of the Birds of the World, Vol. 4. Lynx Edicions, Barcelona, Spain (Plate 33, Nr. 57, page 363).

Fig. 10.1 (**1**) After Bütschli and Schewiakoff. From Prenant, M. 1935. Leçons de Zoologie, Protozoaires. Hermann Ed., Paris, France (Fig. 13, page 21). (**2**) After Trembley. From Bohn, G. 1934. Les Invertébrés, III (133). Hermann Ed., Paris, France (Fig. 3, page III-12). (**3**) After Will. From Bohn, G. 1934. Les Invertébrés, III (133). Hermann Ed., Paris, France (Fig. 4, page III-12). (**4**) After Döflein. From Prenant, M. 1935. Leçons de Zoologie, Protozoaires. Hermann Ed., Paris, France (Fig. 19, page 24).

Fig. 10.2 (**1**) Francé, R.H. 1943. La Maravillosa Vida de los Animales. Editorial Labor, Barcelona, Spain (Fig. 28, page 179). (**2**) Frisch, K.V. 1938. Noi e la Vita. U. Hoepli Ed., Milano, Italy (Fig. 122, page 234). (**3**) Pope, J. 1986. Do Animals Dream? Natural History Museum London, Michael Joseph, London, UK (Fig. page 53).

Fig. 10.3 (**1**) del Hoyo, J. *et al.* (Editors) 1996. Handbook of the Birds of the World, Vol. 3. Lynx Edicions, Barcelona, Spain (Plate 18, Nr. 126). (**2**) del Hoyo, J. *et al.* (Editors) 1997. Handbook of the Birds of the World, Vol. 4. Lynx Edicions, Barcelona, Spain (Plate 3, Nr.

1, page 113). (**3**) Av Raveesh, M.S. — Eget arbete, CCBY-SA 4.0, https://commons.wikimedia.org/w/index.php?curid=49257617 (**4**) From Chapman, R.F. 1969. The Insects. The English Universities Press, London, UK (Fig. 377, page 566). (**5**) From Wells, M. 1983. Cephalopods do it differently. New Scientist 100: 332-337.

Fig. 10.4 (**1**) From Niklitschek, A. 1943. Tecnica de la Vida. Iberia, Joaquin Gil Ed., Barcelona, Spain (Fig. 111, page 378). (**2**) From Katzir, G. 1993. Visual mechanisms of prey capture in water birds. In: Vision, Brain and Behaviour in Birds. Zeigler, H.P. and Bischof, H.-J. (Editors) MIT Press: 301-315, London, UK (Fig. 17.1, page 303).

Fig. 10.5 Garb, J.E. *et al.* 2006. Silk genes support the single origin of orb webs. Science 312: 1762 (Fig. 1, page 1762).

Fig. 10.6 (**1**) From *Flickr*, Photographer Toshiyuki Imai. (**2**) From *Flickr*, Photographer Wayne Hoffsinger.

Fig. 10.7 (**1**) After Errera and Laurent. From Guilliermond, A. and Mangenot, G. 1941. Precis de Biologie Vegetale. Masson Ed., Paris, France (Fig. 177, page 406). (**2**) After Errera and Laurent. From Guilliermond, A. and Mangenot, G. 1941. Precis de Biologie Vegetale. Masson Ed., Paris, France (Fig. 177, page 406). (**3**) From file:///C:/Users/Bodil%20Enoksson/AppData/Local/Microsoft/Windows/Temporary%

Fig. 11.1 (**1**) Sandars, E. 1937. A Beast Book for the Pocket. Oxford University Press, London, UK (Fig. page 98). (**2**) Wilson, D.E. and Mittermeier, R.A. (Editors) 2015. Handbook of the Mammals of the World, Vol. 5. Lynx Edicions, Barcelona, Spain (Plate 13, Nr. 1, page 218). (**3**) Sandars, E. 1937. A Beast Book for the Pocket. Oxford University Press, London, UK (Fig. page 117). (**4**) Wilson, D.E. and Mittermeier, R.A. (Editors) 2015. Handbook of the Mammals of the World, Vol. 5. Lynx Edicions, Barcelona, Spain (Plate 15, Nr. 2, page 290). (**5**) Grassé, P.P. 1977. Larousse Animal Portraits (From 1806). Hamlyn, London, UK (page 24). (**6**) Wilson, D.E. and Mittermeier, R.A. (Editors) 2015. Handbook of the Mammals of the World, Vol. 5. Lynx Edicions, Barcelona, Spain (Plate 31, Nr. 10, page 562).

Fig. 11.2 (**1**) Boulenger, E.G. 1937. World Natural History. Batsford, B.T., London, UK (Plate 47, page 102). (**2**) Wilson, D.E. and Mittermeier, R.A. (Editors) 2015. Handbook of the Mammals of the World, Vol. 5. Lynx Edicions, Barcelona, Spain (Plate 14, Nr. 0, page 231). (**3**) Photo by James St. John https://www.flickr.com/photos_2015-08-10 (**4**) Wilson, D.E. and Mittermeier, R.A. (Editors) 2015. Handbook of the Mammals of the World, Vol. 5. Lynx Edicions, Barcelona, Spain (Plate 24, Nr. 1, page 432). (**5**) Wilson, D.E. and Mittermeier, R.A. (Editors) 2009. Handbook of the Mammals of the World, Vol. 1. Lynx Edicions, Barcelona, Spain (Plate 21, Nr. 1, page 413). (**6**) Wilson, D.E. and Mittermeier, R.A. (Editors) 2015. Handbook of the Mammals of the World, Volume 5. Lynx Edicions, Barcelona, Spain (Plate 0, Nr. 0, page 27). (**7**) Wilson, D.E. and Mittermeier, R.A. (Editors) 2009. Handbook of the Mammals of the World, Vol. 1. Lynx Edicions, Barcelona, Spain (Plate 6, Nr. 14, page 145). (**8**) Wilson, D.E. and Mittermeier, R.A. (Editors) 2015. Handbook of the Mammals of the World, Vol. 5. Lynx Edicions, Barcelona, Spain (Plate 16, Nr. 21, page 302).

Fig. 11.3 (**1**) Cabrera, A. 1937. Historia Natural. Geologia, Vol. 4. Instituto Gallach, Barcelona, Spain (page 150). (**2**) Medenbach, O. and Sussieck-Fornefeld, C. 1983. Minerais. Ed. Publica, Lisboa, Portugal. (**3**) Medenbach, O. and Sussieck-Fornefeld, C. 1983. Minerais. Ed. Publica, Lisboa, Portugal.(**4**) Photograph in: Phys. Rev. B 59: 308-

321, 1999. https://commons.wikimedia.org/wiki/File:Ho-Mg-Zn_Quasicrystal.jpg (**5**) Medenbach, O. and Sussieck-Fornefeld, C. 1983. Minerais. Ed. Publica, Lisboa, Portugal. (**6**) Cabrera, A. 1937. Historia Natural. Geologia, Vol. 4. Instituto Gallach, Barcelona, Spain (page 118). (**7**) Medenbach, O. and Sussieck-Fornefeld, C. 1983. Minerais. Ed. Publica, Lisboa, Portugal.(**8**) Cabrera, A. 1937. Historia Natural. Geologia, Vol. 4. Instituto Gallach, Barcelona, Spain (page 198). (**9**) Cabrera, A. 1937. Historia Natural. Geologia, Vol. 4. Instituto Gallach, Barcelona, Spain (page 107). (**10**) Dana, E.S. 1955. A Textbook of Mineralogy with an Extended Treatise on Crystallography and Physical Mineralogy, 4th Edition. John Wiley and Sons, New York, U.S.A., Chapman and Hall Ltd, London, UK (page 191).

Fig. 11.4 (**1**) Baur, E. 1930. Einführung in die Vererbungslehre. Borntraeger, Berlin, Germany (Tafel III, page 94). (**2**) Lindman, C.A.M. 1926. Bilder ur Nordens Flora (Plate on page 484). (**3**) Coutinho, A. 1906. Atlas de Botanica. A Editora, Lisboa, Portugal (Plate XVII, Nr. 7). (**4**) Coutinho, A. 1906. Atlas de Botanica. A Editora, Lisboa, Portugal (Plate XVIII, Nr. 11). (**5**) Shewell-Cooper, W.E. 1951. The ABC of Bulbs and Corms. English University Press, London, UK (page 128). (**6**) Lindman, C.A.M. 1926. Bilder ur Nordens Flora (Plate on page 419). (**7**) Heywood, V.H. 1978. Flowering Plants of the World. Oxford University Press, International Projects, Oxford, UK (Nr. 1, page 280). (**8**) Kihara, H. 1971. Trees in Hakone. Hakone Arboretum, Japan. (**9**) Coutinho, A. 1906. Atlas de Botanica. A Editora, Lisboa, Portugal (Plate XI, Nr. 2). (**10**) Strasburger, E. 1943. Tratado de Botanica. Manuel Marin, Barcelona, Spain (Fig. 656, page 585).

Fig. 12.1 null-Encyclopedia Britanica Online. http://academic.eb.com/bps/media-view/124131/0/1/0

Acknowledgments

Med. Kand. Johan Essen-Möller is to be thanked for an unfailing contribution in converting the manuscript into computer work which he carried out with the utmost accuracy.

Special thanks are due to the staff of the Biology Library, Lund University, for the retrieval of scientific works: Librarian Ph.D. Kristina Arnebrant, Librarian Ph.D. Bodil Enoksson and Librarian B.Sc. Johnny Jönsson.

"Permission Forms" were sent by Air Mail to all Publishers to request permission to use the figures included in this work. Most publishers did not demand a fee, some kindly made a 60% reduction. Other publishers demanded a fee which was payed immediately.

Appendix

CHARTS OF CHEMICAL AND BIOLOGICAL PERIODICITY

PERIODIC TABLE OF CHEMICAL ELEMENTS

Increasing grade of complexity	ALKALI METALS	ALKA-LINE EARTHS	NONMETALS				HALO-GENS	NOBLE GASES
								HELIUM 2 4.0
1	LITHIUM 3 6.9	BERYLL-IUM 4 9.0	BORON 5 10.8	CARBON 6 12.0	NITRO-GEN 7 14.0	OXYGEN 8 15.9	FLUORINE 9 18.9	NEON 10 20.1
2	SODIUM 11 22.9	MAGNES-IUM 12 24,3	ALUMI-NUM 13 26.9	SILICON 14 28.0	PHOS-PHORUS 15 30.9	SULFUR 16 32.0	CHLO-RINE 17 35.4	ARGON 18 39.9
3	POTASS-IUM 19 39.1	CALCIUM 20 40.0	GALLIUM 31 69.7	GERMAN-IUM 32 72.5	ARSENIC 33 74.9	SELENIUM 34 78.9	BROMINE 35 79.9	KRYPTON 36 83.8
4	RUBIDIUM 37 85.4	STRON-TIUM 38 87.6	INDIUM 49 114.8	TIN 50 118.6	ANTI-MONY 51 121.7	TELLUR-IUM 52 127.6	IODINE 53 126.9	XENON 54 131.3
5	CESIUM 55 132.9	BARIUM 56 137.3	THALL-IUM 81 204.3	LEAD 82 207.1	BISMUTH 83 208.2	POLON-IUM 84 (210)	ASTATINE 85 210	RADON 86 (222)
6	FRANC-IUM 87 (223)	RADIUM 88 226.0	113	114	115	116	117	118

Increasing grade of complexity →

LANTHANIDE SERIES

LANTHA-NUM	CERIUM	PRAESE-ODYM-IUM	NEO-DYMIUM	PROME-THIUM	SAMAR-IUM	EUROPI-UM	GADO-LINIUM	TERBIUM	DYS-PROS-IUM	HOLM-IUM	ERBIUM	THULIUM	YTTER-BIUM
57	58	59	60	61	62	63	64	65	66	67	68	69	70
138.9	140.1	140.9	144.2	(145)	150.4	151.9	157.2	158.9	162.5	164.9	167.2	168.9	173.0

ACTINIDE SERIES

ACTIN-IUM	THOR-IUM	PROTAC-TINIUM	URAN-IUM	NEPTU-NIUM	PLUTO-NIUM	AMERI-CIUM	CURIUM	BERKE-LIUM	CALI-FORN-IUM	EIN-STEIN-IUM	FERM-IUM	MEN-DELE-VIUM	NOBEL-IUM
89	90	91	92	93	94	95	96	97	98	99	100	101	102
(227)	232.0	231.0	238.0	237.0	(244)	(243)	(247)	(247)	(251)	(254)	(257)	(256)	(254)

Lima-de-Faria, A. 2017. Periodic Tables Unifying Living Organisms at the Molecular Level. The Predictive Power of the Law of Periodicity. World Scientific.

PERIODIC TABLE OF LIVING ORGANISMS

Increasing grade of complexity	LUMINESCENCE	VISION	REGENERATION (WHOLE BODY)	PLACENTA	PENIS (SINGLE)	FLIGHT (IN AIR)	PLANT CARNIVORY (TRAP)	MENTAL ABILITY
1	MINERALS EUBACTERIA	MINERAL REFRACTION CYANOBACTERIA	CRYSTALS PROTOZOA	ONYCHOPHORANS BRYOZOANS	FLUKES	INSECTS	*Drosophyllum* Leaves slightly curved at top	ANTS TERMITES BEES
2	DINOFLAGELLATES	PLANT CELLS CNIDARIANS	ALGAE SPONGES	SPONGES WORMS FLIES	GNATHOSTOMULIDS GASTROTRICHA	SQUIDS	*Sarracenia* Tube with a lid	SPIDERS OCTOPUSES
3	FUNGI	CRUSTACEANS MOLLUSCS INSECTS	CNIDARIANS	ASCIDIANS FERNS	MOLLUSCS INSECTS	FISH	*Nepenthes* Pitchers of two types	ARCHERFISH
4	STARFISH JELLYFISH	FISHES AMPHIBIANS	FLATWORMS	SHARKS TOADS LIZARDS	CAECILIANS LIZARDS	PTEROSAURS	*Dionaea* Leaves bilobed and movable	MONITOR LIZARDS
5	MOLLUSCS INSECTS	REPTILES BIRDS	ECHINODERMS	FLOWERING PLANTS BANDICOOTS	OSTRICHES DUCKS	BIRDS	*Genlisia* Spirally twisted branches with claws	WEAVERBIRDS
6	NEARSHORE AND ABYSSAL FISH	MARSUPIALS PLACENTALS	FLOWERING PLANTS	EARLY PLACENTALS LATE PLACENTALS	MARSUPIALS PLACENTALS	BATS	*Utricularia* Bladder with movable door	CHIMPANZEES

Increasing grade of complexity →

MARSUPIAL SERIES

POUCHED MOLE	MULGARA MOUSE	PHASCOGALE	OPOSSUM	ANTEATER NUMBAT	GREATER GLIDER	SUGAR GLIDER	WATER OPOSSUM	POUCHED HYENA FOSSIL	TASMANIAN WOLF	MARSUPIAL LION FOSSIL	SABRETOOTH TIGER FOSSIL

PLACENTAL SERIES

MOLE	MOUSE	SHREW	TARSIER	ANTEATER	SQUIRREL	COLUGO	OTTER	SPOTTED HYENA	WOLF	LION	SABRETOOTH TIGER FOSSIL

Lima-de-Faria, A. 2017. Periodic Tables Unifying Living Organisms at the Molecular Level. The Predictive Power of the Law of Periodicity. World Scientific.

See next page

Periodic Table of Chemical Elements

Mendeleev's original Table contained only 61 chemical elements. At present the Table extends to 118 elements (Gray 2009). Over 700 different graphic representations of the periodic system have been published putting in evidence particular properties of the atoms (Mazurs 1974). In the present Table are included the groups of atoms that exhibit the highest degree of regularity. Hydrogen (with its variable location) and the Transition Metals were not included. Most columns consist of elements which have the same basic property. This property resurfaces at intervals independently of whether the atoms are simple or most complex. The intervals are not all of the same length. Columns exhibiting properties that are related, are located forming natural groups. All elements are derived from hydrogen being produced by combination of this simple element. This combination is dominated by order and coherence since the regular location of the electrons is the decisive component of the atom properties.

The well established periodicity of the complete Table of the elements contains many irregularities (see Chapter 2). These are usually not mentioned since the Table is one of the icons of science. No phenomenon is totally regular.

Periodic Table of Living Organisms

The structures and functions studied at the molecular level in animals and plants show also an impressive periodicity which is characterized by the following: (1) The properties analysed show an increase in complexity starting with simple forms and terminating in highly complex ones. These result from the combination of additional coherent components. (2) The same property re-emerges as much as 40 times. (3) It suddenly appears in the simplest as well as the most complex organisms. (4) The sudden recurrence occurs in organisms that are not phylogenetically closely related. (5) The property arises without previous announcement. (6) The novel structures appear "ready made" being functional and coherent from the start. (7) Some properties, such as luminescence, emerge already at the mineral level before DNA arrived in evolution. Luminescence is known to be a pure atomic event. This supports the notion that biological properties have an atomic ancestry.

In the same way as with the atoms, the columns containing properties closely related, were put together in groups. The column *Plant Carnivory* includes more details than Table 12.1 to make it better understood by those less familiar with this phenomenon. *Plant Carnivory* is placed at the side of *Mental Ability* because both represent devices by which an organism improves acquisition of food resources.

The present table, like the initial one of Mendeleev, represents only a first stage in the establishment of periodicity in living organisms. Further research will lead to its enlargement, as more properties will be added. These will be better defined not only at the molecular but also at the atomic level.

Irregularities are expected to be more common at the biological level, than at the atomic one, since one is dealing with a more complex organization.

The emergence of two independent series, in the lower part of the Charts, reinforces the connection between the two types of periodicity. The equivalence found within the Mammals (Marsupials and Placentals) is present as well within the Rare Earths (Lanthanides and Actinides). Only future investigation, at the atomic level, will elucidate properly this surprising similarity that at present may appear fortuitous.